DIGITAL SIGNAL PROCESSING

Efficient Convolution
and
Fourier Transform Techniques

DIGITAL SIGNAL PROCESSING

Efficient Convolution
and
Fourier Transform Techniques

Douglas G. Myers

School of Electrical and Computer Engineering
Curtin University of Technology
Perth, Western Australia

Prentice Hall

New York   London   Toronto   Sydney   Tokyo   Singapore

Typeset by Keyboard Wizards, Harbord, NSW.

Cover design by Kim Webber.

Printed and bound in Australia by Impact Printing, Brunswick, Victoria.

2 3 4 5 93 92 91 90

ISBN 0 13 211814 9.

National Library of Australia
Cataloguing-in-Publication Data

Myers, Douglas Graham
   Digital signal processing

   Includes index.
   ISBN 0 13 211814 9.

   1. Signal processing - Digital techniques. I. Title.

621.3822

Library of Congress
Cataloguing-in-Publication Data

Myers, Doug.
   Digital signal processing / Doug Myers.
      p.   cm.
   Includes bibliographical references and index.
   ISBN: 0-13-211814-9

   1. Signal processing--Digital techniques. I. Title.

TK5102.5.M89   1990   90-7840
621.382'2--dc20   CIP

Prentice Hall, Inc., *Englewood Cliffs, New Jersey*
Prentice Hall Canada, Inc., *Toronto*
Prentice Hall Hispanoamericana, SA, *Mexico*
Prentice Hall of India Private Ltd, *New Delhi*
Prentice Hall International, Inc., *London*
Prentice Hall of Japan, Inc., *Tokyo*
Prentice Hall of Southeast Asia Pty Ltd, *Singapore*
Editora Prentice Hall do Brasil Ltda, *Rio de Janeiro*

PRENTICE HALL

A division of Simon & Schuster

Contents

Part 2 : Digital Convolution on Infinite Fields

Part 3 : Digital Convolution in Finite Structures

Part 4 :   Digital Convolution in Finite Polynomial Structures

PREFACE

Among the numerous uses made of prefaces in books, one is some explanation of why the book was written. In many instances, that explanation begins by justifying why another book should be written on the given subject. This preface, though, belongs to that minority where the object is to explain why any book at all should be written on the subject.

Digital signal processing has a remarkably short history. The early work on speech, image processing, radar signal processing and the like — topics still of considerable interest and still the objects of active research — began less than two decades ago. The rapid development of the subject can be appreciated by scanning the issues of the leading journals. Yet the application of digital signal processing has not been without its problems. It is a field which demands sophisticated technology. Consequently, the significant industrial and applications advances have largely followed major technological innovations such as microprocessors.

Three recent developments have brought new freedoms to designers. One, the digital signal processing chip, is, in effect, a special-purpose microprocessor with enhanced arithmetic capabilities and extensive pipelining. The second is devices for array and wavefront processing such as the transputer. The third, in the area of design technology, is powerful software operating on low-cost systems for the design of large-scale Application Specific Integrated Circuits (ASICs).

With these developments, design approaches for implementing systems once regarded as belonging to the avant-garde of digital signal processing can, and now should, move into the mainstream of the field. The objective of this book is to present the core of this material in an integrated and coherent manner to an audience with some familiarity with digital signal processing and an interest in advanced techniques. At first glance, the material may seem unusual. New tools, though, are not necessarily difficult tools and these are presented and studied because they can be used to great effect.

The particular issue within digital signal processing on which this book focuses is the range of efficient algorithms for implementing digital convolution. That is, of the many and diverse algorithms which can be employed to implement systems described, for example, by:

$$y(n) = \sum_{m=0}^{\infty} h(m) \cdot x(n-m) \qquad\qquad n = -\infty, \infty$$

where $x(n)$ is some input data sequence and $h(n)$ is a processor response. This book is concerned with those which best meet some criteria related to performance, design or operation. How $h(n)$ is derived is not considered in any way and, indeed, it is assumed the reader is already aware of how to solve this particular problem.

There are two broad approaches to implementing digital convolution. One, the direct approach, focuses directly on the equations themselves and attempts to locate symmetries which can be exploited. The other seeks to transform the problem into a mathematical domain where convolution is an easier task. There are a number of transform approaches to examine, including the various fast Fourier transforms (FFTs).

For whom is this book written? The perceived requirements of three groups influenced its construction. One is senior engineering undergraduates continuing on from an introductory course in digital signal processing. A second is beginning graduate students, in particular, students in engineering, geology, physics, mathematics, computer science and similar disciplines seeking a knowledge of advanced signal processing methods. Finally, given the rapid rise to prominence of digital signal processing, an expected readership is professionals engaged in private study or continuing education programs.

What is expected of the reader? Remarkably little. Some background in digital signal processing is, of course, assumed; at least an understanding of z-transforms, basic digital signal processor types, some design methods for them and the elements of digital frequency analysis. Although not frequently employed, the ideas of matrix and vector analysis are certainly alluded to and a good understanding of the rudiments of both is essential for effectively appreciating the techniques employed. No understanding of programming is required (although some problems will require it). However, it would be very unusual for any reader of this book not to be experienced in this.

Some comments on why no programs appear in this book. A major reason is that the book is on understanding algorithms not on the fabrication of software. Another is that many programs are readily available and it is really not necessary to list them once again. A third is that the book interprets implementation broadly, covering both hardware and software realizations.

What demands are placed on the reader? The philosophy of this book is not to demand but to encourage. The desire is to impart knowledge and to encourage the development of skill, not to set an obstacle course. The text is therefore deliberately expansive, perhaps too explicit and elementary for some minds. However, this is done to effectively communicate ideas, to clearly signal significant issues and in so doing, to instill the fundamentals and a desire to use them. The subject matter covered may be 'unusual' but, if approached in a systematic and logical fashion, there is no need for it to be any more difficult than any other of interest to the reader.

Of course, a book like this is valueless without challenges but these (hopefully) are challenges to be tackled with relish, not tests of endurance. They are provided through the medium of exercises and problems. Exercises seek to draw out some of the finer points of the text. For that reason, they are included in the sections labeled 'Further Study'. Problems, on the other hand, aim at honing and developing skills and are framed with applications in mind. Be warned! Some are far from trivial. Some, too, can be quite time consuming. Others can have a variety of solutions or can be worked to a variety of levels. The author is not a believer in neatly packaged doses of computation and clearly delimited thinking but an enthusiast for problems which go in part to simulate problems of a real environment. That means some vagueness, possibly some incoherence and some passing of the responsibility for framing all of the problem's specifications. Part of the skill expected to have been already developed by the reader is the ability to assess the effort needed to solve a problem. A preliminary sketching of possible solutions would therefore be an advisable step in tackling most of the problems of this book.

Encouragement of the reader was also a reason for the slightly unusual structure of this book. It has four Parts. The first sets the framework for the endeavor and reviews and introduces some general mathematical results. The second concentrates on the various fast Fourier transform techniques plus some related methods. It is assumed that the reader has encountered the common base-2 algorithm and this section takes a more advanced viewpoint. The third looks at number theoretic approaches to digital convolution and the fourth at polynomial. All are linked by an underlying abstract algebraic structure. That structure unifies and integrates what may otherwise be seen as merely a collection of useful techniques.

The more unusual aspects of the format lie within the Parts themselves. This may be an eccentricity but it reflects a belief that books aimed at an audience interested in practical application need to consciously stress the desire for means to an end, not the profundity of those means themselves no matter how intriguing they may be. Therefore, the initial emphasis is on tools — mathematical techniques which can be employed. Next theory is considered — the working base of knowledge created with the tools and aimed at the broad spectrum of applications of interest. Third, developments are considered — specific tailoring of the theory to particular problems of interest. Finally, implementation issues are considered — the practical concerns in engineering a solution.

What should the reader gain from this book? It is expected it would form the basis of one or two semesters of formal coursework and at the end of that time, two important results are anticipated. One is clearly to apply this knowledge in different fields and effectively. The other, most important given the rapid development of the field, is the ability to read and judge the advanced literature within digital signal processing.

How should the book be approached? That should be evident from earlier comments. Reading informs. Some effort has been devoted to making that easier through both the text and layout of the book. But reading does not provide skill nor can it be wholly successful in conveying a proper sense of balance with respect to problem solving. Skill is only achieved through application and that gives the exercises and problems a key role in this book. While the author can strenuously urge, even harangue through the strictures of the printed word to tackle these and as well to search out others in the general literature, ultimately it is the reader's choice.

PART

Foundations

THE PROBLEM FOUNDATIONS

The problem addressed in this book is efficient methods of digital convolution. It is concerned purely with the mechanics of digital convolution, not on devising what that convolution should be in order to achieve some given processing action. That is admirably covered by numerous other books. Rather, the domain of interest extends from when the convolution system is created to the point where it is implemented and used. That is, the process of converting the logical structure to an implementation.

Digital convolution is at the very heart of digital signal processing. In abstract terms, it arises when an input signal is provided to a mathematical model in order to gain an output signal. Thus it describes applications as simple as frequency shaping of signals, to as complex as echo canceling processors used in telecommunications. Similarly, it is a natural part of mathematical models defining national economies, power systems operations, industrial plant operations, weather systems, aircraft flight, agricultural production and numerous environmental problems.

As the theoretical foundations of this subject expand, and as the available technology required for implementation both falls in price and rises in performance, so the variety and sophistication of applications grows. To quite a high degree, a book such as this tends to address the more sophisticated end of this spectrum. Once rather limited, now there is a pleasing diversity of applications. For example, digital signal processing techniques are rapidly becoming of critical importance in seismology. The cost of mineral exploration, particularly offshore, demands more sophisticated processing of the available geophysical data. Seismologists, too, are adapting techniques such as tomography to their work to great effect. Another emerging area is beam forming. Phased arrays have existed for many years but sophisticated digital signal processing techniques enable more flexible and efficient systems to be created. These techniques will find increasing application in antenna systems, sonar systems, medical systems and geological systems.

What might be termed the traditional interests of digital signal processing continue to develop. Audio engineering, for example, has yet to exhaust the imagination of engineers, and now they are delving into such complex problems as the elimination of room effects from audio signals. The swing to digital replay systems, and so the conversion of recording studios, radio stations, television stations and similar facilities in the broad audio area to digital, is certain to accelerate this.

Medical applications continue to grow, especially in the imaging area as CAT scanners, MRI systems and ultrasound systems become increasingly more common. As these penetrate the mainstream of diagnosis and surgical planning, the demands are for services such as the better presentation of the data gathered, some interactive control over that presentation, and for a variety of feature-extraction capabilities. Outside of imagery, the low cost of digital signal processing technology now permits a wide range of monitoring and biomedical signal analysis systems to be created. Rapidly increasing health care costs, especially in intensive health care, make this likely to be one of the most commercially significant areas for digital signal processing in coming years.

Two other traditional areas continue to burgeon. Sonar signal processing has always been one of the mainstays of digital signal processing and, unlike some other areas, has been more restricted by the pace of theoretical advances than by technological. The marine

3

environment is particularly complex and demands a very sophisticated approach to signal processing. Sonar, in consequence, can be expected to remain an area of active research and a fertile source of ideas and techniques. Radar, too, has long been important in digital signal processing, but here it is the technology which has set the limits. Now that some of these restrictions are being removed, quite sophisticated units are becoming more common in applications ranging from air traffic control to radars for pleasure boats. The freedom these technological advances has brought, however, has just fueled the demand for even more sophisticated units and so better technology. Imaging radars, for example, could be widely applied if the cost of processing the 300 Mbit per second or so outputs of these units could be made reasonable. Radar's contribution to the future of digital signal processing, then, seems most likely to be as a technological spur.

Image processing tends to reflect many of the same concerns as radar. Applications abound but, until recently, costs meant any implementations were largely confined to the laboratory. Now plug-in boards for personal computers are common and adequately cater for the needs of most basic applications, but not for the more sophisticated sector of the market. The pressures of increasing labor costs and increased demand for quality control in manufacturing industry therefore form another spur for further technological improvement.

An area that needs to be mentioned but which will not be covered here is cryptography. Convolution has an important role here and optimal algorithms are of considerable interest. Finite mathematical structures are also the norm, hence the relevant sectors of this book to this particular problem are Parts 3 and 4.

The book will examine various efficient methods for digital convolution. 'Various methods' because efficient implies optimal which in turn implies that some sets of performance criteria are involved and these will not be uniform from application to application. Further, here 'optimal' will be predominantly used in the sense of achieving ease and economy of design and minimal complexity in implementation. Given the key role of compromise in design, then efficiency or optimality must also mean there exists a choice.

The book focuses on two approaches to convolution. One of these will be referred to as direct approaches and concentrates on a direct implementation of the convolution equations. The other will be referred to as transform approaches. Here, the two signals to be convolved are mathematically transformed to a domain where convolution becomes a very much simpler operation, operated upon in this simpler fashion and then reverse transformed.

In this section of FOUNDATIONS, we look at the problem in some detail. We examine digital convolution in its different forms but only briefly. The object is just to review the material, to clearly establish in the reader's mind the scope of the book and to introduce the terminology that will be used. Next, we examine this question of optimality, ponder on the solutions likely and the problems they may present for implementation. That leads into a small but crucial topic: efficient methods of multiplying complex numbers. Immediately following, we consider one important solution — the Toom-Cook algorithm — which very neatly shows what can be achieved and how some of these problems are actually manifested in a real example.

DIGITAL CONVOLUTION

Terminology

A **discrete signal** is simply a sequence of numbers. A **multidimensional discrete signal** is an n-dimensional array of numbers. If the set of numbers from which elements of these sequences or arrays are selected is finite, then these signals are termed **digital**.

These definitions require no further embellishment. If the signal should represent a time-varying waveform — and this is very common in engineering applications — then it is quite a simple task to relate time to position in the sequence. Similarly, for signals representing images there is no difficulty relating sequence position to some set of spatial coordinates. Equally, the numbers involved in the signals can represent any number of physical or abstract quantities and, again, there is no significant problem in relating the two. Therefore, we shall concentrate on these simple definitions and leave the interpretation of the representation to specific applications.

Consider Figure 1.1. This shows the standard terminology that will be used throughout this book for systems. That is, x(n) will refer to an input sequence applied to a system and y(n) to the corresponding output. The index refers to the position of the particular value within the sequence. For many applications, it is related to time. The system's characteristics will be described by h(n).

Figure 1.1 Terminology

Although they will not be introduced for some pages, it is as well to comment on the terminology to be used for some other concepts. Transformed values will be represented in upper case, such as X(k), where k, for example, identifies spectral line numbers in the frequency domain. Where z-transforms are used, they will also be denoted by capitals but in the form of X(z). For mathematical convenience, z-transforms will be taken as:

$$X(z) = \sum_{n=0}^{\infty} x(n) \cdot z^n$$

rather than with negative powers of z. It needs to be stressed that this is for mathematical convenience but the conversion to the correct form is trivial in most instances.

In keeping with most of the literature on digital signal processing, computer engineering and the like, we shall only distinguish two arithmetic operations. Addition will be assumed to encompass subtraction and multiplication, division.

A comment on the terms 'digital' and 'discrete'

A reader of the digital signal processing literature cannot help but notice that terms like digital, discrete and sampled appear to be used interchangeably. From our definitions, that does not seem justifiable in spite of the fact they all refer to number sequences. There is, though, some justification for this practice. Sampling an analog signal will give a sampled or discrete signal. Quantizing that will give a digital signal. The difference between the two is naturally termed the **quantization error**. Although signal-dependent, it is convenient to treat this as a random effect termed **quantization noise**. This permits a **signal to quantization noise ratio** to be defined as a measure of the difference between the two. To be realistic given the effect is not, in fact, random, means the definition needs to be in the form of an upper bound. It is common to take it as the ratio of the maximum signal level to the root mean square quantization error where that error is taken to be a random signal with a uniform distribution. Thus:

$$\frac{S_{max}}{\sigma_N} = 6.02n + 10.79 \quad \text{dB}$$

For n equal to 16 or 32 bits, the quantization noise is 107.12 dB or 203.45 dB respectively. Signal to thermal noise ratios would be far smaller for real signals. Consequently quantization noise can, in practice, be irrelevant and so a discrete or sampled and digital signal can be effectively the same entity. To be strictly correct, as the various operations described in this book lead to discrete signals, that is the term which should be used. In part for the reason outlined and in part because of the technology used in implementation, the common practice, though, is to use the term digital. This book will conform to that.

Digital signal processors

Only a particular category of **digital signal processors** will be examined in this book. This class has three principal properties, namely:

1.   Linearity

Assume a signal $x_1(n)$ when applied to a given processor generates an output signal $y_1(n)$ and that another input signal $x_2(n)$ when applied to the same processor generates an output $y_2(n)$. This system is **linear** if an input signal ($\alpha x_1(n) + \beta x_2(n)$), where α and β are two numbers, generates an output signal ($\alpha y_1(n) + \beta y_2(n)$).

2.   Shift-invariance

Assume a signal $x(n)$ when applied to a processor generates an output sequence $y(n)$. This system is **shift-invariant** if an input signal $x(n+i)$ generates an output $y(n+i)$ where i is some integer. Since sequence position frequently represents time in engineering applications, this property is often referred to as **time-invariance**.

3. Causality

Assume a signal $x_1(n)$ when applied to a given processor generates an output signal $y_1(n)$ and that another signal $x_2(n)$ when applied to the same processor generates an output signal $y_2(n)$. If $x_1(n)$ is identical to $x_2(n)$ for all values of n up to some value j, then the system is **causal** if $y_1(n)$ is also identical to $y_2(n)$ up to that position j. Any digital signal can be regarded as infinite but many are distinguished by all samples up to some given sequence position (and possibly after some other) being zero. This leads to another common expression of this property, namely that a system is causal if there is no output from a system before a (non-zero) input is applied.

Properties of linear, shift-invariant, causal processors

A signal of great importance is the unit sample. It is described by:

$$x(n) = \begin{array}{l} 0 \text{ for } n \neq 0 \\ 1 \text{ for } n = 0 \end{array}$$

If this is applied to some linear, shift-invariant, causal digital signal processor, an output h(n) results that uniquely defines that processor. It is normally termed the **unit sample response** or, following practice for continuous systems, the **impulse response**.

Consider a signal x(n) applied to some digital signal processor with unit sample response h(n). It may be regarded as the sum of an infinite set of unit sample signals each shifted so that the jth such signal is only non-zero at position j and each scaled by the magnitude of the sample of x(j) at position j. The output sequence y(n) of such a system then, following from the linearity and shift-invariant properties, becomes the sum of an infinite set of unit sample responses each scaled and shifted in the same way. Thus the uniqueness of the unit sample response. Consider that output y(n). At a particular sequence position n_o, y(n) consists of the sum of the components:

$$
\begin{array}{ll}
x(n_o) \cdot h(0) & \text{the beginning of the response due to } x(n_o) \\
x(n_o - 1) \cdot h(1) & \text{the contribution of the previous 'unit sample'} \\
x(n_o - 2) \cdot h(2) & \text{the contribution of the sample two positions prior} \\
\quad \cdot & \\
\quad \cdot & \\
\quad \cdot &
\end{array}
$$

More compactly, the output of the processor at any sequence position n is given by:

$$y(n) = \sum_{m=0}^{\infty} h(m) \cdot x(n-m) \qquad n = -\infty, \infty$$

This describes **digital convolution**. Consider the terms $h(m) \cdot x(n-m)$ of this sum. Let:

$$q = n - m \qquad\qquad \text{i.e.} \qquad\qquad m = n - q$$

Thus we may express them as $h(n - q) \cdot x(q)$. This change in notation doesn't alter the fact that there are an infinite number of such terms, thus summing over the parameter q, we derive an alternative form:

$$y(n) = \sum_{q=0}^{\infty} h(n - q) \cdot x(q)$$

The dummy variable doesn't matter, hence:

$$y(n) = \sum_{m=0}^{\infty} h(m) \cdot x(n - m) = \sum_{m=0}^{\infty} h(n - m) \cdot x(m) \qquad n = -\infty, \infty$$

Consider a special case of these two equations. Assume the unit sample response is zero after some position N. Then the equations become:

$$y(n) = \sum_{m=0}^{N-1} h(m) \cdot x(n - m) = \sum_{m=0}^{N-1} h(n - m) \cdot x(m) \qquad n = -\infty, \infty$$

This describes finite impulse response digital filtering. The original form describes infinite impulse response digital filtering. Note again this book will NOT be concerned with the design of such processors, just the most efficient way of implementing the response $h(n)$. The interest is purely in the convolution — the action of filtering or processing the signal — not in the information processing properties of the processor itself.

There is another property real systems need to exhibit, namely stability. A system is **stable** if a bounded input produces a bounded output. A linear, shift-invariant, causal system with unit sample response $h(m)$ is stable if and only if:

$$\sum_{m=0}^{\infty} |h(m)| < \infty$$

Some particular forms of convolution

Five particular cases of digital convolution are of great practical importance. Four of these cases are distinguished by simple attributes of $x(n)$ and $h(n)$, the two signals being convolved:

1. $x(n)$ finite; $h(n)$ finite

If $x(n)$ is a sequence with only N terms and $h(n)$ a sequence with only M terms, then their digital convolution is given by:

$$y(n) = \sum_{m=0}^{n} h(n-m) \cdot x(m) \qquad\qquad n = 0, 1, \cdots, N+M-2$$

with h(m) and x(m) taken to be zero outside their defined range. This is referred to as **linear** or **aperiodic** convolution.

2. x(n) infinite; h(n) finite

If x(n) is an infinite sequence of non-zero terms but h(n) a sequence with only N non-zero terms, then the convolution is given by:

$$y(n) = \sum_{m=0}^{N-1} h(m) \cdot x(n-m) \qquad\qquad n = -\infty, \infty$$

3. x(n) finite; h(n) infinite

If x(n) is a sequence with only N non-zero terms beginning at sequence number 0, but h(n) is an infinite sequence of non-zero terms, then the convolution is given by:

$$y(n) = \sum_{m=0}^{N-1} h(n-m) \cdot x(m) \qquad\qquad n = 0, \infty$$

4. x(n) infinite; h(n) infinite

Assuming a causal system, then:

$$y(n) = \sum_{m=0}^{\infty} h(m) \cdot x(n-m) \qquad\qquad n = -\infty, \infty$$

The fifth form of importance is a particular case of the fourth. Consider two infinite signals x(n) and h(n), both distinguished by being periodic with a period of N samples. **Periodic** or **cyclic convolution** of these two signals is defined as being:

$$y(n) = \sum_{m=0}^{N-1} h(n-m) \cdot x(m) = \sum_{m=0}^{N-1} h(m) \cdot x(n-m) \quad n = 0,1, \cdots, N-1$$

where this is taken as describing one period of an infinite periodic signal. How is cyclic convolution to be interpreted? We could take h(n) to represent the unit sample response of a system. It would uniquely characterize a linear, shift-invariant system but, because h(n) is infinite, that system cannot be causal. Moreover, because this unit sample response is infinite and periodic, the system cannot be stable, thus the output for any x(n) must be infinite. For the moment, then, we can only interpret cyclic convolution as a mathematical contrivance.

We shall show throughout this book that while it may seem a contrivance, cyclic convolution is a remarkably useful tool in the study of digital convolution. We can hint at that by considering some of the possible relationships between it and the first four cases of digital convolution given above. To begin, note that the mathematical structure of aperiodic and cyclic convolution are similar. Therefore, it should be possible to express the former as a cyclic convolution. Let:

$$Q = N + M - 1$$

Now define two Q-point sequences h'(n) and x'(n) derived by **zero-padding** h(n) and x(n) as follows:

$$
\begin{aligned}
h'(n) &= h(n) && \text{for } n = 0, 1, \cdots, M-1 \\
&= 0 && \text{for } n = M, M+1, \cdots, Q-1
\end{aligned}
$$

$$
\begin{aligned}
x'(n) &= x(n) && \text{for } n = 0, 1, \cdots, N-1 \\
&= 0 && \text{for } n = N, N+1, \cdots, Q-1
\end{aligned}
$$

Then each period of the cyclic convolution:

$$y'(n) = \sum_{m=0}^{Q-1} h'(n-m) \cdot x'(m) = \sum_{m=0}^{Q-1} h'(m) \cdot x'(n-m) \qquad n = 0, 1, \cdots, Q-1$$

equals the desired aperiodic convolution.

Realizable infinite impulse response digital signal processors designed according to the methods outlined in any standard text can be described by an equation of the form:

$$y(n) = \sum_{i=0}^{N-1} a(i) \cdot x(n-i) - \sum_{j=1}^{M-1} b(j) \cdot y(n-j)$$

That is, an infinite impulse response system can be modeled by two finite unit sample response systems. Therefore, cases 3 and 4 above can be computed by any method found for cases 1 and 2. As case 1 has already been covered, this just leaves case 2 to consider.

This case involves a computation of the form:

$$y(n) = \sum_{m=0}^{N-1} h(m) \cdot x(n-m) \qquad\qquad n = -\infty, \infty$$

where h(n) is a finite sequence of length N and x(n) is an infinite sequence. Now x(n) can be expressed as the sum of an infinite set of finite sequences. That is:

$$x(n) = \sum_{p=-\infty}^{\infty} v_p(n)$$

where $\qquad v_p(n) \; = \; x(n) \;$ for $pM \le n \le (p+1)M-1$
$\qquad\qquad\qquad = \; 0 \qquad$ otherwise

Substituting:

$$x(n) \; = \; \sum_{p=-\infty}^{\infty} \left\{ \sum_{m=0}^{N+M-2} h(m) \cdot v_p(n-m) \right\} \qquad\qquad n = -\infty, \infty$$

The second term has the form of a finite aperiodic convolution and it therefore can be determined by cyclic convolution via zero-padding of the sequences.

All four cases, then, can be computed by cyclic convolutions and so efficient means for determining cyclic convolutions are of considerable interest to the general digital convolution problem. Such means exist. Their study is the principal theme of this book.

Sectioned convolution

It would be remiss not to make some additional comments on the **sectioned** or **block convolution** method developed above. The output of this process is the sum of an infinite set of aperiodic convolutions between a processor response h(n) of length N samples and a finite input $v_p(n)$ of length M samples. Each of these convolutions must therefore have $(N+M-1)$ samples and that means there is an overlap of $(N-1)$ points between successive convolutions. As shown in Figure 1.2(a), an implementation is :

1. compute each aperiodic convolution, possibly via a cyclic convolution;
2. add to that result the last (N – 1) points from the previous convolution;
3. save the last (N – 1) points of this convolution for the next stage.

This procedure is termed **overlap and add**. An alternative, termed **overlap and save**, is shown in Figure 1.2(b). Here, the finite sequences $v_p(n)$ are defined as:

$$v_p(n) \; = \; x(n) \;\text{ for } pM-N+1 \le n \le (p+1)M$$

There is now an overlap on the input sequence rather than the output. It is easily shown the required output is constructed by cyclically convolving (with a period (M + N – 1)) each of the sequences $v_p(n)$ with h(n), discarding the first (N – 1) points of those convolutions and passing the rest to the output.

From the practical viewpoint, both sectioned convolution methods share the unfortunate property of an overlap of terms. However, overlap and add requires temporarily storing some output values from each stage, then adding those values to the next set of output values to give the complete output. Overlap and save, though, just requires temporarily storing input values. That makes overlap and save more attractive for most applications.

An important question is the size of M in a sectioned convolution. One way of determining this is to select M so that there are no more operations for a sectioned/cyclic convolution approach than by directly executing the convolution. Assume some algorithm

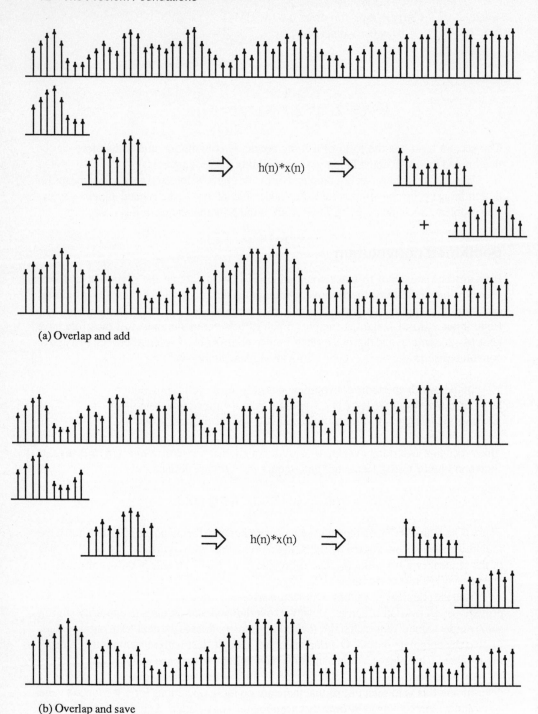

(a) Overlap and add

(b) Overlap and save

Figure 1.2 Sectioned convolution

for a direct convolution requires D(N) operations per output point. The sectioned convolution is executed by 'padding' both h(n) and $v_p(n)$ with zeroes until both are ($M + N - 1$) points in length and then performing the cyclic convolution. Assume this requires C(M+N–1) operations. Because of the overlap, this will only generate M output points. Consequently, the two approaches are the same if:

$$D(N) \cdot M = C(M+N-1)$$

Solving gives the required value of M.

   This formula suggests there can be a value of M which actually gives sectioned convolution an advantage. Consider the figure of merit term:

$$F(M,N) = \frac{C(M+N-1)}{M}$$

For many cyclic convolution algorithms, C(M+N–1) is of the form:

$$C(M+N-1) = c_0 \cdot (M+N-1)^r$$

where c_0 and r are constants. Substituting and taking derivatives gives the optimum value for M as:

$$M_{opt} = \frac{(N-1)}{r-1}$$

ALGORITHMS FOR DIGITAL CONVOLUTION

Introduction

The focus of this book is efficient algorithms for determining cyclic convolution. The two threads of this will be a direct approach which examines the equations for cyclic convolution and a transform approach which maps the convolution into an alternative and simpler mathematical domain. It will seem at first as if these two approaches are quite different. As the book proceeds, it will be increasingly seen that, in fact, a close relationship exists.

   Consider the N equations for cyclic convolution:

$$y(n) = \sum_{m=0}^{N-1} h(n-m) \cdot x(m) \qquad\qquad n = 0,1, \cdots, N-1$$

These completely describe one cycle of the periodic function y(n). As indices –i and (N–i) point to the same sample value, in matrix form these equations may be expressed as:

$$
\begin{bmatrix} y(0) \\ y(1) \\ \cdot \\ \cdot \\ \cdot \\ y(N-1) \end{bmatrix} = \begin{bmatrix} h(0) & h(N-1) & \cdot & \cdot & \cdot & h(1) \\ h(1) & h(0) & & & & h(2) \\ \cdot & & & \cdot & & \cdot \\ \cdot & & & & \cdot & \cdot \\ \cdot & & & & & \cdot \\ h(N-1) & h(N-2) & \cdot & \cdot & \cdot & h(0) \end{bmatrix} \begin{bmatrix} x(0) \\ x(1) \\ \cdot \\ \cdot \\ \cdot \\ x(N-1) \end{bmatrix}
$$

A matrix A is **Toeplitz** if the elements within each of the diagonals are the same. That is, if:

$$
a_{ij} = a_{pq} \qquad \text{for } i - p = j - q
$$

If A is N×N and each row is the preceding row circularly rotated right, then A is termed **circulant**. The matrix in this convolution equation is circulant and Toeplitz. It is also symmetric about the main cross diagonal.

There is another way of expressing convolution. If a sequence y(n) is the convolution of two sequences x(n) and y(n), then a well-known property of z transforms is that:

$$
Y(z) = H(z) \cdot X(z)
$$

where $Y(z)$, $H(z)$ and $X(z)$ are the z transforms of y(n), h(n) and x(n) respectively. Although this describes aperiodic convolution, we shall show later that it may also be related to cyclic. Thus efficient methods for convolving sequences are also efficient methods for multiplying two polynomials, and vice versa.

What we mean by a transform approach is an operation similar to the discrete Fourier transform (DFT). Given a periodic digital signal x(n), then the discrete Fourier transform pair is:

$$
X(k) = \sum_{n=0}^{N-1} x(n) \cdot W^{kn} \qquad\qquad k = 0, 1, \cdots, N-1
$$

$$
x(n) = -\frac{1}{N} \sum_{k=0}^{N-1} X(k) \cdot W^{-nk} \qquad\qquad n = 0, 1, \cdots, N-1
$$

where   N = the number of samples in one period of x(n)

$$
W = \exp(-\frac{j2\pi}{N})
$$

This transform maps the sequence x(n) into a complex frequency domain.

The DFT is a procedure that can be only applied to discrete or digital data to give a discrete output. It therefore has meaning only when applied to periodic signals as only such signals have a discrete spectrum. Cyclic convolution, too, deals only with discrete periodic data. The two are related in that the DFT has the **cyclic convolution property**. That is, given two periodic digital signals, x(n) and h(n), their convolution y(n) has a DFT:

$$Y(k) = H(k) \cdot X(k) \qquad\qquad\qquad k = 0, 1, \cdots, N–1$$

where $H(k)$ and $X(k)$ are the DFTs of $h(n)$ and $x(n)$ respectively. Therefore, an alternative way of determining a cyclic convolution is as:

$$x(n) = -\frac{1}{N}\sum_{k=0}^{N-1} H(k) \cdot X(k) \cdot W^{-nk} \qquad\qquad n = 0, 1, \cdots, N–1$$

$H(k)$ and $X(k)$ can be computed in parallel and so may the N products $H(k) \cdot X(k)$.

What makes this approach different from the direct is the dependence on the cyclic convolution property of the transform. All the transform techniques to be investigated here will have that property. This is an important distinction and, indeed, a property with significant practical implications.

The above suggests one simple relationship between the direct and transform approaches. Consider the z-transform of the periodic sequence $x(n)$:

$$X(z) = \sum_{n=0}^{N-1} x(n) \cdot z^n$$

This is a mapping into the complex plane. Evaluating it on the unit circle, that is, on z equal to $\exp(-j\omega)$ gives:

$$X\left(e^{-j\omega}\right) = \sum_{n=0}^{N-1} x(n) \cdot e^{-j\omega n}$$

If this is evaluated at a set of N points spaced $\dfrac{2\pi}{N}$ apart, the result is the DFT.

A comparison of the direct and transform approaches

Like the direct approach, the transform approach is also essentially a matrix computation. In this case, the matrix is given by:

$$
\begin{bmatrix} X(0) \\ X(1) \\ \cdot \\ \cdot \\ \cdot \\ X(N-1) \end{bmatrix}
=
\begin{bmatrix}
W^0 & W^0 & \cdot & \cdot & & W^0 \\
W^0 & W^1 & & & & W^{N-1} \\
\cdot & & \cdot & & & \\
\cdot & & & \cdot & & \\
\cdot & & & & \cdot & \\
W^0 & W^{N-1} & \cdot & \cdot & & W^{(N-1)(N-1)}
\end{bmatrix}
\begin{bmatrix} X(0) \\ X(1) \\ \cdot \\ \cdot \\ \cdot \\ X(N-1) \end{bmatrix}
$$

Although the direct convolution matrix is real, this is a matrix of complex terms.

Consider these two approaches as computational procedures. The matrix equation given earlier for the direct approach requires N^2 real multiplications and $N \cdot (N - 1)$ real additions to produce the N output points of the cyclic convolution. Also required are N storage cells for the N values of the unit sample response of the processor. No prior calculations are required. At least one additional storage cell is needed for intermediate results but, depending on implementation, up to N can be used.

Each DFT requires $(N^2 - 2N + 1)$ complex multiplications and $N \cdot (N - 1)$ complex additions to produce the N output values. If the usual complex multiplication formula is employed, a DFT requires $(4N^2 - 8N + 4)$ multiplications and $(4N^2 - 6N + 2)$ additions. Using the DFT for convolution requires the forward transformation to be applied to both signals x(n) and h(n), for the resulting transformed values X(k) and H(k) to be pointwise multiplied and then for the result to be reverse transformed. Calculation of H(k) can be done prior to any processing and the scaling value $\frac{1}{N}$ included with these terms. Thus, the number of essential operations is $(8N^2 - 12N + 8)$ real multiplications and $(4N^2 - 10N + 4)$ real additions. The storage required for the unit sample response of the processor is now 2N cells as the values are complex. Also needed are 2N cells to hold intermediate results.

These results do not suggest the transform approach is superior to the direct. For small N, both have a comparable number of operations, but as N grows, the transform approach asymptotically tends to eight times the number of multiplications and additions. Note that both operations for both algorithms grow as N^2. That makes both quite unattractive for large N. In terms of the storage required to hold intermediate and processor unit sample response values, the transform approach fares badly. Indeed, as the evidence currently stands, all that can be said of the transform approach is that it does derive the digital spectrum and there may be circumstances where that is needed as well as the convolution.

The question of optimality

The engineering of digital signal processors must be concerned with the design of those processors, their implementation and their operation. The concerns of each of these areas differ. Furthermore, the balance of those concerns varies across the different applications areas. The question of what is efficient or optimal — the best practice or the best approach — is, as a result, a complex question.

Algorithms with optimal design characteristics, in general, mean algorithms which allow designs to be easily created and modified. This is achieved through the designer's understanding of good design techniques and employment of effective design tools. The former is the role of texts and experience and not the province of the algorithms themselves. The latter for digital signal processing undoubtedly means interactive software running on a range of computer systems from workstations through to personal computers, but can mean the use of mainframe machines in a non-interactive mode. The common thread with all of these is ease of programming.

A digital signal processor may be implemented as software running on a special or general-purpose computer system, or as dedicated hardware. Algorithms which exhibit some symmetry generally give rise to more compact programs. Symmetry is also desirable for hardware but now because it can lead to simple, repetitive subsystem elements which are easily and economically constructed. The demand for symmetry may not be so great in hardware though, because costs may be set by the number of multipliers or elements used. A greater concern can be **minimal operation** algorithms. Software can also have a general requirement that algorithms be well-suited to the general architecture of Von Neumann machines. The use of specialized programmable digital signal processing chips, or array processors in larger machines, makes this less of a concern nowadays. However, this implies an interest in algorithms more suited to the Harvard architecture of the former and the vector processing structure of the latter.

Minimal operation algorithms are certainly important but, in many instances, the demand is more specific than that. The use of simple microprocessors, or the need to use serial multiplier chips, sees a demand for **minimal multiplication** algorithms. For software, all forms of multiplication must be implemented in the same way. For hardware though, scaling can be very easy to implement, consequently interest can be in just minimizing the number of non-scaling operations. Software and hardware are both concerned with round-off errors and overflows from multiplications and divisions, hence another demand related to this topic is minimization of these **finite register effects.**

Operational requirements can set many optimality factors. An obvious one is **time**. Since the total time taken is dependent on the number of and time required for each operation, this requirement is usually expressed via minimal operations. Another is **minimal memory** in the sense of minimal additional memory for partial results as well as minimal storage for constants used within the process.

Apart from these factors, there is a demand for a range of algorithms due to the differing size of tasks. As the number of points N in the computation grow, so the importance of the different optimality factors within a given computation can vary. For example, minimal memory is rarely going to be a problem with N equal to 10, but if images of say 1024×1024 pixels are to be processed, then it can be the dominant factor. Size also introduces new problems associated with memory. For example, it may not be possible to store all the points to be processed within the available memory and that will demand algorithms which can efficiently work within a memory hierarchy of disks, caches and RAM. Another issue along similar lines is partitioning problems for scheduling on array or multiprocessor systems, that is, a requirement for algorithms to be well-tuned to parallel or pipelined computing structures.

Although optimal algorithms have gained considerable prominence in recent years, they are by no means new. Versions of the fast Fourier transform have been traced back to Gauss. For largely technological reasons, optimal has generally meant minimal multiplications. Until quite recently, it was difficult and expensive providing multipliers in all but larger computers. The myriad small machines dedicated to laboratory use, process control, data acquisition, personal use and so forth, had to employ software-implemented multiplication routines. The time required to multiply two numbers was therefore well over an order of magnitude greater than the addition time and so, provided the algorithm did not make much use of secondary storage, computational effort tended to be determined by the multiplication time.

The study of minimal multiplication algorithms has been very successful. Part 4 examines Winograd's theorem, a result giving the theoretically minimum number of multiplications required for the class of mathematical structures to which convolution belongs. Microprocessors and digital signal processors with integral multipliers are now easily produced with multiplication times close to addition. Consequently, there is now more interest in the minimal operations problem. Theoretical advances, though, have not been quite as forthcoming as for the minimal multiplication algorithms. While Parts 3 and 4 discuss means of reducing the number of additions, the theoretical minimum remains unknown.

Three emerging technology trends suggest interesting times lie ahead for convolution and transform theory. One is the development of the specialized but very powerful digital signal processor (DSP) chips. They can usually be organized into arrays and typically have a Harvard architecture sometimes with two input data ports, a pipelined structure with several stages, a multiplier/accumulator and a barrel shifter. Specialized parallel computing elements such as the transputer, which allow richly interconnected arrays to be easily created and programmed, is another. Finally, there are ASICs — application specific integrated circuits. Advanced design tools for these allow relatively inexperienced users to create sophisticated computing elements at quite modest cost. Again, that is likely to lead to array processing elements. Future systems, then, should have a capability for throughputs of hundreds if not thousands of megabytes per second.

The emerging theoretical questions in convolution and transform theory are focusing on algorithms for parallel structures. The freedom semiconductor technology will bring suggests array structures will dominate this work, particularly structures like wavefront or systolic systems. However, there will still be applications where all that is available is a cluster of conventional machines. In addition, the emerging general purpose parallel machines, such as hypercube and data flow machines, cannot be ignored. Optimization here remains a nebulous question but, initially at least, will probably relate to cost/performance issues or speed-up factors. As well, issues such as the development of fault-tolerant arrays will need to be considered.

The study of parallel algorithms for convolution or transforms has considerable theoretical appeal. The mathematical structure of these problems has significant symmetry. In addition, there are large numbers of repetitive operations where the input data for each is not dependent on other operations within the sequence. Developing a practical parallel algorithm frequently centers on how to partition the problem into a series of independent subproblems where each has a comparable computational burden. That seems inherently easier for convolution and transform problems than for others and for this reason, the study of these problems appears to be a useful avenue to the study of parallel algorithms in general. A typical approach is the following. DFT or convolution problems can be expressed as equations like:

$$H(z) = a_0 \cdot z^n + a_1 \cdot z^{n-1} + \cdots + a_{n-1} \cdot z + a_n$$

In recursive form, this is:

$$H(z) = (\cdots ((a_0 \cdot z + a_1) \cdot z + a_2) \cdot z + \cdots + a_n)$$

As the recursive element is an operation ($a_i \cdot z + a_{i+1}$), a sequence of N stages will solve this particular equation and an array of N×N such elements a DFT or convolution. This operation is easily implemented on standard multiply and accumulate chips.

Complex multiplication

Consider two polynomials:

$$X(z) = x_0 + x_1 \cdot z \qquad\qquad H(z) = h_0 + h_1 \cdot z$$

Their product $Y(z)$ is:

$$Y(z) = x_0 \cdot h_0 + (x_0 \cdot h_1 + x_1 \cdot h_0) \cdot z + x_1 \cdot h_1 \cdot z^2$$

Substituting j for z gives the well-known result that the product $(x_0 + jx_1) \cdot (h_0 + jh_1)$ is:

$$Y(j) = (x_0 \cdot h_0 - x_1 \cdot h_1) + j(x_0 \cdot h_1 + x_1 \cdot h_0)$$

Since this result was derived from a polynomial product, any efficient algorithms for multiplying polynomials can be used in the development of efficient algorithms for multiplying complex numbers.

The problem of multiplying polynomials is a recurring one in this book as many direct methods of implementing convolution rely on improved multiplication algorithms. Complex multiplication is a simple polynomial multiplication problem and as such is an ideal introduction to the more general problem discussed in association with the Toom-Cook algorithm of the next section. It is most important, too, to list common algorithms for future reference. Except for some very specialized machines, complex multiplication is implemented as a sequence of real operations. Therefore, the type of complex multiplication algorithm chosen can greatly affect the computational effort required in the overall implementation of an algorithm.

The direct algorithm, that listed above, requires four real multiplications and two real additions to find one complex value. We can more explicitly describe it via the equations:

$$
\begin{array}{ll}
a_0 = x_0 & \qquad b_0 = h_0 \\
a_1 = x_1 & \qquad b_1 = h_1 \\
a_2 = x_0 & \qquad b_2 = h_1 \\
a_3 = x_1 & \qquad b_3 = h_0
\end{array}
$$

$$m_i = a_i \cdot b_i \qquad\qquad i = 0, 1, 2, 3$$

$$
\begin{aligned}
y_0 &= m_0 - m_1 \\
y_1 &= m_2 + m_3
\end{aligned}
$$

and the result is $(y_0 + jy_1)$. These equations can be succinctly expressed in matrix form as:

$$
\begin{bmatrix} y_0 \\ y_1 \end{bmatrix} =
\begin{bmatrix} 1 & -1 & 0 & 0 \\ 0 & 0 & 1 & 1 \end{bmatrix}
\left(
\begin{bmatrix} 1 & 0 \\ 0 & 1 \\ 1 & 0 \\ 0 & 1 \end{bmatrix}
\begin{bmatrix} x_0 \\ x_1 \end{bmatrix}
\bullet
\begin{bmatrix} 1 & 0 \\ 0 & 1 \\ 0 & 1 \\ 1 & 0 \end{bmatrix}
\begin{bmatrix} h_0 \\ h_1 \end{bmatrix}
\right)
$$

where $\bullet$ represents pointwise multiplication of the elements of the two vectors concerned.

In convolution and transform problems, as indeed in many others, one of the two terms in a multiplication is known. That allows some of the terms in an algorithm to be pre-computed. Consequently, it is necessary to distinguish between total operations and essential operations, the latter referring to the number when one element is known. For this algorithm, all operations are essential.

The next section discusses the Toom-Cook algorithm. It gives rise to the set of equations:

$$a_0 = x_0 \qquad\qquad\qquad b_0 = h_0$$
$$a_1 = x_0 + x_1 \qquad\qquad b_1 = h_0 + h_1$$
$$a_2 = x_0 - x_1 \qquad\qquad b_2 = h_0 - h_1$$

$$m_i = a_i \cdot b_i \qquad\qquad i = 0, 1, 2$$

$$y_0 = 2m_0 - \frac{1}{2} \cdot m_1 - \frac{1}{2} \cdot m_2$$

$$y_1 = \frac{1}{2} \cdot m_1 - \frac{1}{2} \cdot m_2$$

and this requires three real multiplications, but five real additions and five scaling operations by powers of 2. That is hardly attractive. However, consider the output terms. We can express these as:

$$y_0 = 2m_0 - \frac{1}{2} \cdot m_1 - \frac{1}{2} \cdot m_2 \qquad = a_0 \cdot (2b_0) + a_1 \cdot (-\frac{1}{2} \cdot b_1) + a_2 \cdot (-\frac{1}{2} \cdot b_2)$$

$$y_1 = \frac{1}{2} \cdot m_1 - \frac{1}{2} \cdot m_2 \qquad = -a_1 \cdot (-\frac{1}{2} \cdot b_1) + a_2 \cdot (-\frac{1}{2} \cdot b_2)$$

and so rephrase the algorithm to:

$$a_0 = x_0 \qquad\qquad\qquad b_0 = 2h_0$$

$$a_1 = x_0 + x_1 \qquad\qquad b_1 = -\frac{1}{2}(h_0 + h_1)$$

$$a_2 = x_0 - x_1 \qquad\qquad b_2 = -\frac{1}{2}(h_0 - h_1)$$

$$m_i = a_i \cdot b_i \qquad\qquad i = 0, 1, 2$$

$$y_0 = m_0 + m_1 + m_2$$
$$y_1 = -m_1 + m_2$$

If the b_i terms are assumed to be pre-calculated, in matrix form these equations become:

$$\begin{bmatrix} y_0 \\ y_1 \end{bmatrix} = \begin{bmatrix} 1 & 1 & 1 \\ 0 & -1 & 1 \end{bmatrix} \begin{bmatrix} b_0 & 0 & 0 \\ 0 & b_1 & 0 \\ 0 & 0 & b_2 \end{bmatrix} \begin{bmatrix} 1 & 0 \\ 1 & 1 \\ 1 & -1 \end{bmatrix} \begin{bmatrix} x_0 \\ x_1 \end{bmatrix}$$

This form of the algorithm requires just three real multiplications, five essential additions and no essential scalings by 2. If real multiplications are a critical factor in framing an algorithm, then this approach to complex multiplication is superior to the initial algorithm.

A heuristic approach can often give a better result than a systematic procedure when the two polynomials to be efficiently multiplied are of small degree. As:

$$x_0 \cdot h_1 + x_1 \cdot h_0 = (x_0 + x_1) \cdot (h_0 + h_1) - x_0 \cdot h_0 - x_1 \cdot h_1$$

we can devise an algorithm:

$$
\begin{aligned}
a_0 &= x_0 \\
a_1 &= x_1 \\
a_2 &= x_0 + x_1
\end{aligned}
\qquad\qquad
\begin{aligned}
b_0 &= h_0 \\
b_1 &= h_1 \\
b_2 &= h_0 + h_1
\end{aligned}
$$

$$
m_i = a_i \cdot b_i \qquad\qquad i = 0, 1, 2
$$

$$
\begin{aligned}
y_0 &= m_0 - m_1 \\
y_1 &= m_2 - m_0 - m_1
\end{aligned}
$$

This also requires three real multiplications and five real additions. In matrix form, it is:

$$
\begin{bmatrix} y_0 \\ y_1 \end{bmatrix} =
\begin{bmatrix} 1 & -1 & 0 \\ -1 & -1 & 1 \end{bmatrix}
\begin{bmatrix} h_0 & 0 & 0 \\ 0 & h_1 & 0 \\ 0 & 0 & (h_0 + h_1) \end{bmatrix}
\begin{bmatrix} 1 & 0 \\ 0 & 1 \\ 1 & 1 \end{bmatrix}
\begin{bmatrix} x_0 \\ x_1 \end{bmatrix}
$$

The objective of these algorithms is to minimize the operations with the unknown terms. Noting that:

$$
\begin{aligned}
y_0 &= x_0 \cdot (h_0 - h_1) + (x_0 - x_1) \cdot h_1 \\
y_1 &= x_1 \cdot (h_0 + h_1) + (x_0 - x_1) \cdot h_1
\end{aligned}
$$

a second heuristic algorithm is:

$$
\begin{aligned}
a_0 &= x_0 \\
a_1 &= x_1 \\
a_2 &= x_0 - x_1
\end{aligned}
\qquad\qquad
\begin{aligned}
b_0 &= h_0 - h_1 \\
b_1 &= h_0 + h_1 \\
b_2 &= h_1
\end{aligned}
$$

$$
m_i = a_i \cdot b_i \qquad\qquad i = 0, 1, 2
$$

$$
\begin{aligned}
y_0 &= m_0 + m_2 \\
y_1 &= m_1 + m_2
\end{aligned}
$$

or in matrix equation form:

$$
\begin{bmatrix} y_0 \\ y_1 \end{bmatrix} =
\begin{bmatrix} 1 & 0 & 1 \\ 0 & 1 & 1 \end{bmatrix}
\begin{bmatrix} (h_0 - h_1) & 0 & 0 \\ 0 & (h_0 + h_1) & 0 \\ 0 & 0 & h_1 \end{bmatrix}
\begin{bmatrix} 1 & 0 \\ 0 & 1 \\ 1 & -1 \end{bmatrix}
\begin{bmatrix} x_0 \\ x_1 \end{bmatrix}
$$

This requires just three essential real multiplications and three essential real additions. It is no different from the original algorithm in terms of total operations but it is superior in terms of the number of multiplications. In general, this is the best all-round algorithm because it has fewer multiplications and it is very easy to implement.

The Toom-Cook algorithm

A relatively simple algorithm which can be employed to minimize the number of multiplications serves as a useful introduction to some ideas and approaches to constructing optimal algorithms. It is known by several names but here will be referred to as the Toom-Cook algorithm. As will be seen, this algorithm has some flaws but in spite of these has proved very useful as the foundation for other algorithms.

Consider two finite length sequences $x(n)$ and $h(n)$, where $x(n)$ is of length N and $h(n)$ is of length M. Their aperiodic convolution is a sequence $y(n)$ whose elements are given by:

$$y(n) = \sum_{m=0}^{N+M-2} h(n-m) \cdot x(m) \qquad\qquad n = 0, 1, \cdots, N+M-2$$

Determining $y(n)$ from this set of equations requires $N \cdot M$ multiplications and $N \cdot (M-1)$ additions. The z-transform of this sequence is:

$$Y(z) = \sum_{n=0}^{N+M-2} y(n) \cdot z^n$$

A problem long studied in mathematics is the interpolation problem. Given a sequence of numbers y_i located at positions z_i, the interpolation problem is to find a polynomial $Y(z)$ passing through these points. Therefore:

$$Y(z_i) = y_i$$

One of the oldest solutions to this problem is to use Lagrange interpolation polynomials, viz:

$$L_i(z) = \prod_{j \neq i} \frac{(z - z_j)}{(z_i - z_j)}$$

where z_i and z_j are interpolation points and z is the polynomial variable. For $(N+M-1)$ interpolation points as in this case, the Lagrange interpolating polynomial will be of degree $(N+M-2)$. An algorithm for multiplying two polynomials is therefore:

1. Choose a set of $(N+M-1)$ sample points. Label these $z_0, z_1, \cdots, z_{N+M-2}$
2. Determine:

$$Y(z_i) = H(z_i) \cdot X(z_i) \qquad\qquad i = 0, 1, \cdots, N+M-2$$

where $H(z)$ and $X(z)$ are the z-transforms of $h(n)$ and $x(n)$ respectively.
3. Form the Lagrange interpolated polynomial:

$$Y(z) = \sum_{i=0}^{N+M-2} Y(z_i) \cdot L_i(z)$$

where $L_i(z)$ are the Lagrange interpolation polynomials given before.

This describes the Toom-Cook algorithm. There are a number of points to discuss in this algorithm but they are amply illustrated via examples.

Example

We need to consider four examples to fully explore the Toom-Cook algorithm; two where $(N+M-1)$ is odd and two where it is even. In both cases we shall examine a simple problem first and then a more complex case leading to some general conclusions.

Let the two polynomials to be multiplied be:

$$X(z) = x_0 + x_1 \cdot z \quad \text{and} \quad H(z) = h_0 + h_1 \cdot z$$

Thus N and M are 2 and $(N+M-1)$ is 3. The Toom-Cook algorithm proceeds in the following stages:

Stage 1 : Choose interpolation points
We need three and we select:

$$z_i = 0, \pm 1$$

Stage 2 : Determine the constant terms
Substituting the interpolation points z_i into the expressions for $X(z_i)$ and $H(z_i)$, we obtain:

$$
\begin{array}{ll}
X(0) = x_0 & H(0) = h_0 \\
X(1) = x_0 + x_1 & H(1) = h_0 + h_1 \\
X(-1) = x_0 - x_1 & H(-1) = h_0 - h_1
\end{array}
$$

Stage 3 : Form the product terms

$$
\begin{aligned}
m_0 &= X(0) \cdot H(0) \\
m_1 &= X(1) \cdot H(1) \\
m_2 &= X(-1) \cdot H(-1)
\end{aligned}
$$

Stage 4 : Form the interpolation polynomials
The Lagrange polynomials are:

$$L_0(z) = \frac{(z-1) \cdot (z+1)}{(0-1) \cdot (0+1)} = 1 - z^2$$

$$L_1(z) = \frac{1}{2} \cdot z^2 + \frac{1}{2} \cdot z$$

$$L_{-1}(z) = \frac{1}{2} \cdot z^2 - \frac{1}{2} \cdot z$$

Stage 5 : Form the result
The interpolated polynomial is:

$$Y(z) \quad = \quad m_0{\cdot}L_0(z) \quad + m_1{\cdot}L_1(z) \quad + m_2{\cdot}L_{-1}(z)$$

$$= \quad (\quad m_0 \quad + \quad 0 \quad + \quad 0 \quad)$$

$$+ \quad (\quad 0 \quad + \quad \frac{1}{2}{\cdot}m_1 \quad - \quad \frac{1}{2}{\cdot}m_2 \quad){\cdot}z$$

$$+ \quad (\quad -m_0 \quad + \quad \frac{1}{2}{\cdot}m_1 \quad + \quad \frac{1}{2}{\cdot}m_2 \quad){\cdot}z^2$$

which is the desired result (and the second complex multiplication algorithm above).

For a second example, choose polynomials:

$$X(z) = x_0 + x_1{\cdot}z + x_2{\cdot}z^2 + x_3{\cdot}z^3 + x_4{\cdot}z^4 \qquad \text{and} \qquad H(z) = h_0 + h_1{\cdot}z + h_2{\cdot}z^2$$

Thus N is 5, M is 3 and (N + M −1) is 7. Choose interpolation points $z_i = 0, \pm1, \pm2, \pm3$. The constant terms are therefore:

$$
\begin{aligned}
X(0) &= x_0 \\
X(1) &= x_0 + x_1 + x_2 + x_3 + x_4 \\
X(-1) &= x_0 - x_1 + x_2 - x_3 + x_4 \\
X(2) &= x_0 + 2x_1 + 4x_2 + 8x_3 + 16x_4 \\
X(-2) &= x_0 - 2x_1 + 4x_2 - 8x_3 + 16x_4 \\
X(3) &= x_0 + 3x_1 + 9x_2 + 27x_3 + 81x_4 \\
X(-3) &= x_0 - 3x_1 + 9x_2 - 27x_3 + 81x_4
\end{aligned}
\qquad
\begin{aligned}
H(0) &= h_0 \\
H(1) &= h_0 + h_1 + h_2 \\
H(-1) &= h_0 - h_1 + h_2 \\
H(1) &= h_0 + 2h_1 + 4h_2 \\
H(-1) &= h_0 - 2h_1 + 4h_2 \\
H(1) &= h_0 + 3h_1 + 9h_2 \\
H(-1) &= h_0 - 3h_1 + 9h_2
\end{aligned}
$$

The product terms are:

$$
\begin{aligned}
m_0 &= X(0){\cdot}H(0) \\
m_1 &= X(1){\cdot}H(1) \\
m_2 &= X(-1){\cdot}H(-1) \\
m_3 &= X(2){\cdot}H(2) \\
m_4 &= X(-2){\cdot}H(-2) \\
m_5 &= X(3){\cdot}H(3) \\
m_6 &= X(-3){\cdot}H(-3)
\end{aligned}
$$

The Lagrange polynomials are:

$$L_0(z) = \frac{(z-1){\cdot}(z+1){\cdot}(z-2){\cdot}(z+2){\cdot}(z-3){\cdot}(z+3)}{(0-1){\cdot}(0+1){\cdot}(0-2){\cdot}(0+2){\cdot}(0-3){\cdot}(0+3)}$$

$$= -\frac{1}{36}{\cdot}(z^6 \quad - 14z^4 \quad + 49z^2 \quad - 36)$$

$$L_1(z) = \frac{1}{48}{\cdot}(z^6 + z^5 - 13z^4 - 13z^3 + 36z^2 + 36z)$$

$$L_{-1}(z) = \frac{1}{48} \cdot (z^6 - z^5 - 13z^4 + 13z^3 + 36z^2 - 36z)$$

$$L_2(z) = -\frac{1}{120} \cdot (z^6 + 2z^5 - 10z^4 - 20z^3 + 9z^2 + 18z)$$

$$L_{-2}(z) = -\frac{1}{120} \cdot (z^6 - 2z^5 - 10z^4 + 20z^3 + 9z^2 - 18z)$$

$$L_3(z) = \frac{1}{720} \cdot (z^6 + 3z^5 - 5z^4 - 15z^3 + 4z^2 + 12z)$$

$$L_{-3}(z) = \frac{1}{720} \cdot (z^6 - 3z^5 - 5z^4 + 15z^3 + 4z^2 - 12z)$$

The result is therefore:

$$Y(z) = m_0 \cdot L_0(z) + m_1 \cdot L_1(z) + m_2 \cdot L_{-1}(z) + m_3 \cdot L_2(z) + m_4 \cdot L_{-2}(z) + m_5 \cdot L_3(z) + m_6 \cdot L_{-3}(z)$$

$$= (m_0 + 0 + 0 + 0 + 0 + 0 + 0)$$

$$+ \frac{1}{60} \cdot (0 + 45m_1 - 45m_2 - 9m_3 + 9m_4 + m_5 - m_6) \cdot z$$

$$+ \frac{1}{360} \cdot (-490m_0 + 270m_1 + 270m_2 - 27m_3 - 27m_4 + 2m_5 + 2m_6) \cdot z^2$$

$$+ \frac{1}{48} \cdot (0 - 13m_1 + 13m_2 + 8m_3 - 8m_4 - m_5 + m_6) \cdot z^3$$

$$+ \frac{1}{144} \cdot (56m_0 - 39m_1 - 39m_2 + 12m_3 + 12m_4 - m_5 - m_6) \cdot z^4$$

$$+ \frac{1}{240} \cdot (0 + 5m_1 - 5m_2 - 4m_3 + 4m_4 + m_5 - m_6) \cdot z^5$$

$$+ \frac{1}{720} \cdot (-20m_0 + 15m_1 + 15m_2 - 6m_3 - 6m_4 + m_5 + m_6) \cdot z^6$$

Using these two examples for guidance, the number of operations in the Toom-Cook algorithm for $(N + M - 1)$ odd can be calculated. Thus:

1. The constant terms require:

   $(N + M - 2) \cdot (N - 1)$ additions for the $X(z_i)$ terms
   $(N + M - 2) \cdot (M - 1)$ additions for the $H(z_i)$ terms

   or $(N + M - 2) \cdot (N + M - 2)$ in total. If $(N + M - 1)$ exceeds three, then there are also a number of multiplications by constants. In fact:

   $(N + M - 4) \cdot (N - 1)$ for the $X(z_i)$ terms
   $(N + M - 4) \cdot (M - 1)$ for the $H(z_i)$ terms

   or $(N + M - 4) \cdot (N + M - 2)$ in total.
2. The product stage requires $(N + M - 1)$ general multiplications.

3. The final stage involves the product of $(N+M-1)$ coefficients and polynomials, where the polynomials are of order $(N+M-1)$. That requires:

$$(N+M-2)\cdot(N+M-2)-\frac{1}{2}\cdot(N+M-2)\ \text{ additions}$$

$$(N+M-2)\cdot(N+M-2)+\frac{1}{2}\cdot(N+M-2)\ \text{ multiplications}$$

Overall, the algorithm requires:

1. $2(N+M-2)\cdot(N+M-\frac{9}{4})$ additions;

2. $(N+M-2)\cdot(N+M-\frac{3}{2})$ multiplications by constants if $(N+M-1)\leq 3$ and

    $2(N+M-2)\cdot(N+M-\frac{11}{4})$ otherwise;

3. $(N+M-1)$ general multiplications.

If one polynomial is known, some terms can be pre-computed. The number of operations falls to :

1. $(N+M-2)\cdot(2N+M-\frac{7}{2})$ additions;

2. $(N+M-2)\cdot(N+M-\frac{3}{2})$ multiplications by constants if $(N+M-1)\leq 3$ and

    $(N+M-2)\cdot(2N+M-\frac{5}{2})-2(N-1)$ otherwise;

3. $(N+M-1)$ general multiplications.

The second two examples cover $(N+M-1)$ even. This suggests that $(N+M-2)$ of the interpolation points be chosen as:

$$z_i\ =\ 0,\pm1,\pm2,\pm3,\ \cdots,\ \pm\frac{(N+M-3)}{2}$$

and the remaining point be chosen as either $\frac{(N+M-1)}{2}$ or $-\frac{(N+M-1)}{2}$. While there is no reason not to do this, it introduces an asymmetric term with a number of multiplications by constants. It can be avoided by choosing infinity as an interpolation point. How, though, can this be accommodated when the interpolation polynomials are of the form:

$$L_i(z)=\frac{(z-z_0)\cdot\ \cdots\ \cdot(z-z_{i-1})\cdot(z-z_{i+1})\cdot\ \cdots\ \cdot(z-z_{N+M-2})}{(z_i-z_0)\cdot\ \cdots\ \cdot(z_i-z_{i-1})\cdot(z_i-z_{i+1})\cdot\ \cdots\ \cdot(z_i-z_{N+M-2})}$$

Clearly, infinity is a pole of order $(N+M-2)$ of $H(z)\cdot X(z)$ and a zero of the same order of $L_i(z)$. We can solve this problem by taking limits with respect to $Y(z)$. That allows us to define $L_\infty(z)$ as:

$$L_\infty(z) = (z - z_0) \cdot \cdots \cdot (z - z_{i-1}) \cdot (z - z_{i+1}) \cdot \cdots \cdot (z - z_{N+M-2})$$

and $Y(\infty)$ as:

$$Y(\infty) = H(\infty) \cdot X(\infty)$$

where $\quad X(\infty) = \lim_{z \to \infty} \dfrac{X(z)}{z^N} = x_{N-1}$

$$H(\infty) = \lim_{z \to \infty} \frac{H(z)}{z^M} = h_{M-1}$$

As the remaining interpolation polynomials have a simple zero and simple pole due to the infinite term, the limit also extends to these.

For the first example, choose:

$$X(z) = x_0 + x_1 \cdot z \qquad \text{and} \qquad H(z) = h_0 + h_1 \cdot z + h_2 \cdot z^2$$

Thus N is 2, M is 3 and ($N + M - 1$) is 4. The stages of an even algorithm follow those for an odd, thus the first step is to choose interpolation points. Select $z_i = 0, \pm 1, \infty$. The constant terms are:

$$
\begin{array}{ll}
X(0) = x_0 & H(0) = h_0 \\
X(1) = x_0 + x_1 & H(1) = h_0 + h_1 + h_2 \\
X(-1) = x_0 - x_1 & H(-1) = h_0 - h_1 + h_2 \\
X(\infty) = x_1 & H(\infty) = h_2
\end{array}
$$

The product terms are:

$$
\begin{array}{rcl}
m_0 & = & X(0) \cdot H(0) \\
m_1 & = & X(1) \cdot H(1) \\
m_2 & = & X(-1) \cdot H(-1) \\
m_3 & = & X(\infty) \cdot H(\infty)
\end{array}
$$

The interpolation polynomials are:

$$L_0(z) = \qquad\qquad -z^2 \qquad\qquad\qquad + 1$$

$$L_1(z) = \qquad\qquad \tfrac{1}{2} \cdot z^2 \quad + \quad \tfrac{1}{2} z$$

$$L_{-1}(z) = \qquad\qquad \tfrac{1}{2} \cdot z^2 \quad - \quad \tfrac{1}{2} z$$

$$L_\infty(z) = z \cdot (z - 1) \cdot (z + 1) = z^3 \qquad\qquad - \qquad z$$

Hence:

$$Y(z) = m_0 \cdot L_0(z) + m_1 \cdot L_1(z) + m_2 \cdot L_{-1}(z) + m_3 \cdot L_\infty(z)$$

$$= (\ m_0 \quad + \qquad 0 \quad + \qquad 0 \ + \ 0 \ )$$

$$+ (\quad 0 \quad + \ \tfrac{1}{2} \cdot m_1 \quad - \quad \tfrac{1}{2} \cdot m_2 \quad - \ m_3 \ ) \cdot z$$

$$+ (-m_0 \ + \ \tfrac{1}{2} \cdot m_1 \quad + \quad \tfrac{1}{2} \cdot m_2 \ + \ 0 \ ) \cdot z^2$$

$$+ (\quad 0 \quad + \qquad 0 \quad + \qquad 0 \ + \ m_3 \ ) \cdot z^3$$

For a second example, choose polynomials:

$$X(z) = x_0 + x_1 \cdot z + x_2 \cdot z^2 + x_3 \cdot z^3 \qquad \text{and} \qquad H(z) = h_0 + h_1 \cdot z + h_2 \cdot z^2$$

Thus N is 4, M is 3 and (N + M − 1) is 6. Choose the interpolation points as $z_i = 0, \pm1, \pm2, \infty$. The constant terms in this case are:

$$
\begin{aligned}
X(0) &= x_0 & H(0) &= h_0 \\
X(1) &= x_0 + x_1 + x_2 + x_3 & H(1) &= h_0 + h_1 + h_2 \\
X(-1) &= x_0 - x_1 + x_2 - x_3 & H(-1) &= h_0 - h_1 + h_2 \\
X(2) &= x_0 + 2x_1 + 4x_2 + 8x_3 & H(2) &= h_0 + 2h_1 + 4h_2 \\
X(-2) &= x_0 - 2x_1 + 4x_2 - 8x_3 & H(-2) &= h_0 - 2h_1 + 4h_2 \\
X(\infty) &= x_3 & H(\infty) &= h_2
\end{aligned}
$$

The product terms giving the general multiplications are:

$$
\begin{aligned}
m_0 &= X(0) \cdot H(0) \\
m_1 &= X(1) \cdot H(1) \\
m_2 &= X(-1) \cdot H(-1) \\
m_3 &= X(2) \cdot H(2) \\
m_4 &= X(-2) \cdot H(-2) \\
m_5 &= X(\infty) \cdot H(\infty)
\end{aligned}
$$

The interpolation polynomials are:

$$L_0(z) = \tfrac{1}{4} \cdot (z^4 \qquad\quad - \ 5z^2 \qquad\quad + 4)$$

$$L_1(z) = -\tfrac{1}{6} \cdot (z^4 + \ z^3 \ - \ 4z^2 \ - \ 4z)$$

$$L_{-1}(z) = -\tfrac{1}{6} \cdot (z^4 - \ z^3 \ - \ 4z^2 \ + \ 4z)$$

$$L_2(z) = \tfrac{1}{24} \cdot (z^4 + \ 2z^3 \ - \ z^2 \ - \ 2z)$$

$$L_{-2}(z) = \tfrac{1}{24} \cdot (z^4 - \ 2z^3 \ - \ z^2 \ + \ 2z)$$

$$L_\infty(z) = (z^5 \qquad\quad - \ 5z^3 \qquad\quad + \ 4z)$$

Hence:

$$Y(z) = m_0 \cdot L_0(z) + m_1 \cdot L_1(z) + m_2 \cdot L_{-1}(z) + m_3 \cdot L_2(z) + m_4 \cdot L_{-2}(z) + m_5 \cdot L_\infty(z)$$

$$
\begin{aligned}
= \; & (& m_0 & + & 0 & + & 0 & + & 0 & + & 0 & + & 0 &) \\
& + \tfrac{1}{12} \cdot (& 0 & + & 8m_1 & - & 8m_2 & - & m_3 & + & m_4 & + & 48m_5 &) \cdot z \\
& + \tfrac{1}{24} \cdot (& -30m_0 & + & 16m_1 & + & 16m_2 & - & m_3 & - & m_4 & + & 0 &) \cdot z^2 \\
& + \tfrac{1}{12} \cdot (& 0 & - & 2m_1 & + & 2m_2 & + & m_3 & - & m_4 & - & 60m_5 &) \cdot z^3 \\
& + \tfrac{1}{24} \cdot (& 6m_0 & - & 4m_1 & - & 4m_2 & + & m_3 & + & m_4 & + & 0 &) \cdot z^4 \\
& + (& 0 & + & 0 & + & 0 & + & 0 & + & 0 & + & m_5 &) \cdot z^5
\end{aligned}
$$

Again, these results can be generalized and the number of operations for the Toom-Cook algorithm for $(N+M-1)$ even computed. Here:

1. In forming the constants, there are:

    $(N+M-3) \cdot (N-1)$ additions for the $X(z_i)$ terms
    $(N+M-3) \cdot (M-1)$ additions for the $H(z_i)$ terms

    or $(N+M-3) \cdot (N+M-2)$ in total. For $(N+M-1)$ greater than 4, there are also:

    $(N+M-5) \cdot (N-1)$ multiplications by constants to form the $X(z_i)$ terms
    $(N+M-5) \cdot (M-1)$ multiplications by constants to form the $H(z_i)$ terms

    That is, $(N+M-5) \cdot (N+M-2)$ multiplications in all.
2. There are $(N+M-1)$ general multiplications.
3. Forming the final result requires:

    $(N+M-3) \cdot (N+M-3)$ additions
    $(N+M-3) \cdot (N+M-2)$ multiplications by constants

Overall, the algorithm requires:

1. $2(N+M-3) \cdot (N+M-\tfrac{5}{2})$ additions;
2. $(N+M-3) \cdot (N+M-2)$ multiplications by constants if $(N+M-1) \le 4$, and
    $2(N+M-2) \cdot (N+M-4)$ otherwise;
3. $(N+M-1)$ general multiplications.

If one term is known, this reduces to:

1. $(N + M - 3)\cdot(2N + M - 4)$ additions;
2. $(N + M - 3)\cdot(N + M - 2)$ multiplications by constants if $(N + M - 1) \leq 4$, and $(N + M - 3)\cdot(2N + M - 3) - 2(N - 1)$ otherwise;
3. $(N + M - 1)$ general multiplications.

In one very important sense, Toom-Cook is an optimal algorithm for multiplying two polynomials. It requires $(N + M - 1)$ essential multiplications whereas a direct approach requires $N\cdot M$. However, it also requires a large number of additions and multiplications by constants. If it is not possible to distinguish these from essential multiplications as, for example, in software, then Toom-Cook is quite inferior to the direct method. Clearly, this is a substantial disadvantage for many implementations.

It is interesting to observe all four cases in the example can be expressed in matrix form as:

$$\mathbf{y} = C(A\mathbf{x}\bullet B\mathbf{h})$$

where $\bullet$ is pointwise multiplication of vector elements. If, say, $\mathbf{h}$ is known, this becomes:

$$\mathbf{y} = CDA\mathbf{x}$$

where D is a diagonal matrix.

DISCUSSION

The purpose of this section was merely to review but in such a manner as to set the context for the book. The essential problem is one of executing digital convolution in an efficient way. Not all digital convolution, though, but only that exhibiting the key properties of linearity, causality and shift-invariance. It was shown that convolution of this type could be expressed as cyclic convolution and, while to some degree an abstraction, cyclic convolution has some desirable properties for the construction of algorithms.

A key point in the execution of digital convolution is how to view that operation. Here, the two most fruitful approaches are to regard it as either a matrix operation or as a problem of multiplying two polynomials. For the question of how to actually perform the operation, again there are two principal approaches. One is to examine the equations directly, to examine matrix or polynomial multiplication directly to see how optimal algorithms could evolve. The other is to use the cyclic convolution property of some transform and convert the problem into an alternative form.

One of the issues raised in this section which bears repeated comment is the question of optimality. Optimization implies minimizing or maximizing some performance criterion which is itself constructed from a range of measures. These measures will vary from application to application; there is no one optimal problem. At a minimum, optimal algorithms for digital convolution must be grouped into hardware or software-oriented techniques. However, what the section has tried to stress is that optimality covers numerous issues. Interest to date has been in minimal multiplication algorithms but is turning to minimal operation algorithms and is certain to settle into optimal algorithms for parallel computing systems for much of the foreseeable future.

FURTHER STUDY

References

This book assumes a good working knowledge of introductory digital signal processing. It also assumes some experience of digital filter design, an understanding of the z-transform and a knowledge of the DFT at the level discussed in, among others, the following books:

R.D. STRUM and D.E. KIRK, *First Principles of Discrete Systems and Digital Signal Processing*, Addison Wesley, 1988

W.D. STANLEY, G.R. DOUGHERTY, and R. DOUGHERTY, *Digital Signal Processing*, Reston Publishing Company, 2nd edition, 1984

R.A. ROBERTS and C.T. MULLIS, *Digital Signal Processing*, Addison Wesley, 1987

S.D. STEARNS and R.A. DAVID, *Signal Processing Algorithms*, Prentice Hall, 1988

A. BATEMAN and W. YATES, *Digital Signal Processing Design*, Pitman Publishing Company, 1988

Stearns and David also includes a floppy disk of software. Bateman and Yates is interesting in that it delves into DSP technology and associated practical issues.

An excellent article which reviews the history of the FFT is:

M.T. HEIDEMAN, D.H. JOHNSON, and C.S. BURRUS, Gauss and the History of the Fast Fourier Transform, *I.E.E.E. Acoustics, Speech and Signal Processing Magazine*, V 1 N 4 , Oct 1984

It includes 47 references to papers of historical interest.

Exercises

1. Early in the text, digital convolution was described by two summations but, in the four special cases discussed, only one of these was ever used. Derive the other summation for each of these cases.

2. In the Toom-Cook algorithm, interpolation points were chosen as $z_i = 0, \pm 1, \pm 2, \pm 3, \cdots$. Why is this choice preferable to $z_i = 0, 1, 2, 3, \cdots, (N + M - 2)$?

3. The discussion on the number of operations required for the direct and transform approaches assumes a processor acting on a stream of input data and generating a stream of output data. Derive formulae for each case when the processor is a general-purpose computer. That is, where all data and intermediate values must be held in random access memory. Does this significantly alter the balance between the two?

4. Prove that, for aperiodic convolution:

$$Y(z) = H(z) \cdot X(z)$$

5. Show that if one sequence in a digital convolution is reversed the result is digital correlation. Examine the special cases again. Does this result mean that any optimal algorithm for cyclic convolution can also be applied to digital correlation?

6. Consider some random digital signal $x(n)$ where $-\infty < n < \infty$. Assume the mean of this signal is zero. The mean square value is:

$$\sigma_x^2 = \sum_{n=-\infty}^{\infty} x^2(n) - \mu$$

Assume this is applied to a causal, shift-invariant, linear system with a unit sample response of $h(n)$. Prove that the mean square value of the output of this system is:

$$\sigma_y^2 = \sigma_x^2 \cdot \sum_{n=0}^{\infty} h^2(n)$$

7. Sampling the unit circle in the z-plane at equi-spaced intervals given by:

$$z_n = (\frac{2\pi}{N}) \cdot n \qquad\qquad n = 0, 1, \cdots, N-1$$

results in the discrete Fourier transform. What results if the unit circle is sampled at:

$$z_n = (\frac{2\pi}{N}) \cdot (n + \frac{1}{2}) \qquad\qquad n = 0, 1, \cdots, N-1$$

8. (a) Does stability imply that:

$$\lim_{n \to \infty} h(n) = 0$$

   (You may care to investigate Schwartz' inequality.)
   (b) Do the conditions of linearity, shift-invariance and causality demand a changing output? That is, is a system which accepts an input but gives a constant output — say a failed system — linear, shift-invariant and causal?

9. Numbers are usually written as strings of symbols drawn from some finite set. For example, the decimal system uses ten symbols or **digits** and numbers become strings like '1947'. Clearly, numbers can be regarded as digital sequences. Show that multiplication of numbers is a convolution operation.

10. If a continuous signal is to be processed, this can be done by digitizing that signal via an analog-to-digital converter and then processing via some digital signal processor. Indeed, this is a very common practice. However, consider the process of digitization. There are two ways this can be done. In one, given a quantization step q, the quantization error is always $0 \leq \text{error} \leq q$. The process is similar to the real-to-integer conversion process in computers. In the other, the quantization error is always $-\frac{q}{2} \leq \text{error} \leq \frac{q}{2}$. The process is similar to rounding of real numbers. Now consider an input signal $k_0 t$ to the second type of system, where t is just time and k_0 some scaling value. Here, an output can occur before the appropriate input value. That suggests non-causality. Is this the case and does it mean this approach cannot be used for the processing of analog signals?

PROBLEMS

1. Compact audio discs provide a digital output (usually 16 bit). Assume a particular manufacturer wishes to continuously display the signal spectrum on the front of a player as a sales feature. Perhaps this would be done with some sort of bar display with each bar corresponding to one octave. How can frequency be derived to give a realistic (i.e. accurate) result? What do you mean by 'realistic'?

2. The formula quoted in the text assumes quantization noise can be described by a uniform noise distribution of magnitude $\frac{1}{Q}$ over ($-\frac{Q}{2}, \frac{Q}{2}$) where Q is the quantization step.

   (a) Prove the formula given.
   (b) A uniform distribution seems a little unrealistic. (Can it be justified as an upper bound?) Derive signal to quantization noise for a triangular distribution.
   (c) A Gaussian distribution is clearly unrealistic (why) but a Cauchy distribution looks similar and can be used. Derive the signal to quantization noise ratio for this.

   Comment on your results.

3. Applications which use pipelined structures require a high throughput but can tolerate a delay between input and output. Sketch some possible pipelined structures for signal processing. From these, list performance criteria which may apply.

4. Repeat problem 4 for parallel systems.

5. Let T represent the addition time of some processor. Let its multiplication time be αT and its memory reference time be βT. Assume that some optimal algorithm is found that requires:

   N^γ real multiplications
   $N^{\gamma-1} \cdot (N-1)$ real additions

per N output points. (What can be assumed about memory references?) Can you derive a formula which determines the optimum value of N? Optimum here is in the sense of minimizing the computation time per output point. How does this formula differ for overlap and add and overlap and save? Collect data on typical microprocessors and digital signal processing chips and produce plots for different values of γ. What do these suggest about the search for (time) optimal algorithms?

6. Consider the Toom-Cook algorithm. Now the Lagrange polynomials are acting as low-pass filters recovering a continuous signal from the digital $Y(z_i)$ values. Is this a reasonable interpretation? If so, then:

   (a) Lagrange polynomials are not an ideal low-pass filter function and that suggests information is being lost. Is that so, and if so, what information?
   (b) Why use Lagrange polynomials? Window functions as used in FIR digital filter design have a better low-pass response, so can they be used instead? If so, devise a modified Toom-Cook using one of these and discuss the advantages and disadvantages with respect to Toom-Cook. (It may be a good idea to read a mathematics text on interpolation.)

7. A point of some concern in practice is the sensitivity of algorithms to coefficient error. Many optimal algorithms are quite sensitive. Assuming these errors can be modeled by a uniform source of noise, analyze the complex multiplication routines given and comment on the results.

8. A wavefront processor is, as the name suggests, an array of simple computing elements whereby data passes through the structure undergoing much the same operation at e ach stage. Given the form of a Toeplitz matrix, devise a wavefront processor for convolution. Note that a desirable feature of such a structure is a very low level of cross-coupling between elements. Comment on your result with respect to this.

9. Under what circumstances can discrete convolution be used to compute continuous convolutions?

10. For hardware implementations, multiplication by a constant is extremely simple if the constant is a power of two. It is then just a wiring connection. Then:

   (a) Examine the Toom-Cook algorithm for interpolation at points:

   $$z_i = 0, \pm 1, \pm 2, \pm 2^2, \cdots, \pm 2^n$$

   Is this possible? Assuming all multiplications by powers of 2 can be ignored, does this lead to a significant reduction in operations?
   (b) Can you devise a scheme for increasing the number of constants in $Y(z)$ with equal powers of two?

THE MATHEMATICAL FOUNDATIONS

The effective practice of any activity requires both a thorough understanding of the subject and the skilled use of appropriate tools. One of the principal tools of digital signal processing is mathematics. As with all tools, the questions to direct at mathematics must concern utility and flexibility. Here, digital signal processing is pioneering the practical application of some key sectors of mathematics. Underpinning much of the advances of this very new branch is over a century of investigation by mathematicians seeking to find the most abstract and subtle characteristics of mathematics. The tools which we shall employ here are part of abstract and linear algebra, number theory and similar topics still regarded as at the core of pure mathematics. Rarely are these topics considered in applications-oriented fields such as engineering. Courses on matrices usually skip over the linear algebraic foundations in favor of the more practical concerns of inversion, matrix equations and the like. Coding theory and information theory share this concern with abstract algebra and number theory but neither has been at the center of communications engineering.

A first glance at abstract algebra tends to induce thoughts along the lines of it being too abstruse, too 'exotic' for those with a practical bent. The value of a tool, though, is in what it can do not in how it appears. We will show that as a tool these topics are immensely useful. Therefore, it behoves us to learn something of them and how they may be used. A cautionary word. From a mathematical viewpoint, the material presented here is abbreviated. A number of important mathematical concepts are just introduced and proofs, if given, are cursory. The object, though, is simply to introduce an important tool and show how it may be applied, not to meet given standards of scholarship and rigor.

This section is termed mathematical foundations. In keeping with the first section of this part, it looks at the central framework of all the mathematical procedures underlying this book. Although undoubtedly much is familiar to most readers, this section does not assume quite the prior experience of the last section. Let us stress, too, the term 'framework'. There will be further mathematical development but it will be filling in the detail.

ABSTRACT MATHEMATICAL STRUCTURES

Introduction

Around 150 years ago, mathematicians began to study **mathematical structures** — the systems used for mathematical manipulations. These investigations sought to unify the different branches of mathematics and, at the same time, to clarify the various techniques employed. One desirable outcome of such work was to enable useful methods developed in one branch of mathematics to be employed in others. Any process of unification and clarification like this is also a process of **abstraction**. Naturally then, the objects of these studies were defined as **abstract systems** and a key theory evolved to study such systems was termed **abstract algebra**.

Digital signal processing has an interest in abstract systems and abstract algebra for a number of reasons. One to stress is that a theory of abstract structures makes it possible to propose and examine entirely new structures never before contemplated. This is an important freedom. There is no reason to persist with particular operations if, via abstract analysis, more easily implemented but more generalized or abstract versions of addition, subtraction, multiplication or division, or any other mathematical operation for that matter, can be created. Alternatively, abstract analysis may show ways of creating systems with particular desirable properties. This is the sort of role a tool should be playing and because of the practical significance of this, it warrants close attention.

Abstract systems

Set theory plays a most important role in abstract systems theory. Recall a set is merely a collection or aggregation of objects. There is no need to be more explicit than this. Then an **abstract system** comprises:

1. a set E of **elements** which are manipulated within the system;
2. a set R of **relations** between elements which are mainly concerned with how different elements can be distinguished;
3. a set O of **operations** describing how elements can be combined to produce new elements;
4. a set of **postulates** stating properties of the system taken to be true;
5. a set of **theorems** derived from the postulates and constituting a body of knowledge about the system;
6. a set of **definitions** for defining notation and generally describing the system.

Since the key components are the sets E, R and O, abstract systems are often described as just a system {E, R, O}.

Let us consider these various constituents a little more deeply beginning with sets. As mentioned, a set is just an aggregation of elements. An element b belonging to some set B is denoted as $b \in B$. All elements of a set can be denoted by:

1. listing those elements, such as $\{1, 2, 3, 4, 5\}$;
2. stating some common property of the elements as in { x is an integer ; x>0, x<10^6).

The set without any elements at all is the **null set**, denoted {} or Ø. Sets are equal only if they contain exactly the same number of elements. This relationship is denoted by:

$$A = B$$

Sets may also be divided into **subsets**. If a set A is a subset of a set B, then this is denoted A⇐B. Ø and B are always subsets of a set B. A more strict definition is that of a **proper subset** where:

$$A \Leftarrow B \qquad \text{and} \qquad A \neq B$$

That is, while all the elements of A are elements of B, as they should be for a subset, there are some elements of B not members of A.

Given two sets A and B, a new set S may be formed via the three principal operations of:

1. UNION $A \cup B = \{ x \in S : x \in A \text{ or } x \in B \}$
2. INTERSECTION $A \cap B = \{ x \in S : x \in A \text{ and } x \in B \}$
3. COMPLEMENTATION $A' = \{ x \in S : x \notin A \}$

Another set operation of considerable importance is the **Cartesian product**. Given a collection of sets $S_1, S_2, \cdots, S_N$, the Cartesian product, denoted $S_1 \times S_2 \times \cdots \times S_N$, is the set of all ordered n-tuples. That is, the set given by:

$$\{ (s_1, s_2, \cdots, s_N) : s_i \in S_i \text{ for } i = 1, 2, \cdots, N \}$$

The set of relations in an abstract structure usually consists of **binary relations**. A binary relation R from a set A into a set B is a subset of A×B. A very important class of binary relations on the set A are those comprising a subset of A×A. Now the **domain** of a binary relation R from a set A into a set B is the set S of all elements of A related to at least one element of B. More precisely, the domain is the set $\{ a \in S : (a,b) \in R \text{ for } b \in B \}$. Similarly, the **range** of R is the set of all elements of B to which at least one element of A is related by R. That is, the range is the set $\{ b \in S : (a,b) \in R \text{ for } a \in A \}$.

Binary relations are useful for defining abstract equalities. Three of particular concern are:

1. REFLEXIVITY $(a,a) \in R$ for $a \in S$
2. SYMMETRY   if $(a,b) \in R$ then $(b,a) \in R$
3. TRANSITIVITY   if $(a,b) \in R$ and $(b,c) \in R$ then $(a,c) \in R$

Any relation exhibiting all three of these is termed an **equivalence relation**.

In set theory, a **function** or **mapping** F is a binary relation with the two additional properties:

1. the domain of F is not the null set;
2. if $(a,b) \in F$ and $(a,c) \in F$ then b equals c.

It is common to denote the mapping or function as F(a). If several mappings are made, this notation allows the very simple notation for the final result of $F_1(F_2(\cdots (a) \cdots))$. Such a complex grouping of functions is called a **composition**. Now a function maps elements of A into B but it may not be unique. That is, several elements of A may map into the same element of B. If all the elements of A are mapped into all the elements of B, then F is termed **onto**. If each element of A is mapped into just one element of B, then the mapping is **one to one**. If a mapping is both one to one and onto, another mapping F* can be found such that:

$$F^*(F(a)) = a$$

Thus F is called a **reversible** mapping.

Like relations, operations in abstract structures are closely linked with the concept of Cartesian products. However, whereas relations serve to classify given elements, operations combine them in various ways. Operations themselves are categorized as:

1.   UNARY a mapping from A into itself;
2.   BINARY a mapping from A×A into A;
3.   N-ARY a mapping from A×A×A× ⋯ ×A into A;

and they are usually denoted by a special symbol. Some may be **closed** which simply means the domain of the operation is the full set of the Cartesian product.

Binary operations are extremely important. Consider three elements { a, b, c } drawn from a set A. Consider further some binary operation • from A×A onto A. Then this operation can have three key properties. They are:

1.   ASSOCIATIVITY $a•(b•c)= (a•b)•c$

2.   COMMUTATIVITY $a•b = b•a$

3.   DISTRIBUTIVITY   A second operation ° is distributive over • if and only if
$$a°(b•c) = (a°b)•(a°c)$$
$$(b•c)°a = (b°a)•(c°a)$$

Postulates are mainly concerned with the existence of certain elements within a set A. The elements of particular interest are the following:

1.   IDENTITY
With respect to some operation •, the identity i is some element with the property that

$$i•x = x•i = x \qquad\qquad \text{for all } x \in A$$

2.   INVERSE
Given that an identity exists for some operation •, then an inverse x' exists for an element x of A if

$$x•x' = x'•x = i$$

Inverses are often denoted x^{-1}.

3.   IDEMPOTENT
An idempotent element of a set A is an element x with the property that with respect to some operation •

$$x•x = x$$

An example of an abstract system

Boolean algebras are very simple examples of an abstract system. They have:

1. a set of elements;
2. two binary operations $\wedge$ and $\vee$ which are
    (i) commutative;
    (ii) distributive over each other;
    and both have identities;
3. a unary operation * such that:

$$x \wedge x^* = \text{the identity of } \wedge$$
$$x \vee x^* = \text{the identity of } \vee$$

Boolean algebras can be viewed as abstractions of set theory. Given the importance of set theory to abstract systems, Boolean algebras play an important role in all abstract algebra.

The concept of a group

Groups are the simplest abstract structures that can be envisaged. They consist of:

1. a non-empty set of elements G;
2. a binary operation • which is
    (i) associative,
    (ii) possesses a right identity i, namely:

$$g{\bullet}i = g \quad \text{where } g \in G$$

    (iii) possesses a right inverse g^{-1}, namely:

$$g{\bullet}g^{-1} = i;$$

A group where • is also commutative is termed a **commutative** or **abelian** group.

Examples

1. The set of all integers with + is a group.
2. The set of all integers with – is not a group as

$$(a - b) - c \neq a - (b - c)$$

3. Matrices form groups under matrix addition and multiplication.

Groups seem so simple that it is hard to imagine they can be studied to any depth. A quick perusal of the shelves of a good library will dispel this idea and then the question arises of how to condense that theory to a few meaningful pages. However, our interest is not in all of group theory but just a small fraction of it. Here, some key theorems will be identified but only occasionally will their proofs be examined.

THEOREM   Given a group with a set of elements G and an operation •, then if a and b are elements of G, there exist elements c and d such that:

$$a•c = b$$
$$d•a = b$$

This theorem may seem a little obscure but consider an alternative description of it. In effect what it is saying is that if a and b are given elements of a group and if x is some variable representing any other element, then the equations:

$$a•x = b$$
$$x•a = b$$

have unique solutions. Given the properties of a group, those solutions in fact, must be:

$$x = a^{-1}•b$$
$$x = b•a^{-1}$$

The number of elements in the set G is termed the **order** of the set. The order can be infinite but for digital signal processing the most interesting groups have finite order. An important theorem details many interesting properties of such groups.

THEOREM   Let { G, • } be a group. Let g be a member of the set G and let i be the identity of the operation •. Denote g•g as g^2, g•g•g as g^3 and so on. If there is some positive integer N such that:

$$g^N = i$$

where N is the smallest such integer with this property, then G is the finite set

$$\{ i, g, g^2, g^3, \cdots, g^{N-1} \}$$

and for $0 \le j < k < N$,

$$g^j \ne g^k$$

It is useful to sketch a proof of this. First, assume that:

$$g^j = g^k$$

If this is the case, then it implies:

$$g^{j-k} = i$$

and as both j and k are less than N, so meaning their difference is substantially less than N, the condition of the theorem is violated. Now consider some integer M greater than N. This can always be expressed as:

$$M = aN + b$$

where $0 \le b < N$. Consider g^{aN+b}. Clearly:

$$g^{aN+b} = (g^N)^a \cdot g^b$$

$$= i^a \cdot g^b$$

$$= g^b$$

Hence G must be finite with the elements given.

What is interesting about this group is that it maps the infinite set of integers into a finite set of elements. This leads to the concept of **cyclic groups**. The general definition is that a group is cyclic if there exists a given element g of G such that for any other element q belonging to G, an integer n can be found such that:

$$g^n = q$$

Naturally, g is termed the **generator** of the set and the set G can simply be described as { g }. The proof is beyond the scope of this book but there is another important theorem which states that if the order of a finite group is prime, then that group is cyclic.

Studying how groups can be mapped into one another is an important topic. To begin discussion on this, we will define a **homomorphism** as a mapping F between two groups { G, • } and { H, ◊ } such that:

$$F(p \cdot q) = F(p) \lozenge F(q)$$

where p and q are both elements of G. A homomorphism of G into itself is termed an **endomorphism**.

A more strict definition is an **isomorphism**. An isomorphism is a homomorphism which is also one to one and onto. An isomorphism of G onto itself is an **automorphism**. Then isomorphic groups must have the same order. Further, if p and q are both i, the identity of •, then substituting shows F must map these to the identity of ◊. Similarly, the inverse g^{-1} of an element g must map into $F(g)^{-1}$. It also follows that two cyclic groups of the same order must be isomorphic.

Examples

1. Denote the operation modulo 7 as •. Then 3 and modulo 7 form a cyclic group of order 6 as:

    $$
    \begin{aligned}
    1 &= 1 \\
    3 &= 3 \\
    3\bullet 3 &= 2 \\
    3\bullet 3\bullet 3 &= 6 \\
    3\bullet 3\bullet 3\bullet 3 &= 4 \\
    3\bullet 3\bullet 3\bullet 3\bullet 3 &= 5 \\
    3\bullet 3\bullet 3\bullet 3\bullet 3\bullet 3 &= 1
    \end{aligned}
    $$

2. Consider the two groups:

    (a) the set G of odd, positive integers with arithmetic addition as an operation;
    (b) the set H of even, positive integers with arithmetic addition as an operation.

    Define a mapping F between these two groups as 2α where α is any element of G. Let x and y be two elements of G. Then:

    $$
    \begin{aligned}
    F(x + y) &= 2(x + y) \\
    F(x) + F(y) &= 2x + 2y
    \end{aligned}
    $$

    Thus the mapping is at least a homomorphism. Clearly though, it is not one to one, and so cannot be an isomorphism. If G was the set of positive integers though, then F would be an isomorphism.

3. An obvious example of an isomorphism is logarithms. These map the group comprising the set of real numbers and multiplication as an operation into the set of real numbers with addition as an operation. The attraction of logarithms is a result of their being isomorphic.

4. Consider a mapping of numbers given by $(a + b\sqrt{n}\,)$ into $(a - b\sqrt{n}\,)$ where a, b and n are integers. Taking addition as an operation, it is easily shown this mapping is isomorphic. It remains isomorphic for multiplication as an operation.

The concepts of a ring and a field

A ring is a slightly more complex structure than a group in that it incorporates two operations. A strict definition is that a ring $\{ R, \oplus, \otimes \}$ is an abstract structure comprising a non-empty set R of elements and two binary operations $\oplus$ and $\otimes$ where $\{ R, \oplus \}$ is an abelian group, $\otimes$ is associative and for any a, b and c drawn from R:

$$a \otimes (b \oplus c) = (a \otimes b) \oplus (a \otimes c)$$
$$(b \oplus c) \otimes a = (b \otimes a) \oplus (c \otimes a)$$

In a **ring with identity** R, there is some element i such that for any r belonging to R:

$$r \otimes i = i \otimes r = r$$

If the operation $\otimes$ has the property of being commutative, then the ring becomes a **commutative ring**. Let the identity of the group $\{ R, \oplus \}$ be e. An **integral domain** is both a commutative ring and a ring with identity such that any two elements a and b belonging to R for which:

$$a \otimes b = e$$

implies either a is e or b is e. A **division ring** is a ring where the elements of R less the identity e form a group with $\otimes$.

Consider two rings $\{ A, \oplus, \otimes \}$ and $\{ B, \oplus, \otimes \}$. A mapping H between these rings is termed a **ring homomorphism** if:

$$H(a \oplus b) = H(a) \oplus H(b)$$

and:

$$H(a \otimes b) = H(a) \otimes H(b)$$

for a and b belonging to A. If the mapping is one to one, then it is a ring isomorphism.

A subset $\{ S, \oplus, \otimes \}$ of a ring $\{ R, \oplus, \otimes \}$ that is also a ring under the operations $\oplus$ and $\otimes$ is termed a **subring** of $\{ R, \oplus, \otimes \}$, while $\{ R, \oplus, \otimes \}$ is termed a **ring extension** of the ring $\{ S, \oplus, \otimes \}$.

A **field** is more complex. It is strictly a division ring in which $\otimes$ forms a commutative group with the elements of R less the identity e. That is, a field is a structure $\{ F, \oplus, \otimes \}$ where $\{ F, \oplus \}$ is a commutative group with identity e, $\{ F_0, \otimes \}$ is a commutative group with identity i where F_0 is the set F less the element e and operation $\otimes$ is distributive over $\oplus$.

A trivial property of fields is that they must have at least two elements, namely e and i. More significant, only finite fields with a prime number of elements are of practical interest. As with rings, there can be **subfields** and any field which has a subfield is an **extension field** of that subfield.

Examples

Consider the set of numbers of the form $(a + b \sqrt{n})$, where a, b, and n are positive integers, and the two operations of arithmetic addition and multiplication. Then these form:

a group with either arithmetic addition or multiplication

a (commutative) ring under both operations
an integral domain
a field

Now consider matrices of the form:

$$\begin{bmatrix} a & b \\ c & d \end{bmatrix}$$

where a, b, c, and d are positive integers. Together with the operations of matrix addition and multiplication, these form:

a group with either matrix addition or multiplication
a ring, but not a commutative ring in the general case, under both operations

Because matrix multiplication in the general case is not commutative, then neither an integral domain nor a field can exist.

Polynomial rings

In the theory examined so far, there is an implicit assumption the set involved in the structure is constructed of simple elements. Consider, though, sets constructed of elements which are n-tuples. That is, elements such as $(a_0, a_1, \cdots, a_n)$. Abstract structures based on such elements are termed **polynomial rings**.

Polynomials in elementary algebra are functions involving the legendary unknown x such as:

$$F(x) = a_0 + a_1 x + a_2 x^2 + \cdots + a_n x^n$$

It is understood this essentially represents a computational procedure. At some stage a value will be substituted for x so enabling F(x) to be evaluated. The tuple $(a_0, a_1, \cdots, a_n)$ uniquely characterizes the polynomial and the mapping between the two forms is trivial.

In higher algebra, polynomials often have a slightly different meaning. It is very similar, in fact, to the interpretation given to the polynomials formed in a typical z-transform such as:

$$F(z) = a_0 + a_1 z + a_2 z^2 + \cdots + a_n z^n$$

Here, the coefficients $(a_0, a_1, a_2, \cdots, a_n)$ of the polynomial are elements of a sequence and so z is interpreted as a shift operator that defines position within the sequence. A convenient representation of any tuple $(a_0, a_1, a_2, \cdots, a_n)$ is therefore as a polynomial and the different powers of x within that polynomial simply reflect the position of the given element within the tuple. The two sets:

$$\{ (0,0), (0,1), (0,2), (1,0), (1,1), (1,2), (2,0), (2,1), (2,2) \} \text{ and}$$

$$\{\ 0, 1, 2, x, x + 1, x + 2, 2x, 2x + 1, 2x + 2\ \}$$

can therefore be identical; they may only differ in their notation.

The astute reader will have noticed some quite untenable assumptions. The process of forming tuples presents no problems with the preceding definitions of groups, rings or fields. However, the formation of the polynomials is very difficult to reconcile. The polynomials discussed so far involve two operations (taken to be the usual arithmetic addition and multiplication), whereas the tuples involve none. Polynomials, then, cannot be defined for a group. Further, they involve a symbol x but where does that symbol come from? After all, the elements of the sets in groups, rings, and fields are also symbols as the whole point of abstract structures is to focus on the abstract.

It behoves us to be a little formal in our definitions.

DEFINITION   A **polynomial** over a ring $\{\ R, \oplus, \otimes\ \}$ is a sequence $(a_0, a_1, \cdots)$ where the elements a_i are drawn from the set R. A finite number of these elements are different from e, the identity of R under $\oplus$. If a_n is the most distant element from the start of the sequence not equal to e, then the polynomial is represented by the finite sequence $(a_0, a_1, \cdots, a_n)$ and n is termed the **degree** of the polynomial. If a_n equals i, the identity of $\otimes$ over R, then the polynomial is termed **monic**.

This definition describes how to construct elements. That being the case, then a set P can be formed from polynomials of some ring. Further, that set P can be used as the base for constructing a ring but that requires defining two operations. This leads to a second important result.

THEOREM   Let P be the set of polynomials drawn from a ring $\{\ R, \oplus, \otimes\ \}$. Let two polynomials drawn from the set P be $(a_0, a_1, \cdots)$ and $(b_0, b_1, \cdots)$. Then the structure $\{\ P, \Diamond, \bullet\ \}$ forms a ring where $\Diamond$ is defined as:

$$(a_0, a_1, \cdots)\ \Diamond\ (b_0, b_1, \cdots) = (a_0 \oplus b_0, a_1 \oplus b_1, \cdots)$$

and $\bullet$ as:

$$(a_0, a_1, \cdots) \bullet (b_0, b_1, \cdots) = (c_0, c_1, \cdots)$$

where $c_i = a_0 \otimes b_i \oplus a_1 \otimes b_{i-1} \oplus \cdots \oplus a_i \otimes b_0$

It is not difficult proving this theorem. Note that the identity for $\Diamond$ is $(e, e, e, \cdots)$ where e is the identity of $\oplus$ and the identity for $\bullet$ is $(i, e, e, \cdots)$ where i is the identity for $\otimes$.

The conventional notation for polynomials is just too convenient to ignore. In order to overcome the philosophical objections given before though, we shall adopt the following procedure. Consider a polynomial $(a_0, a_1, \cdots, a_n)$. Now from the definition of polynomial rings, we can consider a set of polynomials which we shall label as:

$$a_0 \bullet x^0 = (a_0, e, e, \cdots)$$

$$a_1 \bullet x^1 = (e, a_1, e, e, \cdots)$$
$$a_2 \bullet x^2 = (e, e, a_2, e, \cdots)$$

and so on, where e is the identity of $\oplus$ and then define the notation:

$$a_0 \bullet x^0 \lozenge a_1 \bullet x^1 \lozenge \cdots \lozenge a_n \bullet x^n$$

as meaning:

$$(a_0, e, \cdots, e) \lozenge (e, a_1, \cdots, e) \lozenge \cdots \lozenge (e, e, \cdots, a_n)$$

Clearly, there is some ambiguity here. To add to that, we shall adopt the usual notation of representing $a_0 \bullet x^0$ by a_0 alone and $a_1 \bullet x^1$ as $a_1 \bullet x$. Further, we indicate that $(a_i \bullet x^i) \bullet (b_j \bullet x^j)$ which simply means:

$$(e, e, \cdots, a_i, \cdots) \bullet (e, e, \cdots, b_j, \cdots)$$

can be represented by the symbol $(a_i \bullet b_j) \bullet x^{i+j}$.

There is no need to define x; it is merely part of a symbol used to represent a given polynomial. (Mathematicians, though, frequently refer to it as an **indeterminate** and so a polynomial becomes a function in the indeterminate x over some ring.) Continuing in this vein, the symbol f(x) will be used to represent this new notation in compact form and, from that, R[x] could be used to describe polynomial rings. However, to add further to the confusion of the previous paragraph, consider:

DEFINITION   Consider a ring $\{ R, \oplus, \otimes \}$ and some polynomial $(a_0, a_1, \cdots)$ defined on that ring. A **polynomial function** is a mapping Γ from R into R given by:

$$\Gamma(b) = a_0 \oplus a_1 \otimes b \oplus \cdots \oplus a_n \otimes b^n$$

If $\Gamma(b)$ should equal e, the identity of $\oplus$, then that element b is termed a **zero** or **root** of the polynomial function. Also, if f(x) is the polynomial given by $g(x) \bullet h(x)$, it is not true that:

$$f(b) = g(b) \otimes h(b)$$

This can only be true if the polynomial ring is commutative and that will be so if the original ring is commutative. Most work with polynomials is with commutative polynomial rings.

Polynomial rings are very important in digital signal processing and will be examined in more detail in Part 4. It is appropriate to comment here that while there is some looseness in the definitions given so far, it will nevertheless always be clear what is meant.

Vector spaces and linear algebra

Vectors are one of the more important areas of mathematics. They feature very prominently in fields such as engineering where the interpretation is usually geometrical. However, vectors can also be described by ordered sets of numbers where these sets represent the coordinates of those vectors in n-dimensional space. Now consider some typical vector operations based on this description:

1. $(k_1 + k_2)(a_0, a_1, \cdots)$ $= k_1(a_0, a_1, \cdots) + k_2(a_0, a_1, \cdots)$
    $k((a_0, a_1, \cdots) + (b_0, b_1, \cdots))$ $= k(a_0, a_1, \cdots) + k(b_0, b_1, \cdots)$
2. $(k_1 k_2)(a_0, a_1, \cdots)$ $= k_1(k_2(a_0, a_1, \cdots))$
3. $1 \cdot (a_0, a_1, \cdots)$ $= (a_0, a_1, \cdots)$

The first of these is distributivity as defined earlier. The second is associativity and the third defines an identity. That suggests defining an abstract form of vector spaces.

DEFINITION   Consider a structure $\{ V, F, \&, \bullet, \oplus, \otimes \}$ with the following properties:

       (a) $\{ F, \&, \bullet \}$ is a field where & has an identity e and $\bullet$ has an identity i.
       (b) $\{ V, \oplus \}$ is an abelian (i.e. commutative) group with an identity ε.
       (c) The operator $\otimes$ obeys the following. If a and b belong to F and α and β belong to V, then:
          (i) $a \otimes \alpha$ belongs to V;
          (ii) $(a \& b) \otimes \alpha = (a \otimes \alpha) \& (b \otimes \alpha)$
          (iii) $(a \bullet b) \otimes \alpha = a \otimes (b \otimes \alpha)$
          (iv) $a \otimes (\alpha \oplus \beta) = (a \otimes \alpha) \oplus (a \otimes \beta)$
          (v) $i \otimes \alpha = \alpha$

The elements α and β are termed **vectors** and the structure is termed a **vector space over a scalar field F**.

A number of simple properties of vector spaces follow from this definition. For example:

$$\alpha = i \otimes \alpha = (i \& e) \otimes \alpha = i \otimes \alpha \& e \otimes \alpha = \alpha \& e \otimes \alpha$$

Thus $e \otimes \alpha$ must be ε. If i^{-1} is the inverse of i, then:

$$(i \& i^{-1}) \otimes \alpha = \varepsilon = i \otimes \alpha \& i^{-1} \otimes \alpha = \alpha \& i^{-1} \otimes \alpha$$

which means $i^{-1} \otimes \alpha$ must be the inverse of α. Generalizing this result, we deduce the inverse of $a \otimes \alpha$ is $a^{-1} \otimes \alpha$. From this, we can show $a \otimes \varepsilon$ must be ε. A final comment here is that if a subset of vectors in V exists that also forms a vector space over the operations $\oplus$ and $\otimes$, then that subset forms a **vector subspace**.

There are many examples of vector spaces including, of course, most of those described geometrically. In the context of this work, one is easily formed from finite sets of tuples.

A most important definition is:

DEFINITION   Let $\{ V, F, \&, \bullet, \oplus, \otimes \}$ be a vector space. Let $a_0, a_1, \cdots, a_n$ be elements of F and let $\alpha_0, \alpha_1, \cdots, \alpha_n$ be elements of V. Then a **linear combination** of vectors of V is the structure:

$$a_0 \otimes \alpha_0 \oplus a_1 \otimes \alpha_1 \oplus \cdots \oplus a_n \otimes \alpha_n$$

A set L of linear combinations can be formed for any vector space and such a set is a subspace of that vector space. So important are these in the study of abstract systems that vector spaces are often termed **linear spaces**. Now consider another definition.

DEFINITION   Let $\alpha_0, \alpha_1, \cdots, \alpha_m$ be a set of vectors drawn from some vector space given by $\{ \text{ V, F, \&, } \bullet, \oplus, \otimes \}$. If, for every linear combination:

$$a_0 \otimes \alpha_0 \oplus a_1 \otimes \alpha_1 \oplus \cdots \oplus a_m \otimes \alpha_m = \varepsilon$$

where $a_0, a_1, \cdots, a_m$ are elements of F, the implication is that:

$$a_0 = a_1 = \cdots = a_m = e$$

then the set of vectors is said to be **linearly independent**.

By manipulating the above linear combination it is easily shown that linearly dependent vectors can be described as combinations of other vectors of the set. Hence an alternative definition of linear independence is a set of vectors where it is impossible to express any one of them as a linear combination of the others. This leads to other important ideas. Any vector space must have at least one **maximum linearly independent subspace** called a **basis**. Therefore, a vector v, not equal to the identity ε, belonging to some vector space, can always be expressed as:

$$v = c_0 \otimes \alpha_0 \oplus c_1 \otimes \alpha_1 \oplus \cdots \oplus c_m \otimes \alpha_m$$

where $\{ \alpha_0, \alpha_1, \cdots, \alpha_m \}$ is a basis for that space. All bases of a vector space have the same size m. This is termed the **dimension** of the space. Thus vectors of the space can be represented as tuples of the form $(c_0, c_1, \cdots, c_m)$ and these, naturally, are termed coordinates.

There is an interesting aside to make here. The important finite fields have a prime number of elements. In a vector space, the coordinates are drawn from a set F which forms part of a field. If that is finite, then there is a prime number of elements in that set. Now a vector space is also a field but the elements are m-tuples drawn from F. Thus in the general case, a finite field can have p^m elements, where p is some prime and m some integer and where, for m>1, the elements are interpreted as tuples. A polynomial drawn from a finite vector field, therefore, can have p^m elements.

Consider two vector spaces $\{ \text{ V, F, } \Diamond, \bullet, \oplus, \otimes \}$ and $\{ \text{ W, F, } \Diamond, \bullet, +, \circ \}$ over some field F. A mapping H of V into W is a homomorphism if:

$$H(\alpha \oplus \beta) = H(\alpha) + h(\beta)$$
$$H(a \otimes \alpha) = a \circ H(\alpha)$$

where a is an element of F and α and β are elements of V. Homomorphisms are more commonly called **linear transformations**. If they are one to one and onto, then they are an isomorphism and the two vector spaces are said to be isomorphic. It can be shown that all n-dimensional vector spaces are isomorphic to one another.

Linear transformations can be regarded as elements of a set. Two operations can be defined which combine them and the result can be a field. These structures will not be investigated here but it is useful to recognize that for some α from V, there must be an **identity transformation** I such that:

$$I(\alpha) = \alpha$$

Consider a linear transformation T from a vector space $\{ V, F, \Diamond, \bullet, \oplus, \otimes \}$ to another vector space $\{ W, F, \Diamond, \bullet, +, \circ \}$. If $T(\alpha) = T(\beta)$ implies that $\alpha = \beta$, then T is termed a **non-singular transformation**. This implies there is a transformation T* from W into V such that for some α an element of V:

$$T*(\, T(\alpha)\,) = I(\alpha) = \alpha$$

where I is the identity transformation. In addition, given some vector ω belonging to W, then:

$$T(\, T*(\omega)\,) = I(\omega) = \omega$$

Further, any basis of V is mapped into a basis of W.

Transformation T* is clearly just an inverse and given that linear transformations can form a field, it is convenient to denote it by T^{-1}. As well, $T(\, T^{-1}(\alpha)\,)$ is more easily expressed as $T^{-1}T$. If T and S are transformations, then in this notation:

$$(\, TS\,)^{-1} = S^{-1}\, T^{-1}$$
$$(\, aT\,)^{-1} = a^{-1}\, T^{-1}$$

As another aside, it is useful to define a **linear algebra**. As mentioned earlier, linear transformations form a set of elements and so can be used to construct groups, rings, fields or vector spaces. Consider such a set $\ddot{Y}$ formed from all the transformations from a vector space into itself. Constructing a ring, field or vector space requires two operations. Looking over the previous results, one of these — a form of addition — already exists. What is lacking is a form of product operation. An **algebra** over a field F, then, is simply a structure $\{ \ddot{Y}, F, \Diamond, \bullet, \oplus, \otimes, \square \}$ where $\square$ is a vector product operation. A linear algebra is such an algebra. One of its conditions is that $\{ \ddot{Y}, F, \Diamond, \bullet, \oplus, \otimes \}$ must be a vector space. The important properties relate to $\square$. Given transformations T_1, T_2, and T_3 belonging to $\ddot{Y}$, elements a and b belonging to F and the notation given above, this operation must be:

1. closed, viz. $T_1 T_2$ must belong to $\ddot{Y}$

2. associative, viz. $T_1(T_2 T_3) = (T_1 T_2)T_3$

3. bilinear, viz. $\quad T_1(\, a \otimes T_2 \oplus b \otimes T_3) = a \otimes T_1 T_2 \oplus b \otimes T_1 T_3$
    $$(a \otimes T_2 \oplus b \otimes T_3)T_1 = a \otimes T_2 T_1 \oplus b \otimes T_3 T_1$$

Such algebras will not be considered here.

Returning to linear transformations, consider a transformation T from an n-dimensional vector space $\{ V, F, \Diamond, \bullet, \oplus, \otimes \}$ into an m-dimensional space $\{ W, F, \Diamond, \bullet, \oplus, \otimes \}$. The first will have a basis $\{ \alpha_0, \alpha_1, \cdots, \alpha_n \}$ and the second a basis $\{ \beta_0, \beta_1, \cdots, \beta_m \}$. Consider the transformations of the vectors α_i. As any vector can be described as a linear combination of the basis vectors, then:

$$T(\alpha_1) = a_{11} \otimes \beta_1 \oplus a_{12} \otimes \beta_2 \oplus \cdots \oplus a_{1m} \otimes \beta_m$$
$$T(\alpha_2) = a_{21} \otimes \beta_1 \oplus a_{22} \otimes \beta_2 \oplus \cdots \oplus a_{2m} \otimes \beta_m$$

$$T(\alpha_i) = a_{i1} \otimes \beta_1 \oplus a_{i2} \otimes \beta_2 \oplus \cdots \oplus a_{im} \otimes \beta_m$$

$$T(\alpha_n) = a_{n1} \otimes \beta_1 \oplus a_{n2} \otimes \beta_2 \oplus \cdots \oplus a_{nm} \otimes \beta_m$$

Consider some arbitrary vector v belonging to V. Through this linear combination property, it can be described as:

$$T(v) = T(c_0 \otimes \alpha_0 \oplus c_1 \otimes \alpha_1 \oplus \cdots \oplus c_n \otimes \alpha_n)$$

$$= c_0 \otimes T(\alpha_0) \oplus c_1 \otimes T(\alpha_1) \oplus \cdots \oplus c_n \otimes T(\alpha_n)$$

or more compactly, by the tuple $(c_0, c_1, \cdots, c_n)$. Substituting for the terms for $T(\alpha_i)$ will then derive a second tuple $(d_0, d_1, \cdots, d_m)$ which will be the image of the vector v in W. The transformation itself, then, is uniquely defined by the set of scalars $\{ a_{ij} \}$ drawn from F.

The structure of this set is in the form of an array of n rows with m columns. In short, it is an n×m matrix as the strict definition of a matrix is a rectangular array of elements of some field F. Consider some operations which can be performed on linear transformations. Let A and B be two n×m matrices. A binary operation **+** can be defined:

$$A + B = C$$

where, using standard matrix notation, C is given by:

$$[c_{ij}] = [a_{ij} \oplus b_{ij}]$$

Another binary operation ∘ can also be defined:

$$A \circ B = C$$

where A is n×k and B is k×m and where C is given by:

$$[c_{ij}] = [a_{i1} \otimes b_{1j} \oplus a_{i2} \otimes b_{2j} \oplus \cdots \oplus a_{ik} \oplus b_{kj}]$$

An identity for the first operation is simply the matrix with all elements equal to e, the identity of ⊕. An identity for the second is more difficult to find. If the requirement is an identity I with the property that:

$$A \circ I = I \circ A = A$$

then the matrices and so the identity must be n×n. In this case, that identity will have elements given by δ_{ij} where this is the Kronecker delta:

$$\delta_{ij} = \iota \text{ for } i = j$$

$$= e \text{ for } i \neq j$$

where ι is the identity of ⊗. This is sufficient to form a field or a linear algebra.

A study of matrices is a study of linear transformations in vector spaces. Some aspects of this vast subject are particularly important to later work. Let V be an n-dimensional vector space and T some transformation. If T maps V into itself, then the matrix representing T must be square. If the basis maps into itself, then T equals the identity. A more interesting case is the following. Let v be some vector of V. Let $T^2(v)$ mean $T(T(v))$, $T^3(v)$ mean $T(T(T(v)))$ and so on. They form a set of vectors:

$$\{ v, T(v), T^2(v), T^3(v), \cdots , T^{k-1}(v) \}$$

Assume this set of vectors is linearly independent and also is the minimum set of vectors of this type which are linearly independent. Therefore:

$$T^k(v) = a_0 \otimes v \oplus a_1 \otimes T(v) \oplus \cdots \oplus a_{k-1} \otimes T^{k-1}(v)$$

and by definition, the terms a_i cannot be zero (or in the general case, equal to e). Since the vector set is linearly independent, it forms the basis of a subspace within V. Consider some vector u which belongs to that subspace. Then:

$$u = c_0 \otimes v \oplus c_1 \otimes T(v) \oplus \cdots \oplus c_{k-1} \otimes T^{k-1}(v)$$

Apply the transformation T to this vector. Thus:

$$T(u) = T(c_0 \otimes v \oplus c_1 \otimes T(v) \oplus \cdots \oplus c_{k-1} \otimes T^{k-1}(v))$$

$$= c_0 \otimes T(v) \oplus c_1 \otimes T^2(v) \oplus \cdots \oplus c_{k-1} \otimes T^k(v)$$

But $T^k(v)$ is a linear combination of the basis of the subspace, hence $T(u)$ must be another member of that subspace. A transformation which maps vectors of a subspace into that subspace is an **invariant transformation**. It is represented by the k × k matrix:

$$\begin{bmatrix} 0 & 1 & 0 & \cdot & \cdot & \cdot & & 0 \\ 0 & 0 & 1 & & & & & 0 \\ 0 & 0 & 0 & & & & & 0 \\ \cdot & & & & \cdot & & & \cdot \\ \cdot & & & & & \cdot & & \cdot \\ \cdot & & & & & & \cdot & \\ 0 & 0 & 0 & & & & 0 & 1 \\ -c_0 & -c_1 & -c_2 & \cdot & \cdot & \cdot & -c_{k-2} & -c_{k-1} \end{bmatrix}$$

There are many forms of invariant transformation. One important group are the **nilpotent transformations**. These have the property that:

$$T^j(v) \neq e \qquad\qquad \text{for } j = 0, 1, 2, \cdots, k-1$$
$$T^k(v) = e$$

Another important group are those where:

$$T(v) = \lambda \cdot v$$

where λ is a constant. That is, the class of transformations where the vector undergoing transformation is mapped into a scalar multiple of itself. The constant λ is termed an **eigenvalue**. If the transformation T is represented by a matrix A, then:

$$A \cdot v = \lambda \cdot v$$
$$\text{i.e.} \qquad [A - \lambda \cdot I] \cdot v = E$$

where E is a vector of zeroes (or e in the general case). All the eigenvalues of A are found by solving:

$$P(\lambda) = \det[A - \lambda \cdot I] = 0$$

This is the **characteristic equation** and $P(\lambda)$ is the **characteristic polynomial**.

From the general properties of determinants, it can be shown:

$$P(\lambda) = (-1)^k \cdot (\lambda^k + b_1 \cdot \lambda^{k-1} + \cdots + b_{k-1} \cdot \lambda + b_k)$$

In the general case, the coefficients b_i of this polynomial are complex combinations of the elements of A. In the particular case of an invariant transformation, however, the characteristic polynomial is:

$$P(\lambda) = (-1)^k \cdot (\lambda^k + c_{k-1} \cdot \lambda^{k-1} + \cdots + c_1 \cdot \lambda + c_0)$$

and now the coefficients relate directly to the matrix elements. Therefore, given any characteristic polynomial, we can always define a matrix of this type and it is known as the

companion matrix. Note that as det$[A - \lambda \cdot I]$ is identical to det$[A^T - \lambda \cdot I]$, the companion matrix can also be defined as:

$$
\begin{bmatrix}
0 & 0 & 0 & \cdot & \cdot & \cdot & -c_0 \\
1 & 0 & 0 & & & & -c_1 \\
0 & 1 & 0 & & & & -c_2 \\
\cdot & & & \cdot & & & \cdot \\
\cdot & & & & \cdot & & \cdot \\
\cdot & & & & & \cdot & \cdot \\
0 & 0 & 0 & & & 0 & -c_{k-2} \\
0 & 0 & 0 & \cdot & \cdot & \cdot & 1 & -c_{k-1}
\end{bmatrix}
$$

This is often the more convenient form to use in applications.

The Cayley-Hamilton theorem states an important attribute of characteristic equations. A square matrix A is always a solution to its own characteristic equation. For an invariant transformation:

$$ P(A) = (-1)^k \cdot (A^k + c_{k-1} \cdot A^{k-1} + \cdots + c_1 \cdot A + c_0) = Z $$

where Z is a square matrix with every element zero (or e in the general case). Interpreted another way, the Cayley-Hamilton theorem is stating that A^k is a linear sum of the vectors $\{I, A, A^2, \cdots, A^{k-1}\}$. Note that the theorem is not implying these are independent.

A final comment on companion matrices. From the form of the matrix, the set of matrices C_j formed from a companion matrix C, with $0 \le j < k$, must all be non-zero. Further, this set is, in fact, linearly independent and so can be used as the basis of a subspace.

DIGITAL CONVOLUTION AND ABSTRACT MATHEMATICAL STRUCTURES

In this segment, abstract structures within the forms of digital convolution outlined in the first section of this Part are considered. A general observation is the importance of abstract structures based on numbers as set elements. Further, as those structures employ addition (incorporating subtraction) and multiplication, then again, in general, interest in digital signal processing is in rings and fields rather than groups.

Those rings and fields describe two main sectors of digital signal processors: first and most obvious, the data processing activity of the processor; second, the structural properties of digital signal processors in general.

Input to a digital signal processor is a sequence of numbers and the action of the processor is to produce an output sequence of numbers. These sequences are drawn from particular sets. There is no theoretical reason why input and output should necessarily be the same type of numbers but it would be unusual if they were otherwise. There is also no reason why they need to be the same as the numbers used internally within the digital signal processor. Indeed, as will be seen later in the book, this is not at all uncommon. These numbers, though, form the elements of an abstract structure; a structure which defines the basic processing activity of the processor. The input number sequence must map into this structure and then the series of operations which is the processor produces elements which must be mapped into the output sequence.

A comment is in order on the types of numbers involved in digital signal processors. The broad classes are **integer**, **real**, and **complex** numbers. Real numbers can be **rational**, meaning they are equal to the ratios of two integers. All other forms are naturally **irrational**. Complex numbers can be formed from real numbers or from integers. In the latter case, they are termed **Gaussian numbers**. Which to employ ideally shouldn't be a concern. In practice, the demands of digital equipment means that interest centers on integers, rationals, Gaussian numbers, and complex numbers formed from rationals. Note, too, the standard mathematical terminology for these sets of numbers. The natural numbers — the set of all positive integers — are denoted N. The set of all integers is denoted Z. The rationals are Q, the reals are R, and the set of all complex numbers is denoted C.

Another comment is in order on this question of different number systems for input, output and internal operations of a processor. In some circumstances, this wouldn't matter. However, usually the input to a digital signal processor is digital — a sequence drawn from a finite set of integers — and so is the output, but design theory often assumes an infinite set of real numbers for the processor, either singly or in complex form. That is, it assumes discrete numbers. The mapping is no longer simple. This is one of several reasons why abstract algebra is of interest for application to digital signal processing.

Although mapping is a significant problem, the body of knowledge concerning methods of convolution in infinite structures, particularly those based on complex numbers, is too significant to ignore. All of Part 2 is devoted to this. Since the structures are infinite and since the operations employed are those of conventional experience, abstract issues do not need to be stressed. They are nonetheless there and that is useful to keep in mind.

Topics that are largely concerned with avoiding this mapping problem will also be explored. We wish to examine the basic computational processes of digital convolution on finite rings and fields, and mappings between these and the finite integer number sets of input and output. While the primary concern must be finite integer rings and fields, there is more. We also desire rings and fields that are easily implemented and ideally in both software and hardware. The former sets quite rigid constraints, because it is demanding the operations must be existing operations within general-purpose computers. The latter is more open. However, here there may be a concern with economics and that suggests an interest in avoiding complex rings and fields, as these require two storage cells per datum and more complex operators. These concerns are mainly explored in Part 3 but some are examined in Part 4.

Digital signal processing developed for a number of reasons, but no doubt a significant stimulus was the rediscovery of a number of fast algorithms for the discrete Fourier transform. These algorithms were termed fast as multiplication has been a computationally expensive operation and they required fewer multiplications than the usual algorithms employed. This was achieved by recognizing symmetries within the computation. To give a very trivial example: $ab + ac$ requires two multiplications and one addition but, in the form of $a(b+c)$, it requires only one multiplication and one addition. In the case of the discrete Fourier transform, the significant multiplications are by W^{nk} where W is the complex number

$\exp(-\frac{j2\pi}{N})$ and N is the number of points involved. Symmetries here are largely recognized through the indices nk of W and the sequence of operations in an algorithm reflected through these indices forms an abstract structure. This particular structure is based on a finite set of integers.

There is little new to study in this structure. Most of the significant results were derived many years ago but it is still important to examine them. The early part of Part 3 will do this.

As shown in the first section of this Part, digital convolution can be viewed in a number of lights. It can be seen to be a matrix operation. It can also be seen to be a polynomial multiplication problem. Both are encompassed by abstract structures. Then here what we can do is to consider rings or fields based on a set of elements which encompass all possible digital signal processors. That, as we shall find in Part 4, leads to some very useful algorithms indeed.

TRANSFORMS IN ABSTRACT STRUCTURES

Digital convolution can be developed along two major lines; one we have termed the direct approach, the other the transform. The work on abstract algebra enables us to be a little more specific.

The direct approach, in general, involves a set of **fixed** elements of some ring or field. That is, a matrix or polynomial that does not change insofar as the given computation is termed. It also involves a set of **variable** elements representing the input values to the computation but these are also elements of the same ring or field as the fixed. Finally, there is an operation — either a matrix operation or a polynomial — which produces the required output. This approach then, is confined to the one structure but, as mentioned earlier, there may need to be a mapping between the actual input values and corresponding elements in this field and another from this structure to the output. Note though, that how the operation is viewed will be limited by whether the structure is a ring or field.

The transform approach uses a linear transformation and so can only be defined for a field. That transformation must satisfy certain constraints, namely the cyclic convolution property must hold (or its equivalent). Our object here is to examine in detail what this demands of mathematical structures.

First, some terminology and general comments. Define the objects $\{ V, F, \&, \bullet, \oplus, \otimes \}$ and $\{ W, F, \&, \bullet, \oplus, \otimes \}$ as two N-dimensional vector spaces where N is some finite integer. Let V represent the first of these and W the second. Clearly, $\{ F, \&, \bullet \}$ must be a field but we will not specify it in any way. Vectors in these spaces are represented by tuples such as $\{ y_0, y_1, \cdots, y_N \}$ but for convenience denote these as y. Now linear transformations between these vector spaces are defined by N×N matrices. Let these be denoted by capitals such as A. Let a transformed vector be denoted as Y. Let an operation on a field such as:

$$a_0 \bullet b_0 \ \& \ a_1 \bullet b_1 \ \& \ . \ . \ . \ \& \ a_{N-1} \bullet b_{N-1} \qquad \text{be denoted as} \qquad \sum_{m=0}^{N-1} a_m \bullet b_m$$

Finally, let ¤ define the matrix product operation for fields described earlier.

Consider digital cyclic convolution. It is:

$$Y(n) = \sum_{m=0}^{N-1} h(m) \cdot x(n-m) = \sum_{m=0}^{N-1} h(n-m) \cdot x(m) \qquad n = 0, 1, \cdots, N-1$$

In the infinite field for which they were defined, these equations describe a linear transformation. We can generalize, therefore, that cyclic convolution is a linear transformation of vector spaces over some given field. However, our interest is in particular transformations. Let A be some non-singular transformation from $\mathbb{V}$ to $\mathbb{W}$ and let A^{-1} denote the inverse transformation. We will refer to A as the **forward transformation** and A^{-1} as the **reverse**. Then transforming the convolution terms:

$$\begin{aligned} \mathbf{Y} &= \mathbf{A} \, \square \, \mathbf{y} & \mathbf{y} &= \mathbf{A}^{-1} \square \, \mathbf{Y} \\ \mathbf{H} &= \mathbf{A} \, \square \, \mathbf{h} & \mathbf{h} &= \mathbf{A}^{-1} \square \, \mathbf{H} \\ \mathbf{X} &= \mathbf{A} \, \square \, \mathbf{x} & \mathbf{x} &= \mathbf{A}^{-1} \square \, \mathbf{X} \end{aligned}$$

Transformation A will be defined as having the cyclic convolution property if and only if the elements $Y(k)$ of the transformed tuple $\mathbf{Y}$ are related to the elements of $\mathbf{X}$ and $\mathbf{H}$ via:

$$Y(k) = H(k) \bullet X(k) \qquad\qquad k = 0, 1, \cdots, N-1$$

Our object, then, is to identify what this implies for the transformation and how to locate the set of transformations with this property.

To begin, we note that each of the $Y(k)$ are given by:

$$Y(k) = \sum_{n=0}^{N-1} a(k,n) \bullet y(n)$$

where $a(k,n)$ describes the elements of the transformation A. Now the elements $y(n)$ are given by the convolution, thus:

$$Y(k) = \sum_{n=0}^{N-1} a(k,n) \bullet \left(\sum_{m=0}^{N-1} h(n-m) \bullet x(m) \right)$$

Given the field conditions, this can be rearranged to:

$$Y(k) = \sum_{m=0}^{N-1} x(m) \bullet \left(\sum_{n=0}^{N-1} h(k,n) \bullet h(n-m) \right)$$

Consider the second term of this. Let $\theta = n - m$, or $n = \theta + m$. Thus:

$$\sum_{n=0}^{N-1} a(k,n) \bullet h(n-m) = \sum_{\theta+m=0}^{\theta+m=N-1} a(k, \theta+m) \bullet h(\theta)$$

$$= \left(\sum_{\theta=-m}^{\theta=-1} a(k, \theta+m) \bullet h(\theta) \right) \, \& \, \left(\sum_{\theta=0}^{\theta=N-1-m} a(k, \theta+m) \bullet h(\theta) \right)$$

Although we are dealing with only a finite number of data elements, the concept of cyclic convolution is that these elements are regarded as one period of an infinite periodic function. Therefore, an index $-j$ is interpreted as meaning $(N-j)$. We shall require the same interpretation for indices of the transform. From this last equation, this means:

$$\sum_{n=0}^{N-1} a(k,n) \bullet h(n-m) = \sum_{n=0}^{N-1} a(k,n+m) \bullet h(n)$$

Hence:

$$Y(k) = \sum_{m=0}^{N-1} \sum_{n=0}^{N-1} a(k,n+m) \bullet h(n) \bullet x(m)$$

If the cyclic convolution property is a property of transformation A, then:

$$Y(k) = H(k) \bullet X(k)$$

$$= \left(\sum_{n=0}^{N-1} a(k,n) \bullet h(n) \right) \bullet \left(\sum_{m=0}^{N-1} a(k,m) \bullet h(m) \right)$$

and, again, from the general properties of fields:

$$= \sum_{n=0}^{N-1} \sum_{m=0}^{N-1} a(k,n) \bullet a(k,m) \bullet h(n) \bullet x(m)$$

Comparing this with the previous expression, the cyclic convolution property holds only if:

$$a(k,n+m) = a(k,n) \bullet a(k,m) \qquad\qquad n, k, m = 0, 1, \cdots, N-1$$

holds as a relationship between elements of the matrix describing the transformation A.
   Consider the implications of this:

1.   For the particular case of m zero:

$$a(k,n+0) = a(k,n) \bullet a(k,0) \qquad\qquad n, k = 0, 1, \cdots, N-1$$

If $a(k,n)$ is not equal to e, the identity of the field F under &, then clearly:

$$a(k,0) = i \qquad\qquad k = 0, 1, \cdots, N-1$$

where i is the identity of $\bullet$ over the field F.

2. Consider a given element $a(k,p)$ of the transformation. From the condition:

$$a(k,p) = a(k, (p-1)+1) = a(k, p-1) \bullet a(k,1)$$

$$= \; a(k,p-2) \cdot a^2(k,1)$$

$$= \; .$$

$$= \; .$$

$$= \; .$$

$$= \; a(k,p-p) \cdot a^p(k,1)$$

$$= \; a(k,0) \cdot a^p(k,1)$$

$$= \; i \cdot a^p(k,1)$$

$$= \; a^p(k,1)$$

Consider the particular case of p equal to N. We have already declared the indices for the matrix must be interpreted as being periodic with period N and so a(k,N) must be identical to a(k,0). But a(k,0) is equal to i, the identity of • over F. Thus:

$$a^N(k,1) = i \hspace{4cm} k = 0, 1, \cdots, N-1$$

Therefore, the elements of $\{\, i, \Omega, \Omega^2, \cdots, \Omega^j, \cdots, \Omega^{N-1} \,\}$ are the rows of the matrix A defining the transformation where i is the identity of • over F in the field for which the transformation is defined and Ω is an element with the property that $\Omega^N = i$. From matrix theory (and its extension to fields), the matrix A will only be non-singular if each of the rows is unique. If N unique values can be found to satisfy this equation, that condition will be satisfied. These rows have exactly the same form as the elements of a finite group under •. Therefore, each is a subgroup of $\{\, F, \bullet \,\}$ and uniquely defined by a generator taken from the set F. Let Ω be that cyclic group generator and let $\{\, i, \Omega, \Omega^2, \Omega^3, \cdots, \Omega^{N-1} \,\}$ be the cyclic group which it generates. If each of these are taken as the generators of the rows of the transforming matrix, then those rows will be unique (as position within the matrix is the important factor here). That is, the transforming matrix A is given by $A = [\,\Omega^{kn}\,]$ where Ω^0 is interpreted as being i, the identity under •.

DISCUSSION

Much of this section has been preparation for later Parts and its merit will not become obvious until then. Nevertheless, it concluded with some important results. First, it was shown that the abstract concept of a vector space is at the core of all approaches to digital convolution. Any new endeavors in this field must follow that requirement. Second, it showed that digital convolution algorithms describe linear transformations on such vector spaces. The two distinctions made earlier between direct and transform approaches therefore share much the same mathematical structure and only differ in their finer detail. This reinforces the earlier view that they are not as radically different as has sometimes been assumed. Third, the general conditions for a transform with the cyclic convolution property to exist were examined and lead to a very interesting result. Such a transform exists for some given field { F, &, • } provided at least one element Ω exists belonging to F that generates a cyclic group of order N with respect to •. The forward transform is described by the matrix [Ω^{kn}]. If more than one such generator exists within F, then there is more than one transform with the cyclic convolution property.

FURTHER STUDY

References

There are any number of good introductory texts on linear algebra. For example,

J.R. DURBIN, *Modern Algebra: An Introduction*, Wiley, 1979

R. LIDL and H. NIEDERREITER, *Introduction to Finite Fields and their Applications*, Cambridge University Press, 1986

R.J. MCELIECE, *Finite Fields for Computer Scientists and Engineers*, Kluwer Academic Publishers, 1987

Exercises

1. Prove digital convolution is commutative, associative, and distributive.

2. Can there be a polynomial field with the definition given? Either give an example, or prove it is not possible.

3. How are polynomials and vector spaces linked?

4. Consider finite impulse response digital filtering. Describe this as an abstract structure.

5. Prove a cyclic group is abelian.

6. Why cannot max(a,b) be used as the operation of a group?

7. Show each element of a group has a unique inverse.

8. Consider the set of integers. Consider the binary operation ! where:

   $a!b = \alpha \cdot a + \beta \cdot b + \gamma$

   What conditions apply to this operation and to the set of integers when used to form a group?

9. Consider some polynomial ring with elements ($a_0, a_1, \cdots, a_N$). Is a transformation T, given by:

$$T\{ (a_0, a_1, \cdots, a_N) \} = a_0 + a_1 + \cdots + a_N$$

a linear transformation?

10. Integers form an integral domain but not a field. With regard to what?

PROBLEMS

1. It is very well known that:

$$\log (x*y) = \log (x) + \log (y)$$

If x and y are integers of the form 2^k, then:

(a) Is { { 2^k }, * } a group?
(b) If so, under what conditions is logarithm an isomorphism and to what?

2. Consider the set of polynomials and the two operations of integration and differentiation. Do these form a field or ring, or can the set and either operation form a group? If so, then derive identities and find the inverse of a given element.

3. Determine the inverse transform on an algebraic structure following the derivation given for the forward transform.

4. Consider numbers of the form ($a + b \cdot \sqrt{n}$) where n is some fixed value. If these numbers together with multiplication are to be formed into a group, then:

(a) What is the identity?
(b) What is the inverse?
(c) What conditions, if any, apply to a and b?

5. Consider a pentagon. Let (a, b, c, d, e) describe which apex is at a given point with respect to some fixed coordinate system. Let an operation be a circular rotation about the center of the figure through an angle of 72°. Let a second be a symmetrical rotation about an axis through one apex and the center of the figure. Then:

(a) Can each of these operations form a group?
(b) Can they together form a ring or field and if not, why not? Can some other operation with either of them form a ring or field?

6. The derivation of the general form for a transform assumed cyclic convolution. What results if just aperiodic convolution is assumed? How does this affect the eventual result?

7. Consider the letters of the alphabet. Can you devise a format for these and a closed operator that form a group?

8. Consider a chess board. Let the squares on the board be defined by coordinate pairs (a,b). Describe the movement of pieces as coordinate transformations. Can any of these 'operators' form a group? If so, describe the identity and categorize the group as closed, abelian, etc.

9. There is a common puzzle which consists of 15 numbered tokens in a block with 16 positions. The object is to arrange the tokens to form given patterns. (An electronic form exists as a desktop accessory for the Apple Macintosh personal computer.) Describe this by a group.

10. A gourmet considers that a good meal should consist of an appetizer, a soup, a main course, and then dessert, a cheeseboard, chocolates, and the very best Darjeeling tea. Naturally, he could not foresee the need to drink anything during the meal other than water and wine. He entertains frequently and has built up a large collection of recipes which he categorizes according to their main constituent, viz.:

    (a) Poultry — chicken, duck, and turkey
    (b) Meat — beef, lamb, veal, and pork
    (c) Seafood — fish, scallops, shrimp, lobster, and squid
    (d) Eggs
    (e) Pasta, potatoes, and rice
    (f) Green vegetables — beans, peas, spinach, zucchini, and artichokes
    (g) Mushrooms, onions, and carrots

    He desires a database system by which he can provide one recipe, say Chicken Kiev and then get a listing of other meal items which are suitable. Suitable here means to some given rules which include:

    (a) at least one item is to be cold
    (b) no major constituent is to appear twice
    (c) meat cannot precede fish
    (d) once red wine is served, all subsequent dishes must be compatible with red wine

    Elaborate on this structure and once this is completed, identify the groups, rings, fields, and linear transformations which exist.

PART

2

Digital Convolution on Infinite Fields

INTRODUCTION

Continuous signals in the real-world are invariably band-limited. That, together with the certain presence of noise, means their information content is finite. For this reason, the study of discrete convolution on infinite number fields seems of limited practical interest.

Mathematical tractability, though, is a factor that cannot be overlooked and in the development of a transform the assumption of an infinite field leads to a mathematically simple result. That has resulted in transform approaches on infinite fields having a long and quite illustrious history. Indeed, it would not be entirely amiss to claim that studies of such transforms were the most significant stimulus for the early development of digital signal processing.

The field that commands the greatest interest is C, the infinite field of complex numbers. The transform with the cyclic convolution property in this field is the widely known discrete Fourier transform (DFT). What has made this transform of particular interest has been the discovery of a number of very efficient computational algorithms. These, naturally enough, are collectively known as fast Fourier transforms or FFTs.

This Part is exclusively devoted to the study of these FFTs. TOOLS is relatively brief as familiarity with the discrete Fourier transform and its properties is assumed. TOOLS, therefore, is essentially just a review with the sole purpose of emphasizing certain key concepts. THEORY examines the basis of FFT algorithms. We will show that a single computational section can be used to construct most FFT algorithms of interest. It is worth studying in some depth and this is done. In DEVELOPMENTS, we examine some of the host of FFT algorithms which can be developed from that section, the most important being the base-2 algorithms.

As mentioned, most signals of practical interest are invariably drawn from a finite set of values. Performing transforms in the infinite field of complex numbers will therefore require a mapping of those finite values into the complex number field. Further, it will require a mapping back if the transform should be used as a means of performing discrete convolution. That immediately suggests problems and these are one of the issues considered in IMPLE-MENTATION ISSUES. The long history of the FFT approach to discrete convolution means there is also a long history of techniques for implementation. Rapidly evolving technology means much of this body of knowledge is obsolete but some remains of interest and that, too, is considered. Implementation here means both software and hardware implementation.

TOOLS

Conditions for a transform to exist

There is only one mathematical structure of interest to this Part — the field of complex numbers. Following convention, we shall denote these numbers as (a,b) and interpret this as $(a + jb)$. The conditions for the existence of a transform with the cyclic convolution property in this field are found by applying the theory given in Part 1. This is a simple exercise and only the results will be quoted here.

As shown in Part 1, a transform with the cyclic convolution property will exist if a field has an element G that can generate a finite cyclic subfield of order N. That means $G^N = i$ where i is the identity for the operator •. Here that identity is the number (1.0,0.0) and therefore the element G is an Nth root of unity. There is only one such generator and consequently there is only one transform with the cyclic convolution property.

Consider some periodic sequence x(n) with period N. In full, the transform pair is:

$$X(k) = \sum_{n=0}^{N-1} x(n) \cdot W^{kn} \qquad\qquad k = 0, 1, \cdots, N-1$$

$$x(n) = \frac{1}{N} \sum_{k=0}^{N-1} X(k) \cdot W^{-nk} \qquad\qquad n = 0, 1, \cdots, N-1$$

where W is $\exp(-\frac{j2\pi}{N})$. This is the discrete Fourier transform (DFT). Note that the term $\frac{1}{N}$ must exist within the transform pair but where it appears is a convention. It is customary in the technical literature to include this in the inverse transform. Also, the value W could be $\exp(\frac{j2\pi}{N})$ or indeed some integral power of either of these values but, again, it is the convention to use the form shown.

Properties of the DFT

The discrete Fourier transform (DFT) is a linear, symmetric discrete transform. While it can be derived from the continuous Fourier transform (CFT), the two are different. It is important to note those differences and what this implies for the application of the DFT. The CFT is primarily used for continuous data and derives from that a continuous spectrum. The data is often expressed in algebraic form so allowing the spectrum to be also expressed in a closed form. There is no reason why the CFT cannot be used on discrete data; either in algebraic form or as a sequence of samples. The spectrum derived, though, will not necessarily be discrete. That only occurs when the data is periodic. Figure 2.1 shows the relationship between time and frequency of various waveforms.

The DFT can be regarded as a special case of the CFT. It processes discrete data and it produces a discrete spectrum. That spectrum is meaningful only if the input data is either one period or can be regarded as one period of some periodic discrete waveform.

If the DFT is used to derive the spectrum of non-periodic waveforms, then the result will be in error. Even with periodic waveforms, though, there still can be errors. In general, these arise for two reasons. First, the time interval used to sample the data is not equal to some integral multiple of the period. In this case, the derived spectrum will consist of the correct spectral lines but their magnitudes will be in error. In addition, there will be extra spectral components. These are usually referred to as **leakage**. The situation is illustrated in Figure 2.2. Second, the sampling rate may be too low with the result that Nyquist's criterion is not satisfied. The result is that the repeated spectra now overlap. This interference from adjacent spectra is termed **aliasing**. Figure 2.3 shows this.

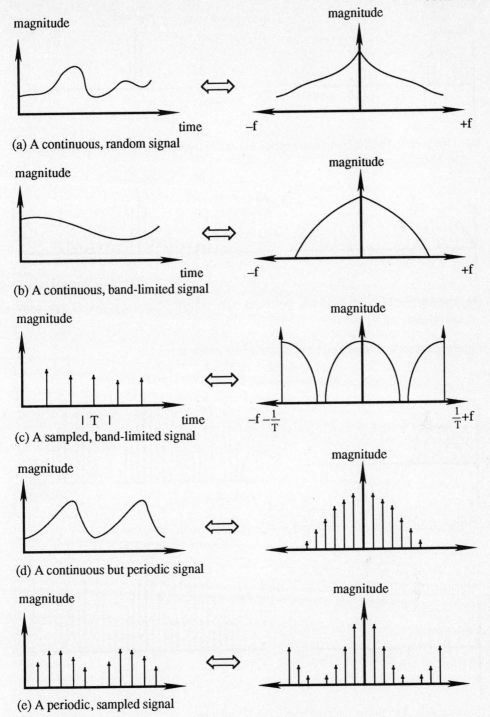

(a) A continuous, random signal

(b) A continuous, band-limited signal

(c) A sampled, band-limited signal

(d) A continuous but periodic signal

(e) A periodic, sampled signal

Figure 2.1 Relationships between the time and frequency domains of different signals

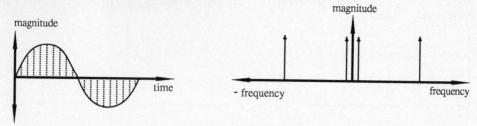

(a) The discrete Fourier transform applied to exactly the number of samples in one period of a periodic waveform

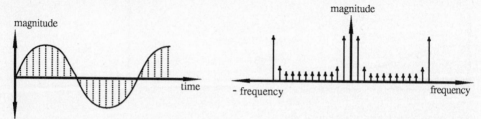

(b) The discrete Fourier transform applied to more samples than one period of a periodic waveform

Figure 2.2 Leakage effects with the discrete Fourier transform

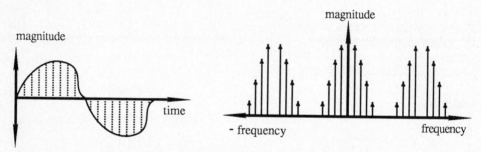

(a) The spectrum of a periodic waveform sampled above the Nyquist rate

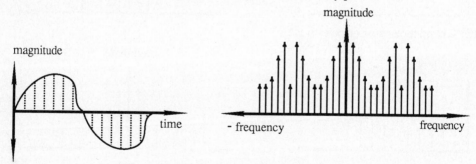

(b) A periodic waveform sampled below the Nyquist rate

Figure 2.3 Aliasing effects in sampled waveforms

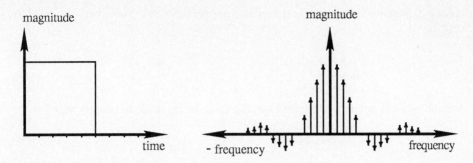

Figure 2.4 The (discrete) sampling function in the time and frequency domain

It would seem relatively easy to avoid both leakage and aliasing but there can be complications. Aliasing is the easier to avoid. In the practical situation, sampling at 20% or so above the Nyquist limit will eliminate it. While leakage is generally easy to avoid, there are circumstances where this may be difficult; for example, measurements taken of very low frequency signals or other situations where for some reason the measurement period is limited. In these instances, as leakage cannot be avoided, the question remains of whether it can be controlled. Fortunately, that is possible and to see why, consider how leakage arises.

A measurement period that is too short is mathematically the same as the correct sample sequence multiplied by a gating function. The spectrum of this signal is therefore the spectrum of the correct signal convolved with the spectrum of a gating function and that spectrum is the well-known sampling function. This is shown in Figure 2.4.

Now this function has the form of a low-pass filter but with rather substantial sidelobes and these are the cause of the leakage. A solution, then, is to choose some other form of low-pass filter where the sidelobes are quite small and to multiply the truncated signal in the time-domain by this function's impulse response. That is, apply one of the many windowing techniques that are commonly employed in designing FIR digital filters.

Consider the spectrum produced by the DFT. As an example, let:

$$x(n) = \sin\left(\frac{2\pi n}{N}\right) \qquad\qquad n = 0, 1, \cdots, N–1$$

Substituting into the equation for the DFT:

$$X(k) = \sum_{n=0}^{N-1} \sin\left(\frac{2\pi n}{N}\right) W^{kn} \qquad\qquad k = 0, 1, \cdots, N–1$$

$$= \sum_{n=0}^{N-1} \frac{1}{2j}\left(\exp\left(\frac{2\pi n}{N}\right) - \exp\left(\frac{-2\pi n}{N}\right)\right) \cdot \exp\left(\frac{-2\pi kn}{N}\right)$$

$$= \sum_{n=0}^{N-1} \frac{1}{2j}\left(\exp\left(\frac{2\pi(1-k)n}{N}\right) - \exp\left(\frac{-2\pi(1+k)n}{N}\right)\right)$$

Since the exponential terms in this summation are periodic, they sum to zero in all cases except for:

$$1 - k = 0 \quad \text{i.e. } k = 1$$
$$1 + k = 0 \quad \text{i.e. } k = -1$$

Given the periodic nature of the transform, the second term needs to be interpreted as:

$$k = -1 = N - 1$$

Substituting, for the first case:

$$X(1) = \frac{N}{2j}$$

and for the second:

$$X(N - 1) = -\frac{N}{2j}$$

The spectrum is shown in Figure 2.5(a). How is this spectrum to be interpreted? The answer is found from Figure 2.5(b) which shows the spectrum derived from the continuous Fourier transform for the same sampled waveform.

Sampling data is mathematically equivalent to multiplying a continuous waveform by a train of impulses. A well-known property of the continuous Fourier transform is that the transform of two functions multiplied in the time domain is the convolution of the two functions' spectra in the frequency domain. Therefore, the spectrum of any sampled waveform is the spectrum of that waveform repeated to infinity about the harmonics of the sampling frequency. What the DFT is deriving is one period of that periodic spectrum and the period is from zero to the first harmonic of the sampling frequency. Thus the lower half of this discrete spectrum consists of all the positive spectral lines and the upper the negative, but these are the negative lines of the first repeated spectrum.

The DFT is a complex to complex transformation and so may be more fully described by:

$$X(k) = X_r(k) + jX_i(k) \qquad\qquad k = 0, 1, \cdots, N-1$$

$$= \sum_{n=0}^{N-1} \left(x_r(n) + jx_i(n) \right) \cdot \left(\cos\left(\frac{2\pi kn}{N}\right) - j \cdot \sin\left(\frac{2\pi kn}{N}\right) \right)$$

$$x(n) = x_r(n) + jx_i(n) \qquad\qquad n = 0, 1, \cdots, N-1$$

$$= \frac{1}{N} \sum_{k=0}^{N-1} \left(X_r(k) + jX_i(k) \right) \cdot \left(\cos\left(\frac{2\pi kn}{N}\right) + j \cdot \sin\left(\frac{2\pi kn}{N}\right) \right)$$

magnitude

0 1 2   N–1

(a) Magnitude of the discrete Fourier spectrum of a sampled sinusoid

magnitude

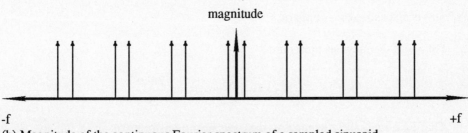

-f   +f

(b) Magnitude of the continuous Fourier spectrum of a sampled sinusoid

Figure 2.5 Magnitudes of the spectrum of a sampled sinusoid derived from the discrete and continuous Fourier transforms

For results which follow, it is also useful to note the fact:

$$X(N-k) = X_r(N-k) + jX_i(N-k) \qquad\qquad k = 0, 1, \cdots, N-1$$

$$= \sum_{n=0}^{N-1} (x_r(N-n) + jx_i(N-n)) \cdot \left(\cos\left(\frac{2\pi kn}{N}\right) + j \cdot \sin\left(\frac{2\pi kn}{N}\right) \right)$$

Now consider some of the symmetry properties of the transform. In particular, those resulting from input data with two forms of symmetry. They are:

(a) conjugate or Hermitian symmetry, where:

$$x(n) = x^*(N-n)$$

(b) skew symmetry:

$$x(n) = -x^*(N-n)$$

where * is complex conjugation. There are four particular cases to note:

1. x(n) is real and symmetric

As $x(n) = x(N-n)$, then from the linearity property of the DFT:

$$X(k) = X(N-k)$$

Therefore, from the formulae for $X(k)$ and $X(N-k)$ above:

$$X(k) + X(N-k) = 2 \cdot \sum_{n=0}^{N-1} x(n) \cdot \cos\left(\frac{2\pi kn}{N}\right)$$

That is, a real, symmetric data sequence transforms to a real, symmetric spectrum or, conversely, a real, symmetric spectrum transforms to a real, symmetric data sequence.

2. $x(n)$ is real and skew symmetric

Following the previous procedure:

$$X(k) = -X(N-k) = j \cdot \sum_{n=0}^{N-1} x(n) \cdot \sin\left(\frac{2\pi kn}{N}\right)$$

That is, a real, skew symmetric data sequence transforms to an imaginary, skew symmetric spectrum or, conversely, an imaginary, skew symmetric spectrum transforms to a real, skew symmetric data sequence.

3. $x(n)$ is imaginary and symmetric

Here:

$$X(k) = X(N-k) = j \cdot \sum_{n=0}^{N-1} x(n) \cdot \cos\left(\frac{2\pi kn}{N}\right)$$

That is, an imaginary, symmetric data sequence transforms to an imaginary, symmetric spectrum or, conversely, an imaginary, symmetric spectrum transforms to an imaginary, symmetric data sequence.

4. $x(n)$ is imaginary and skew symmetric

This situation is given by:

$$X(k) = -X(N-k) = \sum_{n=0}^{N-1} x(n) \cdot \sin\left(\frac{2\pi kn}{N}\right)$$

That is, an imaginary, skew symmetric data sequence transforms to a real, skew symmetric spectrum, or conversely, a real, skew symmetric spectrum transforms to an imaginary, skew symmetric data sequence.

Now consider a more general real data sequence $x(n)$. We can always express this as:

$$x(n) = \frac{1}{2} \cdot (x(n) + x(N-n)) + \frac{1}{2} \cdot (x(n) - x(N-n))$$

$$= x_e(n) + x_o(n)$$

That is, as the sum of a symmetric and skew symmetric part. From the linearity of the DFT, the spectrum must therefore be the sum of the spectra of these two parts. From the above, $x_e(n)$ transforms to a real, symmetric spectrum and $x_o(n)$ to an imaginary, skew symmetric spectrum. Thus we deduce that the DFT of any arbitrary real sequence must exhibit conjugate symmetry.

A similar result applies to purely imaginary data sequences. Here, the DFT has a skew symmetric real part and symmetric imaginary and so, overall, exhibits skew symmetry.

Some transforms related to the DFT

There are three principal transforms related to the DFT. They are:

1. The Discrete Cosine Transform (DCT)
   A transform pair is:

$$X(k) = M(k)\sqrt{\frac{2}{N}} \sum_{n=0}^{N-1} x(n) \cdot \cos\left(\frac{(2n+1)\pi k}{2N}\right) \qquad k = 0, 1, \cdots, N-1$$

$$x(n) = \sqrt{\frac{2}{N}} \sum_{k=0}^{N-1} M(k)X(k) \cdot \cos\left(\frac{(2n+1)\pi k}{2N}\right) \qquad n = 0, 1, \cdots, N-1$$

where $\qquad M(m) = \frac{1}{\sqrt{2}} \qquad$ for $m = 0$

$\qquad\qquad\quad = 1 \qquad$ otherwise

2. The Discrete Sine Transform (DST)
   A transform pair is:

$$X(k) = \sqrt{\frac{2}{N}} M(k) \sum_{n=0}^{N-1} x(n) \cdot \sin\left(\frac{(2n-1)k\pi}{2N}\right) \qquad k = 0, 1, \cdots, N-1$$

$$x(n) = \sqrt{\frac{2}{N}} \sum_{k=0}^{N-1} M(k)X(k) \cdot \sin\left(\frac{(2n-1)k\pi}{2N}\right) \qquad n = 0, 1, \cdots, N-1$$

3. The Discrete Hartley Transform (DHT)
   The transform pair is:

$$X(k) = \frac{1}{\sqrt{N}} \sum_{n=0}^{N-1} x(n) \cdot \left(\sin\left(\frac{2\pi kn}{N}\right) + \cos\left(\frac{2\pi kn}{N}\right) \right) \qquad k = 0, 1, \cdots, N-1$$

$$x(n) = \frac{1}{\sqrt{N}} \sum_{k=0}^{N-1} X(k) \cdot \left(\sin\left(\frac{2\pi nk}{N}\right) + \cos\left(\frac{2\pi nk}{N}\right) \right) \qquad n = 0, 1, \cdots, N-1$$

All three of these transforms are real to real transforms. Further, in each case the inverse transformation has exactly the same form as the forward.

Although the Hartley transform is usually expressed in the form given, there are other possible forms for the DCT and DST. These differ from the expressions here in the range of k and the form of the inverse. Given the relationships between sinusoids, it is possible to define a generalized sinusoidal transform of the type:

$$X(k) = \sqrt{\frac{2}{N}} \sum_{n=0}^{N-1} x(n) \cdot \sin\left(\frac{2\pi}{N}(k+\beta) \cdot (n+\alpha) + \frac{\pi}{4} \right) \qquad k = 0, 1, \cdots, N-1$$

This is known as the discrete W transform.

These discrete transforms exhibit most of the properties of the DFT, the notable exception being the cyclic convolution property. Fast transforms exist for each developed along the same lines as those for the DFT given in this Part. Given the close relationship which exists between these and the DFT, it is quite simple to express a DFT in any of these transforms and vice versa. This does not reduce the number of operations required but it sometimes leads to an easier implementation.

Deriving relationships between these transforms and the DFT follows from the symmetry relations of the last section. For example, if X(k) is the DFT of the data sequence x(n), then (ignoring scale factors):

$$\begin{aligned} DCT\{ x(n) \} &= Re\{ X(k) \} \\ DST\{ x(n) \} &= Im\{ X(k) \} \end{aligned}$$

Similarly, the Hartley transform is given (again ignoring scale factors) by:

$$DHT\{ x(n) \} = (Re\{ X(k) \} - Im\{ X(k) \}) = Re\{ (1 + j) \cdot X(k) \}$$

Although these transforms do not have the cyclic convolution property, it is possible to derive a formula for cyclic convolution expressed in these transforms just by substituting into the DFT equation. So, for example:

$$DHT\{ x(n) * h(n) \} = \frac{1}{2} \cdot [H(k) \cdot X(k) + H(k) \cdot X(N-k) + H(N-k) \cdot X(k) - H(N-k) \cdot X(N-k)]$$

where X(k) and H(k) are the Hartley transforms of x(n) and h(n) respectively.

THEORY

The elemental FFT computational section

To begin, a comment on the terminology used here. The term 'fast Fourier transform' has been widely used to denote algorithms for the DFT which require fewer operations, particularly multiplications, than the direct implementation of:

$$X(k) = \sum_{n=0}^{N-1} x(n) \cdot W^{kn} \qquad\qquad k = 0, 1, \cdots, N-1$$

This terminology will be retained here to collectively describe the algorithms investigated. What is important to stress, though, is that they are all 'fast Fourier transforms' or FFTs and that there is no such thing as 'the' FFT.

Consider this equation for the DFT. Except if it is prime, N can always be factored into:

$$N = M \cdot L$$

Now map x(n) into an array x(u,v) which has M rows and L columns. That is, into:

x(0)	x(1)	$\cdot\ \ \ \cdot$	x(L–1)
x(L)	x(L+1)	$\cdot\ \ \ \cdot$	x(2L–1)
.	.		.
.	.		.
.	.		.
x((M–1)·L)	x((M–1)·L+1)	$\cdot\ \ \ \cdot$	x((M–1)·L+L–1)

Mathematically, this mapping is achieved by transforming the index n according to:

$$n = u \cdot L + v$$

where $0 \le n \le N–1, 0 \le u \le M–1, 0 \le v \le L–1$.

A similar mapping of X(k) into an array X(s,r) can be defined. Here, for reasons which shall become apparent, it is convenient to define this mapping as:

$$k = r \cdot M + s$$

where $0 \le k \le N–1, 0 \le r \le L–1, 0 \le s \le M–1$.

Substituting these transformations into the DFT equation, it becomes:

$$X(s,r) = \sum_{u=0}^{M-1} \sum_{v=0}^{L-1} x(u, v) \cdot W^{(rM+s)(uL+v)}$$

Since W is periodic in N, its notation will be changed to W_N to reflect this. Then:

$$W_N^{(rM+s)(uL+v)} = W_N^{ruML+rvM+suL+sv}$$

$$= W_N^{ruN + (rvM+suL+sv)}$$

$$= W_N^{rvM+suL+sv}$$

Thus:

$$X(s,r) = \sum_{u=0}^{M-1} \sum_{v=0}^{L-1} x(u,v) \cdot W_N^{rvM+suL+sv}$$

This equation can be broken down into three separate actions, namely:

$$X(s,r) = \sum_{v=0}^{L-1} W_N^{rvm} \left\{ W_N^{sv} \cdot \left\{ \sum_{u=0}^{m-1} x(u,v) W_N^{suL} \right\} \right\}$$

Consider this. Its three stages are as follows:

Stage 1

$$P(s,v) = \sum_{u=0}^{M-1} x(u,v) W_N^{suL}$$

where $0 \le s \le M-1, 0 \le v \le L-1$

Now:

$$W_N^{suL} = \left[\exp\left(-\frac{j2\pi}{N} \right) \right]^{suL}$$

$$= \left[\exp\left(-\frac{j2\pi}{ML} \right) \right]^{suL}$$

$$= \left[\exp\left(-\frac{j2\pi}{M} \right) \right]^{su}$$

$$= W_M^{su}$$

Thus:

$$P(s, v) = \sum_{u=0}^{M-1} x(u, v) W_M^{su}$$

and this describes a set of DFTs along the columns of the array x(u,v). Computing each of these as the equation stands requires ($M^2 - 2M + 1$) complex multiplications and $M \cdot (M - 1)$ complex additions. There are L such DFTs to compute, thus this stage requires ($M^2 - 2M + 1$)·L complex multiplications and $M \cdot L \cdot (M - 1)$ complex additions.

Stage 2

$$Q(s, v) = W_N^{sv} \cdot P(s, v)$$

where $0 \le s \le M-1, 0 \le v \le L-1$

The array resulting from the first stage is simply multiplied by a set of factors W_N^{sv}. Since these terms all have unity magnitude but varying phases, they are termed **phase rotation factors**. Another much more poetical term is **twiddle factors** with the stage being termed **twiddling**. Only complex multiplications are involved in this stage and there are at least ($M \cdot L - M - L + 1$) of them.

Stage 3

$$X(s, r) = \sum_{v=0}^{L-1} Q(s, v) \cdot W_N^{rvM}$$

where $0 \le s \le M-1, 0 \le r \le L-1$

Following from the derivation of stage one:

$$W_N^{rvM} = W_L^{rv}$$

Thus:

$$X(s, r) = \sum_{v=0}^{L-1} Q(s, v) \cdot W_L^{rvM}$$

and this is just the computation of L-point DFTs along the rows of the array Q(s,v) resulting from Stage 2 and a mapping of the result into the array X(s, r). It requires in total $M \cdot (L^2 - 2L + 1)$ complex multiplications and $M \cdot L \cdot (L - 1)$ complex additions.

The original N-point DFT equations required ($N^2 - 2N + 1$) complex multiplications and $N \cdot (N - 1)$ complex additions. Summing the operations over the three stages of this algorithm, it requires just:

$$(M + L) \cdot N - 3N + 1 \qquad \text{complex multiplications}$$
$$N \cdot (M + L - 2) \qquad \text{complex additions}$$

There is clearly a smaller number of operations. Indeed, as N grows large this algorithm requires for both complex multiplications and complex additions the fraction:

$$\frac{(M + L)}{N} = \frac{1}{M} + \frac{1}{L}$$

of the number required for the original equations. This is significant. It is approximately $\frac{2}{\sqrt{N}}$ if:

$$M \approx L \approx \sqrt{N}$$

For a value of N as small as 128, this algorithm requires approximately one-eighth as many operations and that is of enormous practical importance. From these results, performing convolutions via a fast Fourier transform approach is far more efficient than via a direct method.

Reviewing the above procedure, we can see there is an alternative formulation. The initial mapping to the array could have been:

x(0)	x(M)	· ·	x(M·(L–1))
x(1)	x(M+1)	· ·	x(M·(L–1) + 1)
·	·		·
·	·		·
·	·		·
x(M–1)	x(2M–1)	· ·	x(M·(L–1)+M–1)

In this case, the mapping could have been chosen as:

$$n = u + v \cdot M$$

where $0 \leq n \leq N{-}1$, $0 \leq u \leq M{-}1$, $0 \leq v \leq L{-}1$, and similarly:

$$k = r + s \cdot L$$

where $0 \leq k \leq N{-}1$, $0 \leq r \leq L{-}1$, $0 \leq s \leq M{-}1$. This results in:

$$X(s,r) = \sum_{u=0}^{M-1} W_M^{su} \left\{ W_N^{ru} \left\{ \sum_{v=0}^{L-1} x(u,v) W_L^{rv} \right\} \right\}$$

Analyzing this shows it requires exactly the same number of operations as the previous form. Consequently, these procedures can be collectively described as an **elemental FFT computational section**.

An important comment to make here is that, in general, desirable algorithms are those which are **in-place**. That is, algorithms where there is no need for any additional storage over that required for the input data. Such algorithms are characterized by computations where at each stage given input and output terms are uniquely linked such as with:

$$X(n) = x(n) + x(n+1)$$
$$X(n+1) = x(n) - x(n+1)$$

The inputs in this calculation can be retrieved, the outputs calculated and then placed in the positions formerly occupied by the inputs. Hence there is no need for intermediate storage but the input data is lost. Creating an in-place algorithm is to a large degree an implementation issue. The elemental section described here is in-place.

The derivation of FFT algorithms

The elemental computational section derived describes a very simple process. Given a need to compute the DFT for a set of N points, the first step is simply to choose a pair of factors M and L of N. Although in fact one-dimensional, the data is treated as two and applying any of the variants of the elemental section involves some combination of:

1. a set of M-point DFTs along either the rows or columns of the structure;
2. a set of L-point DFTs along either the columns or rows of the structure;
3. a set of M×L twiddles over all points of the structure;

and that is all. Figure 2.6 illustrates the first variant. The section places no conditions on N other than it must be possible to factor it and that will be the case if N is composite. Then an **FFT algorithm** can be constructed merely by repeatedly applying the section to these M and L-point DFTs and so achieving further reductions in the number of operations. This continues until a point is reached where the numbers can no longer be factored, meaning they are primes.

The essential structure of an FFT algorithm is clear. Any non-prime integer N can always be expressed as the product of general factors B_i or powers of primes p_i. That is:

$$N = B_1 \cdot B_2 \cdot \ \cdots \ \cdot B_N = p_1^{n_1} \cdot p_2^{n_2} \cdot \ \cdots \ \cdot p_k^{n_k}$$

where the n_i are integers. An N-point FFT is therefore a structure comprising sets of short-length DFTs each based on one of these factors or primes and sets of twiddles. Thus it may be viewed as a procedure by which short-length DFTs can be combined into long-length DFTs via twiddles. An N of this general form gives rise to a **mixed radix** algorithm. If, however, N should be:

$$N = B^n$$

where B is some number and n is an integer, then the algorithm is termed **base-B**.

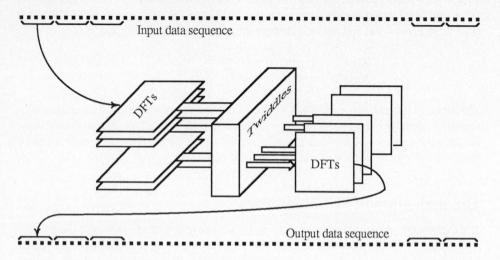

Figure 2.6 The elemental FFT computational section

Traditionally, base-B algorithms have been of more interest than mixed radix; in particular, those where B is a prime. Figure 2.7 gives an example of a mixed radix and base-B algorithm and, as it shows, the latter has greater symmetry than the mixed radix which in turn has implementation advantages. N equal to the power of some prime p gives the minimum possible number of operations in most cases. Because of this, the discussion here will center on these general mixed radix algorithms and on base-p.

FFT algorithms achieve reductions in the number of operations by taking advantage of the symmetries inherent in the periodic function W_N. However, there is a cost for in-place algorithms. In the fundamental section, input data was written into the rows of an array and output data read out from columns or vice versa. Since computer memory is linear, some addressing problems exist because the algorithm fails to preserve the order of the data sequence. To restore this, an additional operation termed **scrambling** is needed.

In the first form of the elemental section discussed, the input data was mapped into the array row by row so requiring the output data to be read out column by column. If the output needs to be in a linear sequence, there needs to be **post-scrambling**. In the second form of the section, the input data was written into columns and output read out from the rows, thus it requires **pre-scrambling**. Examine the equations. It can be seen there is no difficulty reversing these procedures. Thus, there are, in fact, four possible sections; two with pre-scrambling and two with post.

Consider a base-p FFT algorithm with N equal to p^α. The first application of the elemental section could use as parameters:

$$M = p^{\alpha-1} \qquad\qquad\qquad\qquad L = p$$

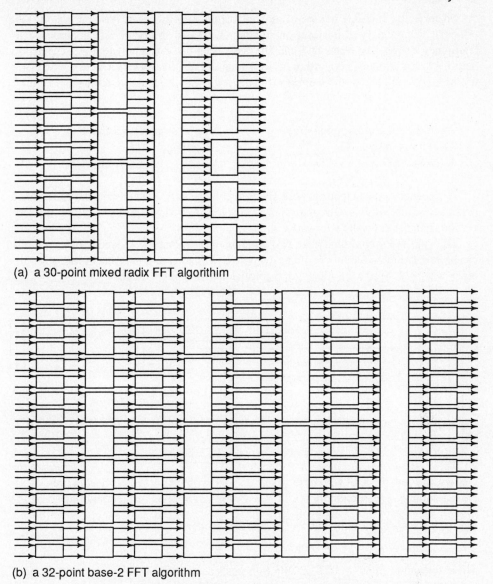

(a) a 30-point mixed radix FFT algorithim

(b) a 32-point base-2 FFT algorithm

Figure 2.7 A 30-point mixed radix and 32-point base-2 FFT algorithm

Equally though, the parameters could be:

$$M = p \qquad\qquad\qquad L = p^{\alpha-1}$$

These, with the two scrambling options, give the four algorithms sometimes refered to as the fundamental algorithms of the FFT.

Historically, the first FFTs were base-2 algorithms and they were developed along two principal lines. One was **decimation in time** (DIT) and the other was **decimation in frequency** (DIF). The terms DIT and DIF have been preserved but are now extended to describe base-p or base-B algorithms as well as base-2. With respect to a section where N is p^{α}, M is $p^{\alpha-1}$, and L is p, these terms mean the following sequence of actions:

DIT	DIF
An initial set of $p^{\alpha-1}$-point DFTs	An initial set of p-point DFTs
Twiddling	Twiddling
A final set of p-point DFTs	A final set of $p^{\alpha-1}$-point DFTs

A DIT algorithm is developed by successively applying the DIT section from the transformed data backwards, while a DIF algorithm is developed by successively applying the DIF section from the data points forwards.

DIT and DIF algorithms have two variants each in that the sequence of operations can be row-twiddle-column or column-twiddle-row. The variation is due to where scrambling is applied. Figure 2.8 illustrates the initial data mappings for each.

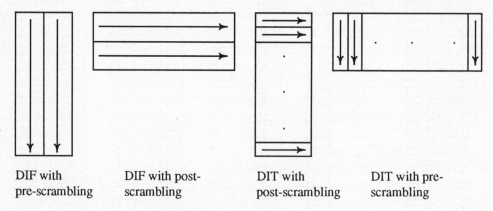

DIF with   DIF with post-   DIT with   DIT with pre-
pre-scrambling   scrambling   post-scrambling   scrambling

Figure 2.8 Initial data mappings for DIT and DIF sections

These four algorithms — DIT or DIF with post- or pre-scrambling — are just the four algorithms of the FFT. There are no conceptual differences between them but there are differences in the ease of implementation. These are only significant for software implementations, though, where they can affect computation time.

The number of operations in a base-p FFT is most easily derived through a recurrence relation and, to find that, it is first necessary to express the number of operations within an elemental section in a suitable form. The number of operations is:

$$M_N = L \cdot M_M + M \cdot M_L + (N - M - L - 1)$$
$$A_N = L \cdot A_M + M \cdot A_L$$

where M_j and A_j are the number of multiplications and additions respectively in a j-point algorithm. The number of real operations depends on the manner in which complex multiplication is implemented.

Assume that N in this base-p FFT is p^α and that M and L are given by:

$$M = p^{\alpha-1} \hspace{6cm} L = p$$

The number of complex multiplications and additions in one application of the section is:

$$M_N = p \cdot M_{\frac{N}{p}} + p^{\alpha-1} \cdot M_p + (p^\alpha - p^{\alpha-1} - p + 1)$$

$$A_N = p \cdot A_{\frac{N}{p}} + p^{\alpha-1} \cdot A_p$$

The number of multiplications and additions in a p-point DFT is $(p^2 - 2p + 1)$ and $p \cdot (p - 1)$ respectively, thus:

$$\begin{aligned} M_{p^2} &= p \cdot (p^2 - 2p + 1) + p \cdot (p^2 - 2p + 1) + (p^2 - 2p + 1) \\ &= 2p^3 - 3p^2 + 1 \\ M_{p^3} &= p \cdot (2p^3 - 3p^2 + 1) + p^2 \cdot (p^2 - 2p + 1) + (p^3 - p^2 - p + 1) \\ &= 3p^4 - 4p^3 + 1 \end{aligned}$$

$$\vdots$$

$$\begin{aligned} M_N &= \alpha \cdot p^{\alpha-1} - (\alpha + 1) \cdot p^\alpha + 1 \\ &= (p - 1) \cdot N \cdot \text{Log}_p N - N + 1 \end{aligned}$$

and:

$$A_N = (p - 1) \cdot N \cdot \text{Log}_p N$$

A comment on the short-length DFTs in FFT algorithms. It is not necessary they are of prime-length. All that ensures is possibly the maximum reduction in the number of operations. However, it is saying nothing of incremental benefit. As was shown earlier, just the application of the section itself can lead to very substantial savings and so, in seeking a balance between savings in operations and complexity in implementation, it may not be desirable to reduce the algorithm to prime length DFTs. There is nothing errant about that; it can be quite a legitimate action in the right circumstances. Further, the algorithm is still quite correctly described as a fast Fourier transform.

An important question here is whether it is indeed the case that extending the algorithm to prime length DFTs achieves maximum reduction in the number of operations. Note the distribution of operations in the three stages of the computational section. All contributed complex multiplications but only the two DFT stages contributed complex additions. This suggests that a prime-length decomposition will achieve maximum reduction in the number of additions but that whether this is true or not of multiplications depends on the twiddling and DFT stages of the algorithm.

One base-p algorithm has a unique property. A two-point DFT:

$$X(0) = x(0) + x(1)$$
$$X(1) = x(0) - x(1)$$

has no complex multiplications at all (or complex additions if $x(0)$ and $x(1)$ are real). A base-2 FFT is therefore unique in that all complex multiplications are due to twiddling.

A 4-point DFT is given by:

$$X(0) = x(0) + x(1) \quad\quad + x(2) \quad\quad + x(3)$$
$$X(1) = x(0) + x(1) \cdot W_4^1 + x(2) \cdot W_4^2 + x(3) \cdot W_4^3$$
$$X(2) = x(0) + x(1) \cdot W_4^2 + x(2) \cdot W_4^4 + x(3) \cdot W_4^6$$
$$X(3) = x(0) + x(1) \cdot W_4^3 + x(2) \cdot W_4^6 + x(3) \cdot W_4^9$$

That is:

$$X(0) = x(0) + \quad x(1) + x(2) + \quad x(3)$$
$$X(0) = x(0) - \quad j \cdot x(1) - x(2) + j \cdot x(3)$$
$$X(0) = x(0) - \quad x(1) + x(2) - \quad x(3)$$
$$X(0) = x(0) + \quad j \cdot x(1) - x(2) - \quad j \cdot x(3)$$

The only complex multiplications here are by j. To perform them just requires complex conjugation of the data followed by swapping of the real and imaginary parts. From the practical viewpoint then, there are no complex multiplications in these terms, only complex additions. As a base-4 algorithm involves fewer twiddles than a base-2, this suggests (and indeed is the case) that a base-4 algorithm requires fewer operations than a base-2. Here is an instance, then, where it is not an advantage to construct an algorithm about prime-length DFTs. Note that the above equations are better expressed as:

$$X(0) = [\, x(0) + x(2) \,] + \quad [\, x(1) + x(3) \,]$$
$$X(1) = [\, x(0) - x(2) \,] - j \cdot [\, x(1) - x(3) \,]$$
$$X(2) = [\, x(0) + x(2) \,] - \quad [\, x(1) + x(3) \,]$$
$$X(3) = [\, x(0) - x(2) \,] + j \cdot [\, x(1) - x(3) \,]$$

While the equations suggest this organization, it is also the result of applying the computational section to a 4-point DFT using factors of M and L equal to 2.

Twiddling involves a set of complex multiplications by the number W_N^{sv} where N is $M \cdot L$ and $0 \le s \le M-1$ and $0 \le v \le L-1$. There are at least ($M \cdot L - M - L + 1$) of these. When N is some power of 2 though, some of the twiddle factors must be j or $-j$. However, as will be seen later, identifying these is difficult and the total number is small.

DEVELOPMENTS

Scrambling and scrambling algorithms

A discrete Fourier transform acts on a linear sequence of data to produce a linear sequence of transformed data. An algorithm based on the section, though, treats that data as multidimensional in order to achieve a significant reduction in the number of operations. However, it does so at a potential cost as the mappings between that multi-dimensional structure and the input and output sequences are not the same. It is a potential cost as problems arise only if the algorithm needs to be implemented for a linear memory system. That is currently true for software but not necessarily for hardware. That cost, too, is a consequence of requiring an in-place algorithm. Algorithms which do preserve the input may not need this mapping but they have an associated cost in the form of the additional storage now required.

Consider an algorithm based on one of the variants of the computational section. It involves a sequence of alternating steps involving short-length DFTs and twiddling. Implementation is far easier if at each stage data is readily accessible in blocks. For half the variants, the input data is in natural or linear order but the output order is non-linear. For these, this requirement is satisfied but there is a need for an additional operation to re-order the output sequence. In the remaining cases, the input sequence is addressed in non-linear order, at each stage data is addressed in non-linear order but the generated output sequence is in natural order. Here, implementation can be difficult and there can be a substantial addressing overhead unless the data can be re-organized prior to application of the algorithm.

Although these two actions may seem different, both require exactly the same operation. In the first case, it converts the output sequence to linear form and, in the second, it groups appropriate terms into blocks. This operation is termed either pre- or post-scrambling depending on where it is applied.

In the first FFT section developed in THEORY, the link between the input sequence index n and the array indices u and v was:

$$n = u \cdot L + v \qquad 0 \leq u \leq M-1 \qquad 0 \leq v \leq L-1$$

and that between the output data sequence index k and the array indices:

$$k = v \cdot M + u$$

A scrambling algorithm is easily determined via these mapping equations. For example:

1. Generate integers n from 0 to N–1.
2. Substitute to find the array coordinates (u,v).
3. Substitute these into the second equation to find the scrambled position k.

What this indicates is that the nth position in the output is in fact the kth transformed value. To unscramble then, that value is moved to the kth position in the output sequence. The symmetry of scrambling means the value at location k should also be moved to position n. That is, scrambling or unscrambling data is in fact a swapping operation.

Where the input needs to be scrambled, exactly the same procedure applies. Integers k from 0 to N–1 are generated and substituted into the second relation above to find the array parameters u and v. They are then substituted into the first relation to find the scrambled input position and the swap made.

This procedure can be viewed in another light. Consider the input sequence x(n). To find the corresponding transformed value:

1. Express n as a number where the least significant digit is in a radix-L number system and the most significant in a radix-M.

2. Reverse those digits and convert to a number k.

The result indicates where the scrambled data should be moved.

Now consider scrambling in a base-B algorithm where N is B^α. For the first section, let:

$$M = B \qquad\qquad\qquad L = B^{\alpha-1}$$

To unscramble data in this first step, the index n is expressed as (b_0, b') where b_0 is a radix-B number system digit and b' is a radix-$B^{\alpha-1}$ digit. The scrambled position is found from the number (b', b_0). The elemental section is applied again and so b' is described by (b_1, b'') where b_1 is a radix-B number system digit and where b" is a radix-$B^{\alpha-2}$ digit. The unscrambled position is given by (b'', b_1). Combining, up to this stage the input is given by (b_0, b_1, b'') and the scrambled output by (b'', b_1, b_0). Continuing this procedure, we eventually find each input position is given by $(b_0, b_1, \cdots, b_\alpha)$ where all digits are within the range of 0 to B and so this is a radix-B number. The scrambled position is simply given by reversing the digits of this number and converting to k.

A mixed radix algorithm is very similar. Input position is now described by some number $(r_0, r_1, \cdots, r_M)$ but each digit is now a number between 0 and R_i where:

$$N = \prod_{i=1}^{M} R_i$$

Hence the term mixed radix. To find the scrambled position, this number is simply reversed and converted. That is:

$$k = r_0 \cdot R_M^{\,m} + r_1 \cdot R_{M-1}^{\,m-1} + \cdots + r_M \cdot R_0^{\,0}$$

A scrambling algorithm generates the scrambled positions from a linear count for software implementations. This enables output values (or input if pre-scrambling is used) to be moved to their correct positions. Since operations in a scrambling algorithm are just simple swaps, there will be no more than $\frac{N}{2}$ such operations.

What must be done is to generate the linear sequence of numbers from 0 to N–1, express each as a base-B or mixed radix number, reverse the digits and convert the result. While an effective way of tackling the problem, it is hardly efficient. Generating the initial sequence requires N real additions in total plus some checks on the magnitude of the digits. However, problems arise with reversing and with generating the scrambled numbers. At least $\frac{M \cdot (N-1)}{2}$ real multiplications and $\frac{(M-1) \cdot (N-1)}{2}$ real additions are required in this last step and these are substantial numbers. Many, though, are unnecessary. In a base-B algorithm, for example, there are B numbers in the sequence such as the first number $(0, 0, \ldots, 0)$ which are their own reverse. None of these numbers need to be reversed, nor does the reversed number need to be calculated. The same applies to any symmetric numbers. Further, if x is a generated reversed number, the next is just $(x + B^{\alpha-1})$. This is often easily generated.

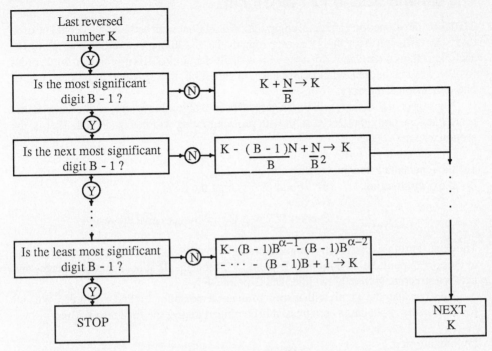

Figure 2.9 A digit-reversal counter scrambling algorithm

A very widely used and quite efficient algorithm, uses a digit reversal counter. Figure 2.9 outlines the algorithm. It requires less than $\frac{(N-B)}{2}$ operations and these are additions and possibly checks on digits. Of these, in $\frac{(B-1)}{B}$ cases only one addition is required, in $\frac{(B-1)}{B^2}$ cases only one addition and a check on a digit, in $\frac{(B-1)}{B^3}$ cases only one addition and two digit checks and so on.

The computation of the scrambled locations plus the swapping operations associated with it are burdens which largely exist for software alone. For hardware implementations, scrambling merely determines the interconnections between circuits. The form of the FFT algorithm, though, has a significant impact on the computational burden for software. Digital computers use a binary number system and therefore the scrambling algorithm for any FFT based on some power of 2 reduces to number reversal alone. In contrast, for mixed radix algorithms all steps in scrambling must be implemented and the digit-reversal counter method becomes quite complex. This together with the other advantages of base-2 FFTs makes it very difficult to find good reasons to use other bases.

The fact that scrambling requires an algorithm with the order of $\frac{N}{2}$ additions is often ignored and not quoted in the total number of operations needed by an FFT. However, this is a substantial number and there are clearly benefits in investigating more efficient techniques. So far, not a great deal of research has been conducted into this question.

The general base-B FFT algorithm

THEORY developed an elemental computational section (with variants) that could be used to develop efficient algorithms for computing the DFT. Although a number of properties of such algorithms were discussed, none were actually developed. It is useful to do this in order to observe some implementation problems. In particular, those created by a linear, that is, one-dimensional, memory.

The object of this study will be the general base-B algorithm with $N = B^\alpha$, where B and α are integers. Only the DIT section with pre-scrambling will be considered. That is, the section where:

1. the input data is mapped to the rows of an array;
2. the operations are: $B^{\alpha-1}$-point DFTs along the rows
                        twiddles
                        B-point DFTs on the columns to give the result.

The extension of this section to an algorithm simply involves repeatedly applying the section to those $B^{\alpha-1}$-point DFTs until only B-point DFTs remain. That occurs after α stages and leaves a structure with only twiddles and B-point DFTs.

Now consider the details with respect to an implementation for linear memory. We will follow how the input data is progressively combined to give the final result. Thus:

Preliminary step
Scramble the data according to the methods outlined in the last section.

Step 1
The data is divided into $B^{\alpha-1}$ groups of B points and each is discrete Fourier transformed. Thus:

$$X_1(k + B \cdot I_1) = \sum_{n=0}^{B-1} x_1(n + B \cdot I_1) \cdot W_B^{kn}$$

where $x_1(n)$ = the scrambled input data
          k $= 0, 1, \cdots, B-1$
          I_1 $= 0, 1, \cdots, B^{\alpha-1} -1$

Step 2
This step will be taken as having two stages. First, the data is twiddled according to:

$$X_2(k + B \cdot I_1 + B^2 \cdot I_2) = (W_{B \cdot B})^{kI_1} \cdot (k + B \cdot I_1 + B^2 \cdot I_2)$$

where $k = 0, 1, \cdots, B-1$
          $I_1 = 0, 1, \cdots, B-1$
          $I_2 = 0, 1, \cdots, B^{\alpha-2} -1$

To complete what is in fact one elemental computational section (and the beginning of the next), the second stage of this step will be a set of B-point DFTs:

$$X_2(k + B \cdot I_1 + B^2 \cdot I_2) = \sum_{n=0}^{B-1} X_2(k + B \cdot n + B^2 \cdot I_2) W_B^{I_1 n}$$

where

$$k = 0, 1, \cdots, B-1$$
$$I_1 = 0, 1, \cdots, B-1$$
$$I_2 = 0, 1, \cdots, B^{\alpha-2} - 1$$

.

.

.

Step L

All subsequent steps are also two-stage. First there is twiddling and then there are B-point DFTs. In keeping with the terminology used so far, the twiddling is given by:

$$X_L(k + B^{L-1} \cdot I_1 + B^L \cdot I_2) = (W_{B^L})^{kI_1} \cdot X_{L-1}(k + B^{L-1} \cdot I_1 + B^L \cdot I_2)$$

where

$$k = 0, 1, \cdots, B^{L-1} - 1$$
$$I_1 = 0, 1, \cdots, B-1$$
$$I_2 = 0, 1, \cdots, B^{\alpha-L} - 1$$

and the DFT stage which follows is given by:

$$X_L(k + B^{L-1} \cdot I_1 + B^L \cdot I_2) = \sum_{n=0}^{B-1} X_L(k + B^{L-1} \cdot n + B_L \cdot I_2) \cdot W_B^{I_1 n}$$

where k, I_1 and I_2 are as above.

Only B-point DFTs exist in this structure and, assuming these are directly implemented, each requires $(B^2 - 2B + 1)$ complex multiplications and $B \cdot (B-1)$ complex additions. Each stage requires $B^{\alpha-1}$ of these and with α stages overall, then the DFTs contribute:

$$\alpha \cdot B^{\alpha-1} \cdot (B^2 - 2B + 1) \qquad \text{complex multiplications}$$
$$\alpha \cdot B^{\alpha} \cdot (B-1) \qquad \text{complex additions.}$$

At the kth stage, each of the $B^{\alpha-k}$ twiddles in that stage has $(B \cdot B^{k-1} - B - B^{k-1} + 1)$ multiplications. Summing over the $(\alpha-1)$ stages where twiddling actually occurs, there are $([\alpha \cdot (B-1) - B] \cdot B^{\alpha-1} + 1)$ in total. As twiddling only has multiplications, then for the complete algorithm there are:

$$\alpha \cdot N \cdot (B-1) \qquad \text{complex additions}$$
$$(\alpha \cdot B - \alpha - 1) \cdot N + 1 \qquad \text{complex multiplications}$$

which is the result derived earlier.

A number of comments need to be made here. As no account has been taken of the computational effort required for scrambling, that will decrease the advantage of the FFT. For large N, the ratio of multiplications in the DFTs to those in twiddling is $(B-1)$. Given that all additions are due to the DFTs as well, the importance of efficient short-length DFTs to any FFT algorithm is very evident. Finally, note that step 1 and step 2 are actually part of the one computational section. Given the repeated application of the section though, it is convenient to express the algorithm as a set of twiddle-DFT steps plus the extra, initial DFT stage.

Consider the constants used by this algorithm. The B-point DFT stages use the set of B complex numbers $\{\ 1, W_B, W_B^2, \cdots, W_B^{B-1}\ \}$. Combined with the values used in twiddling though, the total number of coefficients used must be the set $\{\ 1,\ W_N, \cdots, W_N^2, \cdots, W_N^{N-1}\ \}$. These require another 2N memory cells, or exactly the same number as the data. The original DFT equations also require 2N cells for data and 2N for coefficients. As the number of cells is twice that for direct convolution methods, this is a disadvantage of the Fourier transform approach over direct techniques.

Mixed radix algorithms

We shall only consider the case where N is $p_1 \cdot p_2 \cdot p_3 \cdot \ \cdots\ \cdot p_\alpha$. That is, the product of unique primes p_i. If N has a term p_i^n, then the appropriate part of the decomposition becomes a base-p algorithm and follows the formulation of the base-B algorithm. Thus:

Preliminary step
The initial data set x(n) is scrambled into a sequence $x_1(n)$ according to the algorithm given earlier.

Step 1
The data is discrete Fourier transformed according to:

$$X_1(k + p_1 \cdot I_1) = \sum_{n=0}^{p_1-1} x_1(n + p_1 \cdot I_1) \cdot W_{p_1}^{kn}$$

where $k\ =\ 0, 1,\ \cdots, p_1 - 1$

$I_1\ =\ 0, 1,\ \cdots,\ \dfrac{N}{p_1} - 1$

Step 2
In keeping with the base-B algorithm outlined earlier, each successive stage involves two steps. First, a set of twiddles and then a set of DFTs. The twiddles are given by:

$$X_2(k + p_1 \cdot I_1 + p_1 \cdot p_2 \cdot I_2) = W_{p_1 p_2}^{kI_1} \cdot X_1(k + p_1 \cdot I_1 + p_1 \cdot p_2 \cdot I_2)$$

where $k\ =\ 0, 1,\ \cdots, p_1 - 1$
$I_1\ =\ 0, 1,\ \cdots, p_2 - 1$

$I_2\ =\ 0, 1,\ \cdots,\ \dfrac{N}{p_1 p_2} - 1$

and the DFTs by:

$$X_2(k + p_1 \cdot I_1 + p_1 \cdot p_2 \cdot I_2) = \sum_{n=0}^{p_2-1} X_2(k + p_1 \cdot n + p_1 \cdot p_2 \cdot I_2) \cdot W_{p_2}^{I_1 n}$$

where
$$\begin{aligned}
k &= 0, 1, \cdots, p_1 - 1 \\
I_1 &= 0, 1, \cdots, p_2 - 1 \\
I_2 &= 0, 1, \cdots, \frac{N}{p_1 p_2} - 1
\end{aligned}$$

Step L

Again, there are twiddles and DFTs. The twiddles are given by:

$$X_L(k + S_{L-1} \cdot I_1 + S_L \cdot I_2) = W_{S_L}^{k I_1} \cdot X_{L-1}(k + S_{L-1} \cdot I_1 + S_L \cdot I_2)$$

where
$$\begin{aligned}
k &= 0, 1, \cdots, S_{L-1} - 1 \\
S_L &= p_1 \cdot p_2 \cdots p_L \\
I_1 &= 0, 1, \cdots, p_L - 1 \\
I_2 &= 0, 1, \cdots, \frac{N}{S_L} - 1
\end{aligned}$$

and the DFTs by:

$$X_L(k + S_{L-1} \cdot I_1 + S_{L-1} \cdot I_2) = \sum_{n=0}^{p_L-1} X_L(k + S_{L-1} \cdot n + S_L \cdot I_2) \cdot W_{p_L}^{I_1 n}$$

where I_1, I_2, and, k are as above.

There are α stages so that, over all, there are α sets of DFTs and $(\alpha - 1)$ sets of twiddles.

Computing the number of operations follows from the derivation given for the base-B algorithm but is clearly a much more complex expression.

Base-2 fast Fourier transforms

Apart from being the first to be developed, the most important FFTs are those based on N equal to a power of 2. There are three principal reasons for their popularity. One is that N equal to 2^n is a convenient number for computing. Another is that all complex multiplications in a base-2 FFT are due to twiddling as the 2-point DFT is just:

$$\begin{aligned}
X(0) &= x(0) + x(1) \\
X(1) &= x(0) - x(1)
\end{aligned}$$

Finally, scrambling operations become much simpler.

Expressing the previous base-B algorithm as a base-2 algorithm is simple. The initial set of DFTs are just given by:

$$X_1(2I_1) = x(2I_1) + x(2I_1 + 1)$$
$$X_1(2I_1+1) = x(2I_1) - x(2I_1 + 1)$$

where
$$I_1 = 0, 1, \cdots, \frac{N}{2} - 1$$

All subsequent stages involve twiddling and further base-2 DFTs. For the Lth stage:

$$X_L(k+2^{L-1} \cdot I_1 + 2^L \cdot I_2) = W_{2^L}^{kI_1} \cdot X_{L-1}(k+2^{L-1} \cdot I_1 + 2^L \cdot I_2)$$

where
$$k = 0, 1, \cdots, 2^{L-1} - 1$$

$$I_1 = 0, 1$$

$$I_2 = 0, 1, \cdots, \frac{N}{2^L} - 1$$

describes the twiddling. The DFTs at this stage are given by:

$$X_L(k+2^L \cdot I_2) = X_L(k+2^L \cdot I_2) + X_L(k+2^{L-1}+2^L \cdot I_2)$$

$$X_L(k+2^{L-1}+2^L \cdot I_2) = X_L(k+2^L \cdot I_2) + X_L(k+2^{L-1}+2^L \cdot I_2)$$

where
$$k = 0, 1, \cdots, 2^{L-1} - 1$$

$$I_2 = 0, 1, \cdots, \frac{N}{2^L} - 1$$

The number of two-point DFTs is $\frac{N}{2} \cdot Log_2 N$ and so the number of additions is $N \cdot Log_2 N$. Only twiddling contributes multiplications, thus the total number of these is $(\frac{N}{2} \cdot Log_2 N - N + 1)$.

All algorithms based on powers of 2 have a number of multiplications in the twiddling stages by j alone. Finding and separately executing these introduces an overhead in software that often outweighs the gains. The first stage of twiddling in a base-2 algorithm, though, involves only one multiplication and that is by j. Therefore, there is some advantage in implementing the initial DFTs plus the first stage separately from the rest of the algorithm.

The first base-2 FFT algorithms were developed along the following lines. Divide a set of N data points x(n) into two $\frac{N}{2}$-point sequences $x_1(n)$ and $x_2(n)$ where:

$$x_1(n) = x(2n)$$
$$x_2(n) = x(2n+1)$$

Hence the term decimation in time. (For the decimation in frequency algorithm, the division is just:

$$x_1(n) = x(n) \qquad\qquad n = 0, 1, \cdots, \frac{N}{2} - 1$$

$$x_2(n) = x(n) \qquad\qquad n = \frac{N}{2}, \cdots, N - 1$$

and it is not quite so obvious why the name is chosen.) The DFT equations become:

$$X(k) = \sum_{n=0}^{\frac{N}{2}-1} x(2n) \cdot W_N^{k \cdot 2n} + \sum_{n=0}^{\frac{N}{2}-1} x(2n+1) \cdot W_N^{k \cdot (2n+1)}$$

$$= \sum_{n=0}^{\frac{N}{2}-1} x_1(n) \cdot W_N^{k \cdot 2n} + W_N^k \sum_{n=0}^{\frac{N}{2}-1} x_2(n) \cdot W_N^{k \cdot (2n+1)}$$

As W is periodic in N then W^2 is periodic in $\frac{N}{2}$. Thus this equation becomes:

$$X(k) = X_1(k) + W_N^k \cdot X_2(k)$$

where $X_1(k)$ is the $\frac{N}{2}$-point DFT of $x_1(n)$ and $X_2(k)$ is the $\frac{N}{2}$-point DFT of $x_2(n)$, and it must be interpreted as:

$$X(k) = X_1(k) + W_N^k \cdot X_2(k) \qquad k = 0, 1, \cdots, \frac{N}{2} - 1$$

$$= X_1\left(k - \frac{N}{2}\right) + W_N^k \cdot X_2\left(k - \frac{N}{2}\right) \qquad k = \frac{N}{2}, \cdots, N - 1$$

Given the periodicity of W_N, this can be expressed as:

$$X(k) = X_1(k) + W_N^k \cdot X_2(k) \qquad k = 0, 1, \cdots, \frac{N}{2} - 1$$

$$= X_1\left(k - \frac{N}{2}\right) - W_N^{\left(k - \frac{N}{2}\right)} \cdot X_2\left(k - \frac{N}{2}\right) \quad k = \frac{N}{2}, \cdots, N - 1$$

This elemental computational section describes an in-place algorithm. From an earlier graphical description, it is termed a **butterfly**. The name becomes more obvious when the section is expressed as:

$$X(k) = X_1(k) + W_N^k \cdot X_2(k)$$

$$X(k + \frac{N}{2}) = X_1(k) - W_N^k \cdot X_2(k)$$

for $k = 0, 1, \cdots, \frac{N}{2} - 1.$

Algorithms based on a power of two have a number of multiplications by 1 or –1, or j and –j. The exact formula for identifying these must be left until Part 3 but Table 2.1 shows the number of such operations in a DFT tabulated for different values of N, and the remaining multiplications. As can be seen, it is more worthwhile concentrating on FFT algorithms employing short-length DFTs than identifying such operations in these high-order DFTs.

Table 2.1 Number of multiplications by 1, –j, –1 and j plus the total number of operations in base-2 DFTs and the number in Base-2 and Base-4 FFTs

N	–j	–1	j	1	Remaining operations	Base-2 FFT	Base-4 FFT
4	2	4	2	8	0	0	0
8	9	12	8	20	16	5	–
16	24	32	24	48	128	17	9
32	64	80	64	112	704	49	–
64	160	192	160	256	3328	129	81
128	384	448	384	576	14592	321	–
256	896	1024	896	1280	61420	769	513

Base-3 algorithms

Base-3 algorithms present difficulties in scrambling and addressing due to the radix 3 arithmetic that must be employed and that precludes them from most applications. However, there is a very efficient way of calculating 3-point DFTs that can make base-3 algorithms quite attractive in some cases. Rather than express numbers in the usual complex form, they can be expressed in the form (a + μb) where:

$$\mu^3 - 1 = (\mu - 1) \cdot (\mu^2 + \mu + 1) = 0$$

Hence:

$$\mu = \exp\left(-\frac{j2\pi}{3}\right) = -\frac{1}{2} - j\frac{\sqrt{3}}{2}$$

The conversion between the usual complex format and this one is just:

$$a + jb \rightarrow \left(a - \frac{b}{\sqrt{3}}\right) - \mu\frac{2}{\sqrt{3}} \cdot b$$

$$a + \mu b \rightarrow \left(a - \frac{b}{2}\right) - j\frac{\sqrt{3}}{2} \cdot b$$

Conjugation is given by:

$$(a + \mu b)^* = (a - b) - b\mu$$

Addition is obvious and multiplication is:

$$(a + \mu b) \cdot (c + \mu d) = (ac - bd) + (ad + bc - bd)\mu$$

A 3-point DFT is given by

$$
\begin{aligned}
X(0) &= x(0) + x(1) &+ x(2) \\
X(1) &= x(0) + x(1) \cdot W + x(2) \cdot W^2 \\
X(2) &= x(0) + x(1) \cdot W^2 + x(2) \cdot W
\end{aligned}
$$

where W is $\exp(-\frac{j2\pi}{3})$. Both input and output should be considered complex. Representing them in the new format and defining the input as:

$$x(i) = x_R(i) + \mu \cdot x_I(i)$$

the DFT becomes:

$$
\begin{aligned}
X(0) &= (x_R(0) + x_R(1) + x_R(2) &) + \mu(x_I(0) + x_I(1) + x_I(2) &) \\
X(1) &= (x_R(0) - x_I(1) - x_R(2) + x_I(2)) + \mu(x_I(0) + x_R(1) - x_I(1) - x_R(2)) \\
X(2) &= (x_R(0) + x_I(1) - x_R(1) - x_I(2)) + \mu(x_I(0) - x_R(1) + x_R(2) - x_I(2))
\end{aligned}
$$

These equations involve no multiplications at all; only additions. Therefore, an FFT expressed in this form of complex notation, like the base-2 algorithm, only has multiplications due to twiddling. However, as the base is 3 rather than 2, fewer stages are required than in a base-2 algorithm, hence it is more efficient. The principle can be extended to make quite efficient base-6 and base-9 algorithms.

For software implementations, this algorithm is not entirely attractive given the structure of conventional computers. For hardware, however, these difficulties are relatively minor.

A comparison of different FFT algorithms

To begin, a comment on the different forms of the algorithms. Consider the base-B, mixed radix and base-2 algorithms that have been developed here. All are DIT with pre-scrambling and this gives the very important advantage that all the powers of W in all stages of the algorithm follow in natural order. That makes their generation, or their retrieval from memory, very simple. The DIF section with post-scrambling offers the same advantage. Consequently, most software implementations use one or other of these. For hardware implementations, there is no real differences between any of the forms.

It is sobering to consider the extent of the application of the algorithms developed so far. For example, how many possible base-B FFT algorithms exist up to say, 5000? Expressed another way, how many values of B^α exist for different B and α when $B^\alpha \leq 5000$, $B > 1$ and $\alpha \geq 2$? Table 2.2 lists this data. There are only 106 such algorithms.

Another interesting question is how many possible algorithms are there? To be more specific, how many base-B and mixed radix algorithms are there for n ranging from 2 to 5000. As the methods so far developed cannot treat prime length DFTs, then the answer is 4999 less

Table 2.2 Number of base-B algorithms between 0 and 5000

Bases	Values of α	Total algorithms
2	2, 3, 4, 5, 6, 7, 8, 9, 10, 11, 12	11
3	2, 3, 4, 5, 6, 7	6
4	2, 3, 4, 5, 6	5
5	2, 3, 4, 5	4
6 - 8	2, 3, 4	9
9 - 17	2, 3	18
18 - 70	2	53

the number of primes up to and including 5000. That number is 4330. Base-B algorithms are therefore less than 3% of all possible algorithms. Further, in spite of 5000 being quite a large number, over one half of the base-B algorithms involve just one stage and so are only one application of the computational section. Equally, a very large number of the remaining mixed radix algorithms involve no more than two prime factors and they, too, are just the computational section.

Table 2.2 shows that if an FFT algorithm is needed for arbitrary N, then it will almost certainly need to be a mixed radix algorithm. However, scrambling and addressing become quite complex in software implementations of such algorithms (and in many hardware implementations, too, for that matter). The regularity and symmetry of the base-B appeal but there are not many of these. If used, then almost certainly the data sequence will need to be padded out with zeroes. Since some of the data is now zero, then the algorithm's structure can be **pruned**. Another question here is which base-B algorithm to use? Table 2.2 suggests that for arbitrary N, it is much more likely that a base-2 algorithm will be closer than any other base. These have the further attraction of both addressing and scrambling being considerably easier to implement. Base-3 is not as attractive in this respect but it, too, has a reasonable spread of values and may be worth considering, especially with the improved 3-point DFT given earlier. Since base-2^α algorithms share many of the properties of base-2, they, too, can be examined. In all other cases, though, it is very difficult to see any advantage for software implementation.

Much the same conclusion can be drawn for hardware implementations but for slightly different reasons. Scrambling is just a wiring problem but the computational burden of addressing is a concern, although it is not difficult to ensure it doesn't impinge on performance of the system. What it will do in most instances, though, is impact on the economics. More complex and irregular circuitry is invariably more costly and that can be difficult to justify. Nevertheless, there will be circumstances where a base higher than 2 or a mixed radix algorithm is warranted and for that reason, the hardware designer should not discard the possibility before some preliminary investigation.

Given the importance of base-2^β algorithms, it is useful to examine these in a little more depth. Let N be fixed and given by $(2^\beta)^{\frac{\alpha}{\beta}}$. Now for N equal to B^γ, the total number of multiplications in the base-B algorithm given earlier is:

$$N_M = \left(\frac{N}{B}\right) \cdot \left(\gamma(D+B) - B - \gamma\right) + 1$$

where D is the number of multiplications in the B-point DFTs. For the base-2 algorithm then, where N is $(2^1)^{\frac{\alpha}{1}}$ and D is zero, the number of multiplications is:

$$N_{M2} = \left(\frac{N}{2}\right) \cdot (\alpha - 2) + 1$$

Since the only complex multiplications in a base-4 DFT are by j and so are in fact complex conjugations, D is also zero for this algorithm. Hence the number of multiplications is:

$$N_{M4} = \left(\frac{N}{4}\right) \cdot (1.5\alpha - 4) + 1$$

$$= N_{M2} - \frac{N\alpha}{8}$$

Thus if N is divisible by 4, then the base-4 is superior to the base-2, and for large N this difference is substantial. Consider now a base-8 algorithm, that is β equal to 3. Here, two different approaches can be considered. If an 8-point DFT is closely examined, then by eliminating all operations by j, –j, –1, and 1, only 16 multiplications remain. Thus:

$$N_{M8} = \left(\frac{N}{8}\right) \cdot \left(\frac{23\alpha}{3} - 8\right) + 1$$

$$= N_{M2} + \left(\frac{11}{24}\right) \cdot N\alpha$$

Implementing these 8-point DFTs via an FFT reduces the multiplications to five and so the total number of multiplications in the algorithm is the same as for the base-2 algorithm. Those five multiplications can usually be reduced by heuristic means, so making this algorithm comparable or even better than a base-4. If a base-16 algorithm is examined or indeed any base for higher powers of 2, it is found that none are better than a base-2.

These results do not consider all possible simplifications. For example, they consider no reduction in the number of twiddling multiplications or simplifications which result from the general symmetry of the W^{nk}. A particular class of the latter are the eighth roots of –1 given by $\frac{1}{\sqrt{2}}(\pm 1 \pm j)$ which clearly involve less effort than a full complex multiplication. A thorough study, though, shows the base-8 is slightly better than the base-4 algorithm and both involve significantly fewer multiplications than the base-2 algorithm.

The conclusion is obvious. In very general terms, a base-2 FFT is better on the evidence presented so far than that for any other prime base. However, if it is possible to use it, a base-4 is far better than a base-2 and a base-8 can be slightly better. The base-4 fast Fourier transform, then, can rightly be termed an 'optimal' algorithm in the sense of minimizing multiplications. The success of the base-4 is simply due to the fact there are multiplications by j, –j, –1 or 1 and they are easily eliminated. Other FFT algorithms do not have these and so none of these can be more efficient than the simple routine.

The split radix algorithm

The split radix algorithm is interesting in several respects. It is a base-2 algorithm and based on similar theory to other base-2 algorithms but it has a substantially different structure. This makes it more difficult to implement but, in compensation (using the theory of Part 4), it can be shown it has the minimum number of multiplications possible for an FFT algorithm. It also has fewer additions for small N than any of the other common base-2 algorithms. Both of these properties, though, are dependent on the complex multiplication algorithm used.

Consider the forward DFT equation:

$$X(k) = \sum_{n=0}^{N-1} x(n) \cdot W_N^{kn} \qquad\qquad k = 0, 1, \cdots, N-1$$

This may be divided into even and odd terms, where the even may be grouped into:

$$X(2k) = \sum_{n=0}^{\frac{N}{2}-1} \left(x(n) + x\left(n + \frac{N}{2}\right) \right) \cdot W_N^{2kn} \qquad\qquad k = 0, 1, \cdots, \frac{N}{2}-1$$

$$= \sum_{n=0}^{\frac{N}{2}-1} \left(x(n) + x\left(n + \frac{N}{2}\right) \right) \cdot W_{\frac{N}{2}}^{kn} \qquad\qquad k = 0, 1, \cdots, \frac{N}{2}-1$$

and this describes an $\frac{N}{2}$-point DFT of the sequence $(x(n) + x(n + \frac{N}{2}))$. The odd terms are expressed in the two equations:

$$X(4k+1) = \sum_{n=0}^{\frac{N}{4}} \left\{ \left(x(n) - x\left(n + \frac{N}{2}\right) \right) - j\left(x\left(n + \frac{N}{4}\right) - x\left(n + \frac{3N}{4}\right) \right) \right\} \cdot W_N^n \cdot W_N^{4kn}$$

$$= \sum_{n=0}^{\frac{N}{4}} \left\{ \left(x(n) - x\left(n + \frac{N}{2}\right) \right) - j\left(x\left(n + \frac{N}{4}\right) - x\left(n + \frac{3N}{4}\right) \right) \right\} \cdot W_N^n \cdot W_{\frac{N}{4}}^{kn}$$

and:

$$X(4k+3) = \sum_{n=0}^{\frac{N}{4}} \left\{ \left(x(n) - x\left(n + \frac{N}{2}\right) \right) + j\left(x\left(n + \frac{N}{4}\right) - x\left(n + \frac{3N}{4}\right) \right) \right\} \cdot W_N^{3n} \cdot W_N^{4kn}$$

$$= \sum_{n=0}^{\frac{N}{4}} \left\{ \left(x(n) - x\left(n + \frac{N}{2}\right) \right) + j\left(x\left(n + \frac{N}{4}\right) - x\left(n + \frac{3N}{4}\right) \right) \right\} \cdot W_N^{3n} \cdot W_{\frac{N}{4}}^{kn}$$

for $k = 0, 1, \cdots, \frac{N}{4} - 1$. These two terms describe $\frac{N}{4}$-point DFTs but of much more complicated data sequences than for the even terms equation.

This decomposition describes the basic computational procedure of the split radix algorithm. Compare it to the earlier base-2 algorithms. In those algorithms, at the last stage the required N spectral terms are taken from a series of B identical DFTs of size $\frac{N}{B}$. In turn, each of these derives from B simpler DFTs and so on. Here, however, the N output terms come from an $\frac{N}{2}$-point DFT and two $\frac{N}{4}$-point DFTs. The $\frac{N}{2}$-point DFT can found from the output terms of an $\frac{N}{4}$-point and two $\frac{N}{8}$-point DFTs, and each of the two $\frac{N}{4}$-point DFTs from an $\frac{N}{8}$-point and two $\frac{N}{16}$-point DFTs. Continuing, the eventual result must be a set of terms which prepare data for base-2 DFTs. Hence while this split radix algorithm must be a base-2 algorithm, it has a unique structure.

The elemental computational section in this algorithm can be expressed as a two-stage process with the second only involving half the points. It is:

1.
$$x_0(n) = x(n) + x\left(n + \frac{N}{2}\right) \qquad n = 0, 1, \cdots, \frac{N}{2} - 1$$

$$x_0\left(n + \frac{N}{2}\right) = x(n) - x\left(n + \frac{N}{2}\right) \qquad n = 0, 1, \cdots, \frac{N}{2} - 1$$

2.
$$D(n) = x_0\left(n + \frac{N}{2}\right) - jx_0\left(n + \frac{3N}{4}\right) \qquad n = 0, 1, \cdots, \frac{N}{4} - 1$$

$$x_1(n) = D(n) \cdot W_{N^n} \qquad n = 0, 1, \cdots, \frac{N}{4} - 1$$

$$x_1\left(n + \frac{N}{4}\right) = D*(n) \cdot W_{N^{3n}} \qquad n = 0, 1, \cdots, \frac{N}{4} - 1$$

where * denotes complex conjugation. This structure complicates implementation. The problem is illustrated by the split radix algorithm for 16 points shown in Figure 2.10. As it suggests, for N equal to 2^α all data undergoes α stages of processing. However, that processing cannot be organized in the uniform way achieved by other base-2 algorithms with the result that the algorithm lacks their symmetry. This is a complication for implementation but not an insurmountable one and several exist.

The name 'split radix' arises from the division into $\frac{N}{2}$ and $\frac{N}{4}$ DFTs. Strictly, this is a 2/4 split. Other splits have been investigated but none have proven to be of any practical value.

Consider the number of operations in this algorithm. To begin, we calculate the number of operations in the section alone. The only multiplications are in the second stage of the section and there are nominally two sets of $\frac{N}{4}$ of these. However, one index is always zero and another $\frac{N}{8}$. Thus there are actually just $2\left(\frac{N}{4} - 2\right)$ (complex) multiplications in all, plus two multiplications by an eighth root of one. We distinguish the latter as it is a simpler operation than a general multiplication. Additions only occur in the first part of the section and there are N of these. However, as half the points in this stage are just the conjugation of the other half, this can be reduced to $\frac{N}{2}$. In total then there are $\frac{N}{2} - 4$ general multiplications, 2 multiplications by an eighth root of one and N additions.

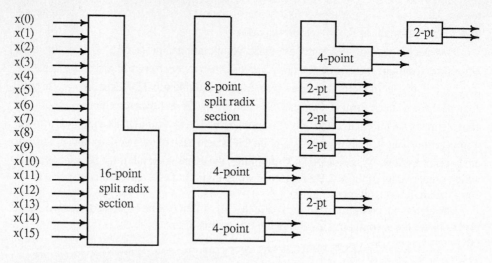

Figure 2.10 A 16-point split radix algorithm

The complex multiplications required here could be performed by the standard four real multiplications/two real additions algorithm. However, given that these multiplications are by the known terms W_N^n and W_N^{3n}, this is inefficient and a better choice is the three real multiplications/three real additions algorithm. An eighth root of one is of the form $(\cos(\frac{\pi k}{4}) - j\sin(\frac{\pi k}{4}))$ which only requires two real multiplications and two real additions. Thus in total the section requires:

number of real multiplications $\qquad M_{es} = 3(\frac{N}{2} - 4) + 2 \cdot 2 = \frac{3}{2} \cdot N - 8$

number of real additions $\qquad A_{es} = 2N + M_{es}$

The number of (real) operations in an N-point algorithm where N is 2^α, is easily found. For multiplications, we observe that a recurrence relation is:

$$M_\alpha = M_{\alpha-1} + 2M_{\alpha-2} + M_{es} = M_{\alpha-1} + 2M_{\alpha-2} + 3 \cdot 2^{\alpha-1} - 8$$

Since both M_1 and M_2 are zero, we derive:

$$M_3 = M_2 + 2M_1 + 3 \cdot 2^2 - 8 = 4 = 0 \cdot 8 + 4$$
$$M_4 = M_3 + 2M_2 + 3 \cdot 2^3 - 8 = 20 = 1 \cdot 16 + 4$$
$$M_5 = M_4 + 2M_3 + 3 \cdot 2^4 - 8 = 68 = 2 \cdot 32 + 4$$

.

.

.

$$M_\alpha = (\alpha - 3) \cdot N + 4$$

The total number of additions is the sum of M_α plus those due to the complex additions. Since there are α stages in an algorithm, there must be $\alpha \cdot 2N$ of these. Hence in total:

real multiplications $\qquad M_\alpha = (\alpha - 3) \cdot N + 4 \qquad = N \cdot \mathrm{Log}_2 N - 3N + 4$
real additions $\qquad\quad A_\alpha = \alpha \cdot 2N + (\alpha - 3) \cdot N + 4 = 3N \cdot \mathrm{Log}_2 N - 3N + 4$

These figures compare very favorably with the earlier base-2 algorithms.

The Rader-Brenner algorithm

The Rader-Brenner algorithm has a very similar structure to those given earlier. Its special significance is that all multiplications are either purely real or purely imaginary. Hence it offers the practical advantages of a substantially reduced component count in hardware realizations or simpler and faster arithmetic in software. Like the algorithms given earlier, it has an elemental form and so is primarily a decomposition technique. It is not specifically aimed at real-valued sequences but has some intrinsic advantages for transforming these.

The advantage of the Rader-Brenner algorithm is in the simplicity of its arithmetic. The disadvantage is its very high sensitivity to coefficient error. That creates problems in realizations, sometimes to the extent of demanding double precision arithmetic. Another disadvantage is that it has far more additions than other algorithms. Nevertheless, it is quite a useful algorithm and of some practical importance. Some of the disadvantages can be overcome using the advanced techniques to be introduced in the next two Parts.

The Rader-Brenner algorithm was originally developed as a base-2 algorithm. It has both a decimation in frequency and a decimation in time form. The latter, like the split radix algorithm, begins with the division of a set of samples x(n) to produce a sequence:

$$a(n) = x(2n+1) - x(2n-1) + C$$

where C is some constant. Taking the Fourier transform of this sequence:

$$A(k) = \mathrm{DFT}\{ x(2n+1) \} - \mathrm{DFT}\{ x(2n-1) \} + C\delta(k) \qquad\qquad k = 0, 1, \cdots, \frac{N}{2} - 1$$

where $\mathrm{DFT}\{\cdot\}$ refers to the Fourier transform. From the linear shift property of the DFT:

$$A(k) = \mathrm{DFT}\{ x(2n+1) \} - W^{2k} \cdot \mathrm{DFT}\{ x(2n+1) \} + C\delta(k)$$
$$\quad\;\, = (1 - W^{2k}) \cdot \mathrm{DFT}\{ x(2n+1) \} + C\delta(k)$$

The constant C can only affect A(0) and so for all other values:

$$A(k) = (1 - W^{2k}) \cdot \mathrm{DFT}\{ x(2n+1) \} \qquad\qquad k = 1, 2, \cdots, \frac{N}{2} - 1$$
$$\quad\;\, = W^k \cdot (W^{-k} - W^k) \cdot \mathrm{DFT}\{ x(2n+1) \}$$
$$\quad\;\, = (W^{-k} - W^k) \cdot \{ W^k \cdot X_{2n+1}(k) \}$$

Now:
$$(W^{-k} - W^k) = 2j\sin(\frac{2\pi k}{N})$$

thus:

$$W^k \cdot X_{2n+1}(k) = -j \cdot \frac{1}{2} \cdot \mathrm{cosec}\left(\frac{2\pi k}{N}\right) \cdot A(k)$$

The DFT can be expressed (in DIT form) as:

$$X(k) = DFT\{ x(2n) \} + W^k \cdot DFT\{ x(2n+1) \} \qquad\qquad k = 0, 1, \cdots, N-1$$

$$= X_{2n}(k) + W^k \cdot X_{2n+1}(k)$$

Substituting the expression just found:

$$X(k) = X_{2n}(k) - j \cdot \frac{1}{2} \cdot \mathrm{cosec}\left(\frac{2\pi k}{N}\right) \cdot A(k) \qquad k = 0, 1, 2, \cdots, \frac{N}{2}-1$$

$$X\left(k + \frac{N}{2}\right) = X_{2n}(k) + j \cdot \frac{1}{2} \cdot \mathrm{cosec}\left(\frac{2\pi k}{N}\right) \cdot A(k) \quad k = 0, 1, 2, \cdots, \frac{N}{2}-1$$

In this form, the equation has a multiplication by a purely imaginary term. For large N and small k though, that multiplication is by a number with large magnitude. If it cannot be represented accurately, then significant errors can occur. Note, too, that C is:

$$C = \frac{2}{N} \sum_{n=0}^{\frac{N}{2}-1} x(2n+1)$$

There is an alternative form for the algorithm. If d(n) is made:

$$d(n) = x(2n+1) + x(2n-1) + C$$

then:

$$X(k) = X_{2n}(k) + \frac{1}{2} \cdot \sec\left(\frac{2\pi k}{N}\right) \cdot D(k)$$

$$X\left(k + \frac{N}{2}\right) = X_{2n}(k) - \frac{1}{2} \cdot \sec\left(\frac{2\pi k}{N}\right) \cdot D(k)$$

The basic principles of the Rader-Brenner algorithm can be extended to other bases. Rather than examining a base-p algorithm though, it is more useful to review the technique in similar terms to the elemental computational section. One version of the second form of the section given earlier with respect to a base of 2 is:

1. map the data into an array with two elements per row and M columns;
2. perform the three-stage operation:

$$X(s,r) = \sum_{u=0}^{M-1} W_M^{su}\left\{ \sum_{v=0}^{1} \{x(u,v)W^{sv}\}W_2^{rv} \right\}$$

which involves twiddling, then 2-point DFTs on the rows and finally, M-point DFTs on the columns.

The Rader-Brenner algorithm simplifies the twiddling stage in this algorithm. It replaces $x(u,v)$ by an alternative and that changes W^{sv} into a purely imaginary value.

Rather than map $x(n)$ into $x(u,v)$ as before, $x(u,v)$ is now formed as:

$$x(u,0) = x(u \cdot 2)$$
$$x(0,1) = x(1) - x(N-3) + C$$
$$x(u,1) = x(u \cdot 2 + 1) - x((u-2) \cdot 2 + 1)$$

where
$$C = \frac{2}{N} \sum_{u=0}^{\frac{N}{2}-1} x(2u + 1)$$

Given the form of the algorithm, clearly operations are commutative. Hence together with the linear shift property of the DFT, this mapping means the second column's elements are equivalent to:

$$x(u,1) = (1 - W^{2s}) \cdot x(u \cdot 2 + 1)$$
$$= (W^{-s} - W^s) \cdot W^{-s} \cdot x(u \cdot 2 + 1)$$

Consequently, the twiddling parameters W^{sv} can be replaced by $T(s,v)$ where:

$$T(u,v) = 1 \qquad\qquad \text{for } u = 0 \text{ or } v = 0$$

$$= -j \cdot \frac{1}{2} \cdot \operatorname{cosec}(\frac{2\pi v}{N}) \qquad \text{otherwise}$$

Extending to other bases requires some variation for the other columns introduced but is otherwise quite straightforward.

The fast Hartley transform

As with the DFT, there is not just one fast Hartley transform (FHT) but many. The close relationship between the discrete Hartley transform (DHT) and the DFT ensures that for every FFT algorithm there exists an equivalent FHT. Nevertheless, the FHT is usually presented as a base-2, DIT algorithm and occasionally as a DIF algorithm. We, somewhat informally, develop that in-place DIT algorithm.

The scale factor doesn't affect the algorithm, so for simplicity, let the DHT equation be:

$$X(k) = \sum_{n=0}^{N-1} x(n) \cdot \left(\cos\left(\frac{2\pi kn}{N}\right) + \sin\left(\frac{2\pi kn}{N}\right) \right) \qquad k = 0, 1, \cdots, N-1$$

where $x(n)$ is a real data sequence. Like the base-2, DIT FFT, we separate this into two terms. That is:

104 Digital Convolution on Infinite Fields

$$X(k) = \sum_{n=0}^{\frac{N}{2}-1} x(2n) \cdot \left(\cos\left(\frac{2\pi k(2n)}{N} \right) + \sin\left(\frac{2\pi k(2n)}{N} \right) \right)$$

$$+ \sum_{n=0}^{\frac{N}{2}-1} x(2n+1) \cdot \left(\cos\left(\frac{2\pi k(2n+1)}{N} \right) + \sin\left(\frac{2\pi k(2n+1)}{N} \right) \right)$$

$$= X_E(k) + X_O(k) \qquad\qquad k = 0, 1, \cdots, N-1$$

Now since it is clear any N-point DHT is periodic in N, the $\frac{N}{2}$-point DHT of the sequence $x(2n)$ is:

$$X_E(k) = \sum_{n=0}^{\frac{N}{2}-1} x(2n) \cdot \left(\cos\left(\frac{2\pi kn}{\frac{N}{2}} \right) + \sin\left(\frac{2\pi kn}{\frac{N}{2}} \right) \right)$$

Using standard trigonometric identities, the second equation becomes:

$$X_O(k) = \sum_{n=0}^{\frac{N}{2}-1} x(2n+1) \cdot \left(\cos\left(\frac{2\pi k(2n+1)}{N} \right) + \sin\left(\frac{2\pi k(2n+1)}{N} \right) \right)$$

$$= \cos\left(\frac{2\pi k}{N} \right) \sum_{n=0}^{\frac{N}{2}-1} x(2n+1) \cdot \left(\cos\left(\frac{2\pi kn}{\frac{N}{2}} \right) + \sin\left(\frac{2\pi kn}{\frac{N}{2}} \right) \right)$$

$$+ \sin\left(\frac{2\pi k}{N} \right) \sum_{n=0}^{\frac{N}{2}-1} x(2n+1) \cdot \left(\cos\left(\frac{2\pi kn}{\frac{N}{2}} \right) + \sin\left(\frac{2\pi kn}{\frac{N}{2}} \right) \right)$$

$$= H(k) \cdot \cos\left(\frac{2\pi k}{N} \right) + H(N-k) \cdot \sin\left(\frac{2\pi k}{N} \right)$$

where H(k) is the $\frac{N}{2}$-point DHT of the data sequence $x(2n+1)$. Thus the original N-point DHT has been decomposed to the combination of two $\frac{N}{2}$-point DHTs and that defines an elemental computational section which can be used to construct a fast algorithm.

The process can be continued. If N is 2^α, then there will be α stages in the algorithm. Note there is scrambling but it is exactly the same as for a base-2 FFT. All operations are real and the end result of the decomposition is a set of two-point DHTs of the form:

$$X(0) = x(0) + x(1)$$
$$X(1) = x(0) - x(1)$$

which also describes 2-point DFTs. The second last stage of decomposition must involve 4-point DHTs and these are, in terms of the FHT section:

$$X(0) = X_E(0) + H(0)$$
$$X(1) = X_E(1) + H(1)$$
$$X(2) = X_E(0) - H(0)$$
$$X(3) = X_E(1) - H(1)$$

Two and 4-point DHTs are the only ones that require no multiplications.

Given the symmetries evident in the section, there are opportunities to reduce the number of operations. An initial observation is that the section may be described as:

$$X(k) = X_E(k) + H(k) \cdot \cos\left(\frac{2\pi k}{N}\right) + H\left(\frac{N}{2} - k\right) \cdot \sin\left(\frac{2\pi k}{N}\right)$$

$$X\left(\frac{N}{2} + k\right) = X_E(k) - H(k) \cdot \cos\left(\frac{2\pi k}{N}\right) - H\left(\frac{N}{2} - k\right) \cdot \sin\left(\frac{2\pi k}{N}\right)$$

where $k = 0, 1, \cdots, \frac{N}{4} - 1$

and this suggests several ways of conserving operations. However, there is a problem. These two equations describe how three input values generate two output values. Consequently, they do not describe an in-place algorithm. To overcome that, we need to modify the section slightly. By expanding the above equations and using standard trigonometric identities, we can show:

$$X(k) = X_E(k) + H(k) \cdot \cos\left(\frac{2\pi k}{N}\right) + H\left(\frac{N}{2} - k\right) \cdot \sin\left(\frac{2\pi k}{N}\right)$$

$$X\left(\frac{N}{2} - k\right) = X_E\left(\frac{N}{2} - k\right) - H\left(\frac{N}{2} - k\right) \cdot \cos\left(\frac{2\pi k}{N}\right) + H(k) \cdot \sin\left(\frac{2\pi k}{N}\right)$$

$$X\left(\frac{N}{2} + k\right) = X_E(k) - H(k) \cdot \cos\left(\frac{2\pi k}{N}\right) - H\left(\frac{N}{2} - k\right) \cdot \sin\left(\frac{2\pi k}{N}\right)$$

$$X(N - k) = X_E\left(\frac{N}{2} - k\right) + H\left(\frac{N}{2} - k\right) \cdot \cos\left(\frac{2\pi k}{N}\right) - H(k) \cdot \sin\left(\frac{2\pi k}{N}\right)$$

where $k = 1, \cdots, \frac{N}{4} - 1$.

The four output terms of this set of equations draw on only four unique input terms, namely

$X_E(k), X_E(\frac{N}{2} - k), H(k),$ and $H(\frac{N}{2} - k)$. Hence this forms an in-place section. It requires four real multiplications and with appropriate groupings of terms, just six additions. The equations where the index k is zero need to be treated separately. For this value:

$$X(0) \quad = X_E(0) + H(0)$$

$$X\left(\frac{N}{4}\right) \quad = X_E\left(\frac{N}{4}\right) + H\left(\frac{N}{4}\right)$$

$$X\left(\frac{N}{2}\right) \quad = X_E(0) - H(0)$$

$$X\left(\frac{3N}{4}\right) = X_E\left(\frac{N}{4}\right) - H\left(\frac{N}{4}\right)$$

These terms only require four additions to compute.

Consider the number of operations in an FHT algorithm with $N = 2^\alpha$ based on this in-place section. The first and second stages have no multiplications, thus from a recursion formula:

$$\text{total number of multiplications} \quad = \frac{3}{2} \cdot N \cdot Log_2 N - \frac{3}{2} \cdot N + 2$$

$$\text{total number of additions} \quad = N \cdot Log_2 N - 3N + 4$$

Some multiplications are redundant as they correspond to indices giving rise to 1 or −1 and so further savings are possible. However, locating these is difficult. A more useful approach to reducing operations is the following. Given the form of the in-place section, we can express the first two equations above as:

$$X(k) \quad = X_E(k) + A \cdot \cos\theta + B \cdot \sin\theta$$

$$X\left(\frac{N}{2} - k\right) = X_E\left(\frac{N}{2} - k\right) - B \cdot \cos\theta + \sin\theta$$

The complex multiplication of $(A - jB)$ and $(\cos\theta + \sin\theta)$ is:

$$(A - jB) \cdot (\cos\theta + \sin\theta) = (A \cdot \cos\theta + B \cdot \sin\theta) + j(-B \cdot \cos\theta + A \cdot \sin\theta)$$

An efficient algorithm for performing this computation was derived in Part 1 and by using that algorithm the in-place section can be computed with just three multiplications and eight additions. This changes the number of operations in an algorithm to a total of:

$$\text{multiplications} = 2N \cdot Log_2 N - 5N + 4$$

$$\text{additions} \quad = \frac{3}{4} \cdot N \cdot Log_2 N - \frac{9}{4} \cdot N + 3$$

which compares favorably against the base-2 FFT algorithms.

As has been mentioned, the DHT and DFT are closely related. The equations linking the two (ignoring the scale factor) are:

$$DHT[\, x(n)\,] \quad = Re\{\, DFT[\, x(n)\,]\,\} - Im\{\, DFT[\, x(n)\,]\,\}$$

and

$$Re\{DFT[x(n)]\} = \frac{1}{2} \cdot \{DHT[x(N-n)] + DHT[x(n)]\}$$

$$Im\{DFT[x(n)]\} = \frac{1}{2} \cdot \{DHT[x(N-n)] - DHT[x(n)]\}$$

It was thought at one time that it would be possible to compute DFTs via an FHT more efficiently than any FFT algorithm. This is not so. The split radix algorithm, when fully optimized for real sequences, has the same number of multiplications as the equivalent FHT but a slightly smaller number of additions. In spite of that, a fast Hartley transform is still an attractive approach to computing DFTs and so convolutions. FHT algorithms have a very regular structure, transforms are their own inverse and they are very efficient. Other factors such as scrambling, generation of sinusoids and so forth, are common to both FHTs and FFTs.

IMPLEMENTATION ISSUES

Efficient use of DFT algorithms for real-valued data

The fact that a DFT is a complex-to-complex transformation creates some inefficiencies when the input data sequence is real. While real-valued FFT algorithms can be devised like those discussed earlier, there remains one important practical problem. The input data requires only N storage locations but the complex transformed values require 2N. Hence, while savings in operations may be made by manipulating the algorithm, memory is still inefficiently used.

Means of overcoming this rely on the conjugate symmetry property of the spectrum of a real sequence. That is, on the fact that $X(k) = X^*(N-k)$ (* equals complex conjugation). The simplest way to take advantage of this is to recognize that half the output terms can be obtained by conjugating the other half. Thus only half the output terms need to be calculated and so only half the number of memory locations is needed. In-place algorithms can be developed along these lines but they tend to be complicated. Another approach is the following. Let u(n) and v(n) be two real-valued sequences. Let:

$$x(n) = u(n) + jv(n) \qquad\qquad n = 0, 1, \cdots, N-1$$

The DFT of x(n) is:

$$X(k) = X_r(k) + jX_i(k) \qquad\qquad k = 0, 1, \cdots, N-1$$

$$= U(k) + jV(k)$$
$$= (U_r(k) - V_i(k)) + j(U_i(k) + V_i(k))$$

where the subscripts r and i refer to the real and imaginary parts respectively. From the conjugate symmetry property of the spectrum of a strictly real data sequence and from the conjugate skew symmetry property of a strictly imaginary data sequence:

$$U_r(k) = \frac{1}{2}(X_r(k) + X_r(N-k)) \qquad\qquad k = 0, 1, \cdots, N-1$$

$$U_i(k) = \frac{1}{2}(X_i(k) - X_i(N-k))$$

$$V_r(k) = \frac{1}{2}(X_i(k) + X_i(N-k))$$

$$V_i(k) = \frac{1}{2}(X_r(k) - X_r(N-k))$$

Using an FFT algorithm then, we can compute the DFTs of two real data sequences at a cost of only an extra 2N real additions above the FFT operations. That is, effectively each data sequence requires half the number of operations of a complex FFT plus N real additions. Given that this requires no change to the FFT algorithm, it is quite an attractive approach.

Another possibility is to take an N-point real data sequence, divide it into two $\frac{N}{2}$ -point data sequences and construct a complex data sequence with one of these declared as the real part and the other the imaginary. This is a much more complex situation but there are several ways of solving it. One follows from the base-2 DIT algorithm. Consider some N-point data sequence y(n) with corresponding DFT Y(k). We can express Y(k) as the sum of two $\frac{N}{2}$ -point DFTs U(k) and V(k) via

$$Y(k) \quad = U(k) + W_N^k \cdot V(k) \qquad\qquad k = 0, 1, \cdots, \frac{N}{2} - 1$$

$$= U\left(k - \frac{N}{2}\right) - W_N^k \cdot V\left(k - \frac{N}{2}\right) \qquad k = \frac{N}{2}, \frac{N}{2} + 1, \cdots, N-1$$

The division for the DIT algorithm resulted in U(k) being the DFT of the even terms of x(n) and V(k) being the DFT of the odd. Thus here we express the N-point sequence x(n) as the complex data sequence:

$$y(n) = x(2n) + jx(2n+1) \qquad\qquad n = 0, 1, \cdots, \frac{N}{2} - 1$$

The DFT is found via some FFT algorithm and this will give an $\frac{N}{2}$ -point DFT Y(k). From the conjugate and skew symmetry properties of purely real and imaginary sequences:

$$U_r(k) = \frac{1}{2}\left(Y_r(k) + Y_r\left(k - \frac{N}{2}\right)\right) \qquad U_i(k) = \frac{1}{2}\left(Y_i(k) + Y_i\left(k - \frac{N}{2}\right)\right)$$

$$V_r(k) = \frac{1}{2}\left(Y_i(k) + Y_i\left(k - \frac{N}{2}\right)\right) \qquad V_i(k) = \frac{1}{2}\left(Y_r(k) + Y_r\left(k - \frac{N}{2}\right)\right)$$

where $k = 0, 1, \cdots, \frac{N}{2} - 1$.

Thus the final desired spectrum is:

$$
\begin{aligned}
X(k) \quad &= U_r(k) + jU_i(k) + \left(\cos\left(\frac{2\pi k}{N}\right) - j\sin\left(\frac{2\pi k}{N}\right)\right) \cdot \left(v_r(k) + jV_i(k)\right) \\
&= \left(U_r(k) + V_r(k) \cdot \cos\left(\frac{2\pi k}{N}\right) + V_i(k) \cdot \sin\left(\frac{2\pi k}{N}\right)\right) \\
&\quad + \left(U_i(k) - V_r(k) \cdot \sin\left(\frac{2\pi k}{N}\right) + V_i(k) \cdot \cos\left(\frac{2\pi k}{N}\right)\right)
\end{aligned}
$$

This algorithm certainly reduces the number of operations but the separation process is much more complex than the last algorithm and that makes it less attractive.

An excellent algorithm for real-valued sequences is the split radix algorithm. Here the conjugate symmetry property can be used very effectively. Recall that in this algorithm, the decomposition is to an $\frac{N}{2}$-point DFT giving transformed terms $X(2k)$ and two $\frac{N}{4}$-point DFTs giving transformed terms $X(4k+1)$ and $X(4k+3)$. From the conjugate symmetry property of the DFT of a real-valued sequence:

$$
\begin{aligned}
X(4k+1) &= X^*(N-4k-1) \\
&= X^*(\{N-4k-4\}+3) \\
&= X^*(4\{2^{n-2}-1-k\} + 3) \\
&= X^*(4m+3)
\end{aligned}
$$

Now k ranges over $0, 1, \cdots, \frac{N}{4} - 1$ and so does m. Consequently, all the transformed terms $X(4k+1)$ can be found by conjugating the terms $X(4k+3)$ or vice versa. That reduces the number of multiplications by a half. Hence the number of real multiplications becomes:

$$2^{\alpha-1} \cdot (\alpha - 3) + 2 = \frac{N}{2} \cdot \mathrm{Log}_2 N - \frac{3}{2} \cdot N + 2$$

The number of additions is more difficult to calculate. The recurrence relation:

$$A_\alpha = A_{\alpha-1} + A_{\alpha-2} + A_{es}$$

holds for this situation. Now $A_{\alpha-2}$ refers to the number of additions in the $\frac{N}{2}$-point DFT, which is a complex data calculation. We already have the formula for that, namely:

$$A_{\alpha-2} = 3(\alpha - 3) \cdot 2^{\alpha-2} + 4$$

The number of the additions in this section is given by:

$$A_{es} = 2^\alpha + 3 \cdot 2^{\alpha-2} - 4$$

hence:

$$\begin{aligned} A_\alpha &= A_{\alpha-1} + 3(\alpha - 3) \cdot 2^{\alpha-2} + 4 + 2^\alpha + 3 \cdot 2^{\alpha-2} - 4 \\ &= A_{\alpha-1} + (3\alpha - 2) \cdot 2^{\alpha-2} \end{aligned}$$

Since A_1 is 2, then:

$$\begin{aligned} A_2 &= 2 + 4 &= 1 \cdot 2 + 4 \\ A_3 &= 6 + 14 &= 4 \cdot 4 + 4 \\ A_4 &= 20 + 40 &= 7 \cdot 8 + 4 \end{aligned}$$

$$\cdot$$
$$\cdot$$
$$\cdot$$

$$A_\alpha = (3\alpha - 5) \cdot 2^{\alpha-1} + 4$$

$$= \frac{3}{2} \cdot N \cdot Log_2 - \frac{5}{2} \cdot N + 4$$

The generation of coefficients for FFT algorithms

Fast Fourier transforms are often compared purely on the basis of the number of multiplications required in the actual processing of data. When used for convolution, that puts them into a very favorable light with respect to a direct implementation. However, we have already seen that scrambling can create a substantial computational burden unless care is shown in formulating the algorithm and so can the computation of the coefficients needed by the algorithm. Further, as multiplication becomes easier to implement and so, as multiplication times approach those of addition, the number of additions and also the number of memory references become more significant.

The coefficients are the terms W^{kn} which are:

$$W^{kn} = \cos\left(\frac{2\pi kn}{N}\right) - j \cdot \sin\left(\frac{2\pi kn}{N}\right)$$

Apart from being complex, these coefficients are expressed in terms of two sinusoids and sinusoids are transcendental functions. Most computers use polynomial approximations to generate these, requiring of the order of six real multiplications and two real additions for single precision, and 15 real multiplications and five real additions for double. (Many microprocessor co-processors include hardwired trigonometric function generators. These are obviously much faster than software routines and usually generate a value in about the same time as a floating point multiplication.) If a convolution algorithm is created from the DFT with every coefficient calculated as required, then the total computation time of that algorithm will be dominated by the time taken to determine these coefficients. Further, it is quite likely that direct methods of convolution would be much faster.

There are two common approaches to overcoming this problem. One is to use a recurrence relation such as:

$$W^{kn} = W^k \cdot W^{k(n-1)}$$

This reduces the computation to one complex multiplication per coefficient (but that still means the number of complex multiplications in the FFT increases by ($N^2 - 2N + 1$)). However, it is impossible to compute sine and cosine precisely on a digital computer as they are transcendental. Errors will therefore accumulate in the calculation of the coefficients and so introduce errors into the computation of the transform. Using double precision arithmetic can correct this but it is computationally costly. Less overhead is involved in checking and then 'resetting' the calculation whenever it should be $-j$, -1 or j.

The alternative is to pre-generate the coefficients and store them in some look-up table. The computational cost now reduces to a memory reference. However, the table needs at least $\frac{N}{2}$ complex entries and that is comparable to the storage needed for the input data. Making use of further symmetries in W^{kn} certainly reduces storage but at a cost of additional overhead in retrieving the coefficients.

Finite register effects

Finite register effects refer to the errors which arise in using finite word sizes for calculations. Since only hardware designers have precise control over word sizes, truncation or round-off of products and so forth, this subject has traditionally been deemed part of hardware implementation techniques.

In the early days of digital signal processing, finite register effects were a major consideration in design as economic pressures demanded the smallest possible word size in systems. Now, with low-cost 32-bit microprocessors commonplace, and with co-processing units for performing floating point arithmetic and computing many common functions readily available, the need to consider finite register effects is diminishing. The literature, for example, has shown a marked decline in papers on this topic since the late 1970s. In spite of this, some comments on finite register effects are worthwhile.

From Parseval's relation:

$$\sum_{n=0}^{N-1} |X(n)|^2 = N \sum_{k=0}^{N-1} |x(k)|^2$$

That is, the average magnitude of the spectral values in a DFT computation is N times the average value of the input data. The increase in levels can cause overflow problems in systems using fixed point arithmetic and that suggests scaling values to limit this. Given the current state of digital electronic technology, that is most easily done by just scaling the input, or if not possible, scaling each stage by one of the factors of N.

The usual causes of finite register effects — truncation or rounding of products — plus most other sources of errors introduce an RMS error to RMS output proportional to 2^{-b} where b is either the word size in fixed point systems, or the mantissa length in floating point. The constant of proportionality can often be taken as $\sqrt{N}$. For 16-bit systems, that is an effective signal to noise ratio (approximating these errors as random disturbances) of the order of 50 dB. For 32-bit systems, this increases to around 100 dB. Apart from very high quality digital audio and some similar applications, these are quite acceptable.

Calculating the influence of finite register effects is far from trivial. Any FFT algorithm has so many combination steps that each error can influence each of the outputs in many ways. It is very tedious having to calculate all the different paths involved and even then, because errors have to be treated as random when they are patently not, the results can only be an upper or lower bound. For these reasons, often the best way of examining finite register effects is simply to simulate the desired system and observe its behavior with sets of test signals. The question arises then, of what test signals? Two are generally favored. First, sinusoids. Because they are transcendental, they are a signal that is very prone to error. However, they are also a signal where the precise output is easily calculated and where there should only be two output spectral values. A very good test with sinusoids, with the imaginary part as zero and the real part as a sinusoid, is to take the Fourier transform, conjugate the result (which introduces no errors) and transform again to obtain the original signal. Measuring the resulting errors, particularly in the imaginary part, is a very good indicator of performance.

The other commonly used test signal is a randomly generated signal. (More accurately, a pseudo-randomly generated signal.) This signal has no input representational error as does a sinusoid and is more representative of real signals. The measured error is entirely due to the computation.

The inverse DFT transform

The forward transformation of the DFT pair is in terms of W^{kn} while the reverse is in terms of W^{-nk}. In addition, the reverse transformation has the additional factor $\frac{1}{N}$ and that introduces a significant number of multiplications. It is clearly undesirable to have two distinct implementations for each part of the pair. For that reason it is important to examine some means of expressing the reverse transformation or **inverse DFT transform** in terms of the forward.

It must be immediately noted that nothing can be done about the scale factor. It introduces an additional step and that must be accepted. Whether this step is executed prior to the transformation or after though, is an implementation issue. If knowing absolute values is not essential, it can be omitted but this isn't the case with convolution problems. While scaling cannot be avoided, it can be ameliorated by a suitable choice of N. A value which is a power of 2 permits a very simple implementation of this stage and that is a further point in favour of base-2 algorithms.

There are three approaches which can be taken to simplify the inverse DFT problem. Two of these are based on some simple properties of complex numbers. Consider a complex number C given by (a + j·b). Its conjugate C* is therefore (a – j·b). Then:

$$(C^*)^* = a + j \cdot b = C$$

In addition, note that:

$$jC^* = b + j \cdot a$$

Thus:

$$(jC^*)^* = b - j \cdot a$$

$$j(jC^*)^* = a + j \cdot b = C$$

What might be termed the traditional approach to expressing the inverse DFT via a forward DFT algorithm is based on the first of these relations. Applying it, it is easily shown the inverse transform is:

$$x(n) = \frac{1}{N} \left[\sum_{k=0}^{N-1} X^*(k) \cdot W^{nk} \right]$$

Therefore, the forward transformation can be used for the reverse transformation simply by conjugating the spectral data beforehand and conjugating the results afterwards. If the output data is known to be real, that second conjugation is omitted. This approach has considerable appeal as conjugation is a simple operation.

A second approach uses the second of the two relations above. Applying it, the reverse transformation can be expressed as:

$$x(n) = \frac{j}{N} \left[\sum_{k=0}^{N-1} (jX^*(k)) \cdot W^{nk} \right]$$

From the above relations, $jX^*(k)$ is formed by swapping the real and imaginary parts of $X(k)$. The forward transformation is applied and then the real and imaginary parts of the output data swapped. Swapping is a very easy operation in hardware and involves much less effort than conjugation. Similarly, it becomes very simple to implement in software if separate arrays of the real and complex parts are formed. Hence this approach is often a better choice than the first.

The third approach relies on the methods to be developed in the next two parts. We note that W^{-nk} and W^{nk} only differ in the exponent and so would suspect that some operation, similar to the scrambling operations discussed earlier, could map the spectral data into a new sequence for which the forward transformation would apply. That is the case, although it will not be developed here. While conceptually interesting, the problem is that such a mapping can be computationally expensive and certainly isn't as simple as the first two approaches. Further, it achieves no practical advantage over either of those. However, in hardware such a mapping is just a wiring problem and so this approach has some attraction for these implementations.

SUMMARY

This Part has concentrated on the single transform in the infinite complex field C; the discrete Fourier transform. The generating element that produces the appropriate cyclic subfield is the Nth root of unity. That is, the quantity W, or $\exp(-\frac{j2\pi}{N})$. The transform pair is conventionally written:

$$X(k) = \sum_{n=0}^{N-1} x(n) \cdot W^{kn} \qquad\qquad k = 0, 1, \cdots, N-1$$

$$x(k) = \frac{1}{N} \sum_{k=0}^{N-1} X(k) \cdot W^{-nk} \qquad\qquad n = 0, 1, \cdots, N-1$$

These equations require at least $(N^2 - 2N + 1)$ complex multiplications. Since in convolution two such transforms are required (assuming one set of terms can be pre-calculated) plus a set of N complex multiplications in the transform domain, a DFT approach compares poorly against a direct computation.

Consider the two approaches. The direct convolution is:

$$y(n) = \sum_{m=0}^{N-1} h(m) \cdot x(n-m) \qquad\qquad n = 0, 1, \cdots, N-1$$

This requires N^2 real multiplications and $N \cdot (N-1)$ real additions per N output points, and N storage cells for the unit sample response. At least $(8N^2 - 12N + 8)$ real multiplications and $(8N^2 - 10N + 4)$ real additions are required for the DFT approach. This assumes the scale factor $\frac{1}{N}$ in the inverse transform is included with the pre-calculated terms and the standard complex arithmetic algorithms. As well, it needs storage for the N complex values of H(n) (previously calculated), the N complex values of the input plus another N for the complex values of W^{kn}. That is, 6N cells in total. If the values of W^{kn} are not stored but generated as needed, then the number of complex multiplications increases by an amount $(N^2 - 2N + 1)$. There is also scrambling to consider. Clearly, the only advantage of this transform approach is that it generates the spectrum and that may be useful in some applications.

This Part examined improved techniques for computing the DFT. Collectively known as fast Fourier transforms, or FFTs, they are so named because they have significantly fewer multiplications than the above transform equations. Most of these techniques are characterized by being a combination procedure for constructing long-length DFTs from short-length. Those short-length routines are usually of prime-length but in the particular case of 2, there are advantages in using lengths based on 2^2, i.e. 4.

Algorithms are developed by considering an elemental computational section which just requires that N can be factored. The section has four variants, each involving a three-stage process following the mapping of the data into a two-dimensional array. That process involves a set of DFTs on the rows of that array, a set of complex multiplications termed twiddles and a second set of DFTs on the columns of the array. The differences between the variants is the order in which these actions apply. If the factors of N can also be factored, the variant selected can be repeated to define an algorithm. The four variants have the same number of operations and storage requirements.

These fast Fourier transforms achieve success by utilizing symmetry within the DFT computation. However, one effect of this is that the order of the data is not preserved. Consequently, the data is said to be scrambled. In terms of the array of the computational section, data is written into the rows of the array and read out via the columns or written into the columns and read out from the rows. The first approach leads to the need for a post-scrambling operation. That is, the output data must be unscrambled after application of the section. The second approach requires the data to be pre-scrambled before application of the section. Given the symmetry of the FFT, these modes can be reversed leading to four distinct FFT algorithms.

Using the computational section to create FFTs leads to two principal types. The first, for those values of N where $N = B^\alpha$ (where B and α are both integers), are termed base-B algorithms. They are entirely expressed in terms of the combination of B-point DFTs and so have great symmetry. The other, termed mixed radix, applies to those values of N which are a product of different integer factors. Since most numbers are of this type, then the appropriate algorithm for an arbitrary N is most likely to be mixed radix.

The symmetry of base-B algorithms and especially those where B is prime has made them very popular. Of these, one group stands out. They are the algorithms based on 2 and powers of 2. These algorithms have a significant number of complex multiplications by j, −1, j and 1. Multiplications by j or −j are just complex conjugations, hence the actual number of multiplications is substantially less than other base-B algorithms. The particular characteristic of base-2 algorithms is that all multiplications are due to twiddling alone. This is also true for base-4. The conclusion, then, is that base-4 is the preferred FFT algorithm and if that cannot be used for some reason, then base-2 is the next best.

One base-2 algorithm is not only more efficient than even a base-4 but quite different in structure to all other such algorithms. It is the split radix algorithm. It can achieve the theoretical minimum number of multiplications and is very parsimonious in its use of additions. This makes it most attractive for a wide range of applications. The elemental section here is two-stage but the second only applies to half the points. Consequently, an algorithm lacks the symmetry of other algorithms and that can create some implementation difficulties.

The general base-B algorithm as developed here requires $\alpha \cdot N \cdot (B-1)$ complex additions. The number of complex multiplications is $\{ (\frac{N}{B}) \cdot (\alpha(D+B) - B - \alpha) + 1 \}$ where D is the number of multiplications in a B-point DFT. Scrambling requires approximately $\frac{1}{2} \cdot (N-B)$ real additions if a digital reversal counter is used. The base-4 FFT can be considered the most efficient of the base-B algorithms. It requires $\{ 2N \cdot (3\alpha - 4) + 8 \}$ real multiplications and $(N \cdot (16.5\alpha + 4) + 2)$ real additions. There is no significant change in the storage required from the DFT. Again, if the coefficients W^{nk} are generated as needed, then $(8N^2 + (1.5\alpha - 5)N + 16)$ real multiplications and $(6N^2 + (16.5 \cdot \alpha - 10) \cdot N + 8)$ additions are needed. Clearly, it is unrealistic to do this, hence table look-up must be the approach used.

Comparing the direct approach to convolution against the (FFT) transform approach, the direct uses far less storage but the number of operations is of the order of N^2. In the transform approach in contrast, the number of operations approaches $N \cdot \text{Log}_B N$. Examining the equations, the transform approach in fact always requires fewer operations. If the number of operations is the performance criterion, then the transform approach is clearly better.

The DFT is a complex to complex transformation but in many important applications the data to be transformed is real. Purely real data sequences give rise to a DFT with conjugate symmetry, while purely imaginary are skew symmetric. These properties can be used to advantage in a number of ways, from pruning existing FFT algorithms to more efficient use of complex FFT algorithms.

Although the DFT is the only transform in C with the cyclic convolution property, there are related transforms that can be used to compute DFTs and vice versa. They are the discrete Cosine transform, the discrete Sine transform and the discrete Hartley transform. The latter in particular has a number of very efficient algorithms with a highly regular structure and that makes an indirect approach to computing DFTs by using these quite appealing.

DISCUSSION

This Part has seen our first venture into a set of optimal algorithms for digital convolution and, at first glance, it seems very successful. The emphasis given to FFT algorithms in the literature would seem to confirm the importance of this technique. Feeling euphoric is very pleasant but before allowing ourselves that luxury it is as well to closely examine some of the implications of the work of this Part.

Return once more to the direct technique of digital convolution. What does it offer? Simplicity is one obvious attribute. Almost an order of magnitude less storage than transform approaches is another. A third is that all operations are real. However, there is one more worth focusing some attention upon. It places no demands on N and neither does the basic algorithm change for particular values of N.

The FFT algorithms investigated are not as simple as the direct approach but not excessively complex either. They do require a lot more storage and that can be a serious disadvantage. Their operations are with complex numbers and that is a disadvantage, but even so, the equivalent number of real operations is less than with the direct approach and they increase more slowly with N. However, the algorithm does change significantly with N. There are no algorithms of the type investigated for any prime values of N. As was noted in DEVELOPMENTS, the vast majority of the remaining algorithms are mixed radix but these do not have a simple structure. Further, scrambling is not simple and can present a significant computational burden. Where the FFT algorithms investigated come to the fore is when N is of the particular form of B^α and especially when B is 2 or 4. However, for N up to 5000, a number well beyond the range likely for most applications, there are only 16 possible algorithms. The transform approach, then, is certainly better in most respects but only within rather serious limits.

There are other conclusions to draw from the work conducted. One is that short-length prime order DFTs are quite important. A search to find efficient methods of executing these would be worthwhile and greatly improve the algorithms investigated. Further, it could well make some higher bases more attractive and so extend the scope of these algorithms. However, there are more multiplications involved in twiddling and therefore the development of minimum multiplication algorithms requires a much closer study of this aspect. A second important conclusion is that techniques for mixed radix algorithms should be investigated in preference to others and that improved techniques in this area are needed. Given there are so few possibilities for base-2 or base-4 algorithms, then for arbitrary N, their use will require padding out of the data sequence with zeroes. That raises the question of whether there exist optimal techniques of pruning. That is, given it is known some data is zero, can the data be re-organized in such a way to take advantages of further symmetries?

At this point, it is useful to take a broader view of this Part. The DFT transform pair can be succinctly described as:

$$\mathbf{x} = \mathbf{W}\mathbf{x}$$

$$\mathbf{x} = \frac{1}{N}\mathbf{W}^{-1}\mathbf{x}$$

where
$$\mathbf{X} = \text{an N-point column matrix of transformed values}$$
$$\mathbf{x} = \text{an N-point column matrix of data values}$$
$$\mathbf{W} = \text{a complex N}\times\text{N matrix}$$

In matrix terms, the elemental sections discussed are merely a means of factoring the matrix W into two. Repeated application of these factoring techniques results in a fast transform and that is described by:

$$\mathbf{X'} = \mathbf{W}_1 \cdot \mathbf{W}_2 \cdot \; \cdots \; \cdot \mathbf{W}_\alpha \, \mathbf{x}$$

or

$$\mathbf{X} = \mathbf{W}_1 \cdot \mathbf{W}_2 \cdot \; \cdots \; \cdot \mathbf{W}_\alpha \, \mathbf{x'}$$

where the superscript ' refers to a scrambled matrix. Matrix theory, then, is another avenue for exploring fast algorithms. As a trivial example, note that:

$$\mathbf{W} = \mathbf{W}^{\mathrm{T}}$$

and:

$$\mathbf{W} = \mathbf{N} \cdot (\mathbf{W}^*)^{-1}$$

Given any matrix description of a fast algorithm, applying these results is a very simple way of generating a new variant. This will not lead to any reduction in the number of operations but may give an algorithm better suited to particular implementations. In fact, in matrix terms, two of the elemental FFT sections are just transposes of the other two.

Scrambling is also a matrix operation. Here, for example:

$$\mathbf{X'} = \mathbf{S}\,\mathbf{X}$$

The matrix S is square and is the identity with rows or columns interchanged. It is therefore the product of elementary matrices, it can be factored and it is always non-singular. These properties opens a number of opportunities for algorithms, including those where both input and output data are in natural order.

A question here is how optimum are the algorithms developed? Using advanced methods, some of which are given in the following Parts, a bound can be determined for the number of multiplications required in FFT algorithms. Assume the input data is complex.

Then if N is 2^n, the theoretically minimum number of real multiplications is:

$$(4N - 2n^2 - n - 4)$$

If N is p^n, then the bound is:

$$(4N - 2n - 4 - (n^2 + n) \cdot d(p - 1))$$

where $d(M)$ is the number of prime divisors of M. This term is defined in Part 3.

FURTHER STUDY

References

A number of good introductory texts to digital signal processing discuss the general properties of the DFT, the concept of frequency in digital systems and the usual development of the base-2 FFT. For example, those mentioned in Part 1. A recently rewritten early text, and excellent introduction to the DFT and to FFT algorithms, is:

O.E. BRIGHAM, *The Fast Fourier Transform and its Applications*, Prentice Hall, 1988

Transforms for real-valued data sequences are discussed in the following:

Z. WANG, Fast Algorithms for the Discrete W Transform and the Discrete Fourier Transform, *I.E.E.E. Transactions on Acoustics, Speech and Signal Processing*, V 32 N 4, August 1984, pp 803–16

P. DUHAMEL and M. VETTERLI, Improved Fourier and Hartley Transform Algorithms : Application to Cyclic Convolutions of Real Data, *I.E.E.E. Transactions on Acoustics, Speech and Signal Processing*, V 35 N 6, June 1987, pp 818–24

H.V. SORENSON, D.L. JONES, M.T. HEIDEMAN, and C.S. BURRUS, Real-valued Fast Fourier Transform Algorithms, *I.E.E.E. Transactions on Acoustics, Speech and Signal Processing*, V 35 N 6, June 1987, pp 849–63

This last paper includes some FORTRAN programs. One of these has an error. The correction is listed on page 1353 in the September issue of the *Transactions*.

The digit reversal counter technique is recounted in many books but originally appeared in:

B. GOLD and C.M. RADER, *Digital Processing of Signals*, McGraw Hill, New York, 1969

Some more recent work is covered in:

D.M.W. EVANS, An Improved Digit Reversal Permutation Algorithm for Fast Fourier and Hartley Transforms, *I.E.E.E. Transactions on Acoustics, Speech and Signal Processing*, V 35 N 8, August 1987, pp 1120–5

The split radix algorithm is succinctly described in two papers:

H.V. SORENSON, M.T. HEIDEMAN, and C.S. BURRUS, On Computing the Split Radix Algorithm, *I.E.E.E. Transactions on Acoustics, Speech and Signal Processing*, V 34 N 1, February 1986, pp 152–6

P. DUHAMEL, Implementation of Split Radix Algorithms for Complex, Real and Real-Symmetric Data, *I.E.E.E. Transactions on Acoustics, Speech and Signal Processing*, V 34 N 2, April 1986, pp 285–95

The first paper here and the paper by Duhamel and Vetterli above include FORTRAN programs for the split radix algorithm. A paper reporting more advanced forms of the algorithm is:

M. VETTERLI and P. DUHAMEL, Split-Radix Algorithms for Length-pm DFTs, *I.E.E.E. Transactions on Acoustics, Speech and Signal Processing*, V 37 N1, January 1989, pp 57–64

The Rader-Brenner algorithm was first outlined in:

C.M. RADER and N.M. BRENNER, A New Principle for Fast Fourier Transformation, *I.E.E.E. Transactions on Acoustics, Speech and Signal Processing,* V 24, N 2, April 1976, pp 264–6

The discrete Hartley transform and its fast algorithms are discussed in:

R.N. BRACEWELL, The Fast Hartley Transform, *Proceedings of the I.E.E.E.,* V 72 N8, August 1984, pp 1010–8

H.V. SORENSON, D. L. JONES, C. S. BURRUS, and M.T. HEIDEMAN, On Computing the Discrete Hartley Transform, *I.E.E.E. Transactions on Acoustics, Speech and Signal Processing,* V 33 N 4, October 1985, pp 1231–8

This second paper discusses many forms of fast Hartley transforms and gives a program. A more recent reference is:

H.S. HOU, The Fast Hartley Transform, *I.E.E.E. Transactions on Computers,* V 36 N 2, February 1987, pp 147–56

Some general references for this Part are:

L.P. JAROSLAVSKI, Comments on "FFT Algorithm for both Input and Output Pruning", *I.E.E.E. Transactions on Acoustics, Speech and Signal Processing,* V 29 N 3, June 1981, pp 448–9

R.D. PREUSS, Very Fast Computation of the Radix 2 Discrete Fourier Transform, *I.E.E.E. Transactions on Acoustics, Speech and Signal Processing,* V 30 N 4, August 1984, pp 595–607

J.B. MARTENS, Discrete Fourier Transform Algorithms for Real-valued Sequences, *I.E.E.E. Transactions on Acoustics, Speech and Signal Processing,* V 32 N 2, April 1984, pp 390–6

A. NORTON AND A.J. SILBERGER, Parallelism and Performance Analysis of the Cooley-Tukey FFT Algorithm for Stored Memory Architectures, *I.E.E.E. Transactions on Computers,* V 36 N 5, May 1987, pp 581–91

P. DUHAMEL, B. PIRON, and J.M. ETCHETO, On Computing the Inverse DFT, *I.E.E.E Transactions on Acoustics, Speech and Signal Processing,* V 36 N 2, February 1988, pp 285–6

D.J. ROSE, Matrix Identities of the FFT, *Linear Algebra and its Applications,* V 29, 1980, pp 423–43

Jaroslavski not only comments on an earlier paper but gives an elegant treatment of the pruning problem. However, the theory to fully understand his arguments is not covered until

Part 4. This also applies to quite a degree to Martens' paper. He develops an algorithm for N equal to $2^t.3^s$ which is better than the normal base-2 but it is quite difficult to implement. Preuss develops an algorithm which could be described as an in-place version of the Rader-Brenner algorithm for real-valued sequences. Norton and Silberger's paper looks at the general question of the optimum MIMD structure for the base-2 FFT algorithm. Duhamel et al look at the general question of inverting the DFT using a forward transformation algorithm. Finally, Rose gives a very interesting exposition of FFTs using traditional matrix and tensor theory.

Exercises

1. Determine whether there is an optimum value for N for the base-B FFT when used in conjunction with the overlap and add, or overlap and save algorithms. If so, what is it for base-4? What does it imply about the relative importance of base-B and mixed radix algorithms?

2. It was stated that the practical Nyquist limit was about 20% higher than the theoretical. Assume a standard one pole low-pass filter is used, such as an RC type, and calculate a precise value. Name the factors other than the filter characteristics which influence that value. Do you think that should be a phase-corrected filter? If so, what does that imply about using a simple RC low-pass filter?

3. Prove scrambling is a swapping formula for mixed radix algorithms. Are there numbers in this case which do not require reversal? How can they be located?

4. Derive a base-5 FFT. Show all stages of the algorithm and derive the number of operations required in scrambling. How does this algorithm compare against the direct convolution approach?

5. Consider the mixed radix algorithm. Does it matter what order the factors are placed in? Consider both the general case and the specific case of a number with 4 as a factor.

6. Find a formula for the number of operations in the mixed radix algorithm.

7. Derive a base-4 algorithm using the Rader-Brenner technique.

8. The base-3 algorithm discussed introduces an interesting idea. Consider a general complex number of the form $(a + \mu.b)$ where μ is given by $\exp(-\frac{j2\pi}{p})$ where p is some integer. Derive the equations for transforming this format to the conventional format and also the equations for conjugation, multiplication, and addition. Do you think p must be prime? If so, on what grounds? Could this be used for more efficient higher base DFTs?

9. In the last Part, it was pointed out that the DFT could be expressed as a set of equations and these could be expressed in recursive form about an element $(b_i \cdot z + a_i + 1)$. Thus

a DFT could be computed by an array of N×N such elements. Show how this array could be reduced by half by using the conjugate symmetry property.

10. In software implementations, memory references are an important consideration. Derive these for the base-B and mixed radix algorithms. How do they compare when used in convolution algorithms against the direct approach?

11. Generalize the discrete cosine transform to

$$X(k) = \sqrt{\frac{2}{N}} \sum_{n=0}^{N-1} x(n) \cdot L(n) \cdot \cos\left(\frac{2\pi}{N}(k+\beta) \cdot (n+\alpha)\right)$$

where $L(n) = \dfrac{1}{\sqrt{2}}$ for k, n = 0

          = 1   otherwise

(a) For what values of a and b does the transform exist? Is another term needed?
(b) Derive the general inverse for these cases.
(c) Are all the transform pairs symmetric? That is, is the reverse transform also the forward transform?

12. Express the cyclic convolution of two N-point data sequences x(n) and h(n) entirely in terms of discrete cosine transforms.

13. Show that if X(k) is a discrete Hartley transform of an N-point data sequence, then:

$$X(k) = X(k+m \cdot N)$$

where m is an integer.

14. Prove Parseval's relation. That is, if X(k) is the DFT of a data sequence x(n), prove:

$$\sum_{k=0}^{N-1} |X(k)|^2 = N \sum_{n=0}^{N-1} |x(n)|^2$$

15. The DFT is conventionally expressed with W_N as $\exp(-\frac{j2\pi}{N})$.

(a) Is it possible to choose:

$$W_N = \exp\left(-j\frac{2\pi}{N} + C\right)$$

and if so, for what values of C?
(b) What is the inverse transform of a DFT based on such a value? Does it offer any practical advantages?

(c) In some FFT algorithms, data is separated into even and odd points and separate DFTs taken of each. Would a DFT such as this be of any advantage here?

16. Derive the symmetry conditions of the DFT in terms of magnitude and phase.

17. The two-dimensional DFT has the form:

$$X(k_1 k_2) = \sum_{n_1=0}^{N_1-1} \sum_{n_2=0}^{N_2-1} x(n_1, n_2) W_{N_1}^{k_1 n_1} W_{N_2}^{k_2 n_2}$$

Now this can be expressed as:

$$X(k_1 k_2) = \sum_{n_1=0}^{N_1-1} \left\{ \sum_{n_2=0}^{N_2-1} x(n_1, n_2) W_{N_2}^{k_2 n_2} \right\} W_{N_1}^{k_1 n_1}$$

which means the transform is separable. That is, it can be executed by taking one-dimensional DFTs along one axis, then along the other. Are the DCT, DST and DHT separable?

18. Consider the terms W_N^{kn} used in the DFT. In an implementation, at one extreme all values could be pre-computed and stored and in the other just one seed value is stored and values computed as required. The first has the advantage of no computations needed to find the values but 2N storage locations are required. The other has the advantage of only needing two storage locations but a complex multiplication is needed every time a value is used. Devise an intermediate scheme where storage is minimized and so are the average number of operations needed to derive each point. Begin by noting the symmetry of W_N^{kn}.

19. The auto-correlation function of a signal x(n) is:

$$C_{xx}(k) = \frac{1}{N} \sum_{n=0}^{N-k-1} x(n) \cdot x(n+k) \qquad\qquad k = 0, 1, \cdots, N-1$$

and the cross-correlation of two signals x(n) and y(n) is:

$$C_{xy}(k) = \frac{1}{N} \sum_{n=0}^{N-k-1} x(n) \cdot y(n+k) \qquad\qquad k = 0, 1, \cdots, N-1$$

Which algorithm discussed in this Part is best suited to these computations? Why?

20. The two-dimensional transform of a sequence x(n,m), where $0 \le n,m \le N-1$, is X(k,l). This may be divided into four quadrants. Explain how the transformed values in each quadrant relate to the original signal.

PROBLEMS

1. The base-B algorithms investigated only exhibited one form of symmetry. Assume N is $2^{2\alpha}$. Rather than employing a base-2 or a base-4 algorithm, would there be an advantage in having an algorithm with the first stage base-4, the second base-2, the third base-4, and so on. Derive equations starting with the elemental computational section and investigate. (Comment: note the twiddles.)

2. Sequences are assumed to be equi-spaced samples. In two dimensions, though, regular sampling is not just achieved on a Cartesian grid. Sampling at $(n_1, n_2 + f(n_1))$ is another choice where:

    $$f(n_1) \quad = \text{C, a constant for } n_1 \text{ even}$$
    $$= 0 \ \text{ otherwise}$$

    (a) Are there conditions on the value C can take?
    (b) Derive a two-dimensional DFT for this sampling grid.
    (c) Can you relate your DFT to that obtained from a Cartesian grid?

3. Fourier transforms are important in image processing. The DFT in two dimensions is just:

    $$X\left(k_1 k_2\right) = \sum_{n_1=0}^{N_1-1} \sum_{n_2=0}^{N_2-1} x(n_1, n_2) W_{N_1}^{k_1 n_1} W_{N_2}^{k_2 n_2}$$

    and as this is separable, it suggests transforming the rows then the columns. Some images, though, are very large. The imagery which can be acquired at any groundstation from the AVHRR (Advanced Very High Resolution Radiometer) instrument carried on the NOAA series of environmental monitoring satellites is 2048 pixels wide and about 4096 pixels long. The instrument provides five spatially coincident images representing samples of five separate bands in the electromagnetic spectrum. Such a mass of data cannot be held in the main memory of many computers and so must be held on disk. However, that means a fast transpose technique must be employed prior to the second set of DFTs. Also, disk storage is very slow compared to main memory.

    (a) Investigate fast transpose techniques. Begin with:

    J.O. EKLUNDH, A Fast Computer Method for Matrix Transposing, *I.E.E.E. Transactions on Computers*, V 21, N 6, July 1972, pp 801–3

    R.E. TWOGOOD and M.O. EKSTROM, An Extension of Eklundh's Matrix Transposition Algorithm and its Application in Image Processing, *I.E.E.E. Transactions on Computers*, V 25, N 9, Sept. 1976, pp 950–2

    (b) Calculate a new formula for the computational effort required when disk storage and fast transposition must be used.

(c) Compare this result against direct convolution methods. Under what circumstances is the FFT approach competitive with these for digital convolution?

(d) Three-dimensional signal processing is very important in areas like seismology. What do these results suggest about Fourier transform techniques in such an area?

4. The scandal rocked Oodnagalarbie! Some vandal had broken into the Bluey Bonzer Memorial Gallery and School of Art and there sprayed Bluey's masterpiece, 'The Sparkling Port Festival, 49' with black paint. All was not completely lost though. The flywire erected in front of the achievement by Bluey's widow years ago had limited the damage to a fine, black grid over the entire work.

After months of anguished debate, the town fathers decided action had to be taken. Thus, the now sorry picture was duly packaged and sent to the art restorers. They, however, discovered a problem. Bluey was much in favor of the abstract approach to non-cubist art. Given his considerable interest in sparkling port, his lines, where he bothered to use any, showed some flutter. Further, he was not averse to using the natural colors of soil and whatever vegetation he may have located. Paint, to Bluey, also meant house paint he could borrow, or the very pleasant green he acquired on an occasional visit to the railway depot two days ride away. Therefore, what was desecration and what was Bluey's inspiration were difficult to say. What was needed was a photograph of the work to help in the restoration. Amazingly, none had ever been taken.

Young Fred, who had just joined the restorers, was suddenly hit with inspiration while pondering this vexing problem. If the image was scanned and fed in digital form to a computer, it would in effect be a signal $x(n_1, n_2) \cdot y(n_1, n_2)$ where $x(n_1, n_2)$ is Bluey's treasure and $y(n_1, n_2)$ is the desecration. If therefore, the logarithm is taken, the result is:

$$z(n_1, n_2) = \log(x(n_1, n_2)) + \log(y(n_1, n_2))$$

The desecration can be measured, so all that had to be done was calculate the Fourier transform of $z(n_1, n_2)$, calculate the Fourier transform of the desecration, subtract this from $Z(k_1, k_2)$, invert the result and that would be the original!

Is young Fred right? If not, why not? Is he partly right? Explain. If you take this middle road, assume that instead of a grid, the vandal had squirted a blob into the center of this objet d'art. Does that cause you to change or modify your views and if so, in what way?

5. A tedious, but necessary, part of supermarket work is to go around the shelves counting the stock that remains at the end of the day or week. What is needed to alleviate this drudgery is some automated means of identifying objects of different sizes and shapes. Fourier transforms are a convenient way of describing the shape of objects. One approach is to simply step around the edge of the object within the image to generate coordinates as complex numbers and to then take a Fourier transform of those numbers as a shape descriptor.

From the properties of the DFT, what can you discern about orientation of objects and the distance from camera to object? Are other properties relevant in this practical situation. Consider the objects typically encountered in a local supermarket and sketch out a design for such an automated stock checking system. How many samples would you take? Using crude figures for a typical microprocessor, how long would your system take to do a complete stocktake in that supermarket?

6. Derive the DIF form of the discrete Hartley transform.

7. Would you class the split radix algorithm developed as DIT or DIF? Derive the other form and comment on the scrambling variants if any.

8. The 2/4 split radix algorithm has an elemental section based on splitting the computation of the transformed values into a set of values obtained from an $\frac{N}{4}$ -point DFT and a set derived from two $\frac{N}{2}$ -point DFTs. Consider an algorithm derived from a 2/8 split. How does it compare to the 2/4 split?

9. In the development of FFT algorithms during the 1970s, some concern was shown over scrambling. Interest was therefore shown in algorithms where the input and output remained in natural order. A further restriction placed on such algorithms was that each stage should be symmetric to simplify implementation. Hence the name 'constant geometry algorithms'. Derive a base-2 constant geometry algorithm.

10. Higher-dimensional DFTs have more potential for symmetries than one-dimensional. Derive the DFT for all forms of symmetric data sequences of a two-dimensional DFT.

PART

3

Digital Convolution in Finite Structures

INTRODUCTION

Two broad observations can be made of the FFT approaches of the last Part. One pertains to implementation and the other to the nature of the algorithms.

Implementing FFT algorithms means either creating software for programmable digital systems or developing specialized hardware. For both, word size is finite although it can be quite large. The algorithms were developed on the basis of an infinite complex domain. When mapped to the finite implementation domain, then, the result is quantization noise, overflows and possibly round-off errors. Further, the magnitude of these errors is usually dependent on the number of points involved. The implementation domain cannot be changed. Therefore, overcoming this problem requires a study of digital convolution in finite mathematical structures.

The success of FFT algorithms is primarily due to the mapping procedures by which a DFT of some given length is transformed into a combination of short-length DFTs. These mapping procedures, though, are relatively simple and based on reasonably obvious symmetries in the indices of W^{kn}. A deeper study of these and their properties may therefore lead to even better algorithms. Consider these indices. They are integers and, as W^{kn} is periodic, are taken from the finite set $\{0, 1, \cdots, N-1\}$. Operations on these indices permutes them in various ways and so these operations together with this set of numbers usually form a group, ring or field. That is, a finite mathematical structure.

The abstract mathematical link between these two problems makes it convenient to study them together. Nevertheless, it needs to be stressed that in terms of applications they remain apart as one problem concerns the mechanics of convolution while the other concerns the formulation of algorithms.

This Part begins with an examination of number theory in TOOLS. Long one of the most fascinating branches of mathematics, it is becoming increasingly important in digital signal processing. Our study begins with a basic examination of numbers and their properties and then proceeds to the question of residues and congruences. The Euclidean algorithm is given some prominence as it will occur repeatedly in the remainder of the book. Then the very important Chinese remainder theorem is discussed. Number sequences and their properties are considered and this leads to the concept of primitive roots. Quadratic residues follow and, finally, a simple but important application of number theory to the FFT.

THEORY opens with a study of finite mathematical structures. Then Rader's theorem is examined; a theorem showing that in certain circumstances the DFT and convolution are very closely related. While simple, it has very significant implications which will be explored both here and in Part 4. Finally, we examine the conditions under which transforms can exist on finite structures. These transforms are termed Number Theoretic Transforms or NTTs.

DEVELOPMENTS begins by examining a very simple application of number theory — the prime factors algorithm. While it essentially belongs to the last Part, its important use of number theory makes it more appropriate that it should be here. The body of this section is an examination of various NTTs. These have some attraction for hardware implementations but virtually none for software. However, such is the nature of some quite significant practical problems that NTTs are often overlooked. As a result, it would not be amiss to describe NTTs as being in the backwater of transform theory.

In the early part of DEVELOPMENTS, we shall examine a range of NTTs. Each has severe constraints on the size of sequence length that can be chosen and, unlike the FFT, the individual techniques appear to have little underlying commonality except number theory. This is the cause of much of the adverse commentary on NTTs. However, we shall then move to some combination procedures which, when employed with these short-length NTTs, enable a wide range of NTTs to be constructed.

The negative attitudes often expressed about NTTs are not entirely fair. Unfortunately, NTTs suffered an 'image problem' in their early development and it has remained. Such is the current state of development, though, that any designer of digital processors must consider the possibility of using them. Certainly, NTTs are not as convenient to use as FFT techniques but they are not nearly as difficult to use as is sometimes implied. From our perspective, it is not just the potential application that makes them of significance. They are based on many interesting and important ideas and must be considered in any well-rounded study of digital convolution.

Since the interest in finite structures is that they are better tuned to implementation requirements, IMPLEMENTATION ISSUES is relatively short. Nonetheless, there are one or two interesting techniques to explore.

TOOLS

An introduction to number theory

Number theory is one of the most interesting branches of mathematics but was long considered too abstruse for practical application. Its achievements in coding theory and similar fields changed that and in recent years it has achieved great success in digital signal processing. In spite of its name, much of number theory is concerned with only a limited range of numbers. Here, unless otherwise specified, numbers will be assumed to be only the non-negative integers. That is, $0, 1, 2, \cdots$.

At the heart of much of number theory is a very simple concept. Let A and B be two numbers. Then it is always possible to find two other numbers Q and R such that:

$$A = B \cdot Q + R$$

This is an elementary concept of division. Here, Q is termed the **quotient** and R is the **remainder**. If R is zero, then B is termed a **divisor** of A and denoted B|A.

For any number A, there are always at least two divisors, namely 1 and A itself. The number is termed **prime** if these are the only divisors. The prime numbers form an infinite sequence beginning with $2, 3, 5, 7, 11, 13, 17, 19, 23, 29,$ and 31. Note that 1 is not considered a prime and that 2 is the only even prime.

Consider some A that is known not to be prime. By definition, there must be particular numbers B and Q such that:

$$A = B \cdot Q$$

Now B and Q may be prime, or they may not. If not, then each can be similarly treated and if they have divisors which are not prime, this procedure can be repeated again. Continuing, eventually it is found that the number A may be described as:

$$A = \Pi p_i^{e_i}$$

where $\quad$ p_i = the ith prime number

$\quad\quad\quad$ e_i = an integer exponent.

It is obvious from this why non-prime numbers are termed **composite** numbers.

All numbers are either prime or composite. An important theorem, the **fundamental theorem of arithmetic**, states that for any composite the set of exponents is unique. The order in this expression, therefore, does not matter. As most exponents for the primes in this representation are zero, it is common to simply list the non-zero prime divisors.

In digital signal processing, as indeed in mathematics, it is frequently important to know whether a number is prime or composite and if the latter, what are its divisors. In addition, it is often extremely important to be able to locate the nearest prime to a given number. These, unfortunately, are very difficult questions that have taxed the finest mathematicians over the centuries. Given the range of numbers used in digital signal processing, the easiest way of resolving these questions is simply to look up tables. Nevertheless, it is useful to examine some of the mathematical techniques employed.

The ideal solution to problems such as these would be to use a prime number generator. Such a generator does not exist. What is available is a number of usually probabilistic tests for determining whether numbers are prime or composite and a range of methods, also usually probabilistic, that can factor detected composites. There is, though, a simple, deterministic and widely used method for identifying primes that also identifies the factors of composites. It is the **sieve of Eratosthenes**. If all the prime numbers from 2 to N are to be determined, the first step is to list those N–1 numbers. Two is noted as a prime and then all the even numbers are eliminated as they are divisible by the prime number 2. Of the numbers which remain, the smallest is three and that, too, must be a prime. Again, this is noted and then every third number is discarded. The next smallest number remaining is 5, that is noted and then every fifth number is removed and so on. Each step in the sieve process identifies a given prime and eliminates a particular group of composites based on that prime. Only about $\sqrt{N}$ steps are needed to detect all the primes. The sieve is a simple procedure but a number of refinements exist to improve the algorithm's efficiency.

While there is no direct means of locating primes, mathematicians have evolved a number of useful functions which can provide valuable clues on primes and divisors. Some have exact, and invariably exceedingly complex, formulae but almost all have simple approximate formulae which can be used to some advantage.

An interesting question is: how many primes exist up to some number x? This is given by the prime distribution function $\pi(x)$. Given some number x, the average distance to the primes above and below it is approximately:

$$D(x) \approx Ln(x)$$

The distribution of primes can be treated statistically, thus the mean density of primes is the inverse of this. A moderately accurate approximation to $\pi(x)$, usually attributed to Gauss, is:

$$\pi(x) \approx \int_2^x \frac{dx}{Ln(x)} = Li(x)$$

where $Li(x)$ is termed the integral logarithm. The **Prime Number Theorem** states that:

$$\underset{x \to \infty}{\text{Lim}} \frac{\pi(x) \cdot Ln(x)}{x} = 1$$

For relatively small values of x, and certainly within the range most likely to be used for digital signal processing , the approximation:

$$\pi(x) \approx Li(x) - \frac{1}{2} Li(\sqrt{x}) - \frac{1}{3} Li\left(\sqrt[3]{x}\right) - \cdots$$

is a much better estimator of the function than the integral logarithm.

A useful formula derived from the prime number theorem is the following. For large x:

$$\pi(x) \approx \frac{x}{Ln(x)}$$

Taking logarithms:

$$Ln(\pi(x)) = Ln(x) - Ln(Ln(x))$$

which, because x is large, means:

$$Ln(\pi(x)) \approx Ln(x)$$

The original approximation can therefore be expressed as:

$$x \approx \pi(x) \cdot Ln(\pi(x))$$

$\pi(x)$ must be n if x is p_n, the nth prime. Substituting:

$$p_n \approx n \cdot Ln(n)$$

Note that this approximation does depend on x being large.

The number of divisors of a number N are given by a function $d(N)$ where:

$$d(N) = \prod (e_i + 1)$$

and where e_i is the exponent of the ith prime in the composite representation given earlier. This, however, depends on knowing the composite form of the number and that is unlikely when the number is large. A very approximate formula is:

$$d(N) \approx (Ln(N))^{Ln2}$$

and is based on the asymptotic behavior of $d(N)$ but, because of this, it may not be all that useful. A better approximate formula is:

$$\sum_{m=1}^{N} d(m) \approx Ln(N) + 0.154431$$

but this is clearly difficult to use in all circumstances.

Another important function, $\omega(N)$, is a count of the number of distinct prime divisors of a number. That is, the number of p_i terms which have non-zero exponents. Hence:

$$\omega(N) = \sum_{p_i|N} 1$$

Like most number theoretic functions, approximations exist and are sufficiently accurate for most practical purposes. One such formula is:

$$\omega(N) \approx Ln(Ln(N)) + 0.2614$$

Applying this to a few large numbers and comparing the results against tables shows, apart from the fact the formula is quite accurate for most practical purposes, the perhaps surprising result that composites up to extremely large values individually have very few prime divisors. Indeed, it is usually less than 6 for numbers of the order of 10^{80}.

Although these formulae are of some benefit, a great deal of labor remains in finding the factors of any large number. To illustrate, the approximate formula predicts 4.37 prime divisors for:

$$2^{88} - 1 = 309, 485, 009, 821, 345, 068, 724, 781, 055$$

but, from the prime number theorem, there are around 19 million possible candidates for these primes. No formula predicts how to locate these. A number of new and very efficient factoring algorithms exist which can compute the factors of large composites in a time proportional to $\exp(\sqrt{Ln(N) \cdot Ln(Ln(N))})$ but this is still a substantial time. In this case, as 88 is even the number factors into $(2^{44} - 1)$ and $(2^{44} + 1)$ and the answer is found by using the tables in the appendices.

Consider two arbitrary numbers A and B. Let them be represented as:

$$A = \Pi p_i^{\alpha i} \qquad\qquad B = \Pi p_i^{\beta i}$$

Form a third number C from these as follows:

$$C = \Pi p_i^{\min(\alpha_i + \beta_i)}$$

This number is termed the **greatest common divisor** or GCD of A and B and is denoted:

$$(A, B) = C$$

A useful result which can be derived from this definition is that:

$$(B^m - 1, B^n - 1) = B^{(m, n)} - 1$$

for any integers B, m, and n.

If (A, B) is 1, then A and B are said to be **relatively prime** or **coprime** to each other. One property of such numbers is that:

$$d(A \cdot B) = d(A) \cdot d(B)$$

and so the divisor function has a multiplicative property. Similarly, $\omega(n)$ is additive. That is:

$$\omega(A \cdot B) = \omega(A) + \omega(B)$$

Two particular types of numbers are worth mentioning at this juncture. One is the set of **Mersenne numbers** given by:

$$M_N = 2^N - 1 \qquad N = 1, 2, \cdots$$

For N equal to the primes 2, 3, 5, 7, 13, 17, 19, 31, 61, 89, 107, and 127, the Mersenne numbers are prime. Other Mersenne primes are all larger than M_{257} and that is an extremely large number. Other primes such as 11, 23, 37, and so on generate composite Mersenne numbers.

The other type is the set of **Fermat numbers** where these are given by:

$$F_N = 2^{2^N} + 1$$

$$N = 1, 2, 3, \cdots$$

The first four are prime and all Fermat numbers are relatively prime to one another. If a Fermat number has a prime divisor, then that divisor is of the form ($k2^{t+2} + 1$). An interesting property of Fermat numbers is that:

$$F_{N+1} = (F_N - 1)^2 + 1$$

From this:

$$F_{N+1} - 2 = F_N \cdot (F_N - 2)$$

which expressed another way is:

$$F_{N+1} - 2 = F_0 \cdot F_1 \cdot F_2 \cdot \quad \cdots \quad \cdot F_N$$

Two properties of Mersenne numbers are important to note. One is that for N even:

$$2^N - 1 = (2^{\frac{N}{2}} - 1) \cdot (2^{\frac{N}{2}} + 1)$$

That is, an even Mersenne number is the product of a low order Mersenne and Fermat number. This greatly simplifies the construction of tables of Mersenne numbers. The other derives from the fact:

$$M = 2^N - 1 = 2^{(N-R)+R} - 1 = 2^R \cdot (2^{N-R} - 1) + (2^R - 1)$$

If ($2^{N-R} - 1$) has either ($2^R - 1$) or any of the divisors of ($2^R - 1$) as a factor, then so must ($2^N - 1$). Extending this:

$$
\begin{aligned}
2^N - 1 \quad &= 2^R \cdot (2^{N-R} - 1) + (2^R - 1) \\
&= 2^{2R} \cdot (2^{N-2R} - 1) + 2^R \cdot (2^R - 1) + (2^R - 1)
\end{aligned}
$$

$$
\cdot
$$
$$
\cdot
$$
$$
\cdot
$$

$$
= 2^{\alpha R} \cdot (2^{N-\alpha R} - 1) + 2^{(\alpha-1)R} \cdot (2^R - 1) + 2^{(\alpha-2)R} \cdot (2^R - 1) + \cdots + (2^R - 1)
$$

If ($N - \alpha R$) equals R, then ($2^R - 1$) must be a divisor of ($2^N - 1$) and R must be a divisor of N. Thus ($2^N - 1$) always has factors ($2^{d_i} - 1$) where the d_i are the divisors of N.

Although we have concentrated on positive integers, the results derived apply to all integers. To conclude this section on number theory, it is useful to examine some properties of particular integers. To lead into this, consider an equation:

$$
a_0 \cdot x^n + a_1 \cdot x^{n-1} + \cdots + a_{n-1} \cdot x + a_n = 0
$$

where $a_0, a_1, \cdots, a_n$ = integers, not all zero.

A number which is a solution to this expression is termed an **algebraic number**. If the smallest polynomial that a number satisfies is of degree n, then the number is said to have degree n. The integers considered so far are algebraic and they have degree one.

If a_0 in this polynomial is one, then the solution is termed an **algebraic integer**. Since $\sqrt{-1}$ (which will be denoted as j) is the solution to the polynomial:

$$
x^2 + 1 = 0
$$

then j is an algebraic integer. So are $\exp(\frac{j2\pi}{3})$ or $\frac{1}{2}(-1 + j\sqrt{3})$ (which will be denoted as ρ) and $\exp(\frac{j4\pi}{3})$, as they are the solutions of:

$$
x^2 + x + 1 = 0
$$

All of these are numbers of degree 2 or **quadratic numbers.** Other numbers of the form $\exp(\frac{j2\pi}{N})$ are also algebraic integers but are of higher degree.

Let δ be an algebraic number. Further, let P(δ) and Q(δ) be two polynomials in δ whose coefficients are integers. Define a mapping which creates numbers R(δ) where:

$$
R(\delta) = \frac{P(\delta)}{Q(\delta)}
$$

Consider the cases when the algebraic numbers are the quadratic numbers j and ρ. For R(j), any number will be of the form:

$$R(j) = \frac{P(j)}{Q(j)} = \frac{a + jb}{c + jd} = \frac{(a + jb)(c + jd)}{c^2 - d^2}$$

$$= A + jB$$

and similarly, for $R(\rho)$:

$$R(\rho) = \frac{P(\rho)}{Q(\rho)} = \frac{a + \rho b}{c + \rho d} = \frac{(a + \rho b)(c + \rho d)}{c^2 + d^2 - cd}$$

$$= A + \rho B$$

A subset of these two sets of numbers is formed by taking only those numbers where A and B are integers. In the case of $R(j)$, those numbers are termed **Gaussian integers** and for $R(\rho)$ they are the **Eisenstein integers**. Both are infinite in extent. Other algebraic numbers have not yet been considered for digital signal processing applications.

Return now but just briefly to the usual integers. However, rather than restricting our attention to the non-negative integers, we shall consider all the integers. We note that the number one in this set has two divisors, namely 1 and –1. Therefore, to generalize the concept of a prime, we need to redefine it as a number N which only has divisors of ε and εN where ε describes each of the divisors of one. Thus in the complete set of integers, a prime has four divisors.

Consider again number sets formed from a mapping $R(\delta)$. We note that unity is a member of each of these sets. Then we shall simply state that for the sets of numbers of interest here, a **prime** is a number N which only has divisors of ε and εN where ε describes each of the divisors of one in that set. Any other number is therefore **composite**. An important result beyond the scope of this book is that any such a composite is the product of primes (as defined here) and that the representation is unique. For Gaussian integers, the divisors of one are $\pm 1, \pm j$, and for Eisenstein, $\pm 1, \pm \rho, \pm(1 + \rho)$. Greatest common divisors can be defined for these numbers, and other results to be developed later also apply.

Gaussian integers are of some importance in digital signal processing but Eisenstein have so far had only limited use. Both, in particular circumstances, can be used as an alternative to the usual complex numbers in digital convolution.

Residues and congruences

Consider once more the simple equation:

$$A = M \cdot Q + R$$

Let M be some fixed number, which we shall term the **modulus**. If we allow A to successively take the value of every (non-negative) integer and derive values for R, then that infinite set will be mapped into the finite set of numbers $\{ 0, 1, 2, \cdots, M{-}1 \}$. These numbers are the only possible remainders and are called the **residues** or, more correctly, the **residues modulo M**. Since the result includes all possible residues, this mapping procedure produces a **complete residue system**. Mathematicians denote the complete residue system modulo p of **Z** as **Z/pZ**.

Residue operations are quite common. For example, time is measured by a process using moduli of 60 seconds, 60 minutes, 12 hours, 7 days, and 365 or 366 days. Angles, too, employ residue techniques where here the modulus is 360 degrees.

All those numbers which generate the same residue are said to be **congruent** modulo M. Consider two such numbers A and B. If they are congruent, then it must be true that:

$$M \,|\, (A - B)$$

This is commonly denoted:

$$A \equiv B \, mod(\, M \,)$$

This can also be used to denote the operation of extracting the residues from some given number. The engineering literature has evolved its own and far more appealing notation for this very important operation in digital signal processing, viz.:

$$< A >_M \, = R$$

In this notation, two numbers can be said to be congruent if:

$$< A >_M \, = \, < B >_M$$

This will be the notation used here.

Computing residues can be greatly simplified by using some elementary properties of congruent numbers. It is easy to show that:

$$< A + B >_M \, = \, << A >_M + < B >_M >_M$$

$$< A - B >_M \, = \, << A >_M - < B >_M >_M$$

$$< A \cdot B >_M \, = \, << A >_M \cdot < B >_M >_M$$

In addition, if d is the largest divisor of A, M, and C where:

$$< A >_M \, = C$$

then the residue operation can be expressed as:

$$\left\langle \frac{A}{d} \right\rangle_{\frac{M}{d}} = \frac{C}{d}$$

If in this equation A is $a \cdot x$, C is $c \cdot x$ and (x, M) is d, then;

$$< a >_{\frac{M}{d}} = c$$

A very useful result which follows from these is that $\{\, k \cdot a_i \,\}$ is a complete residue system modulo M, provided $\{\, a_i \,\}$ is a complete residue system modulo M and (k, M) is one. If this was not true, then two of the values from the set $\{\, k \cdot a_i \,\}$ would be the same. Thus:

$$< k \cdot a_i - k \cdot a_j >_M = 0 \qquad\qquad i \neq j$$

But given that k and M are mutually prime, these results imply it must be true that:

$$< a_i - a_j >_M = 0$$

and this is not the case. Hence the set $\{ k \cdot a_i \}$ forms a complete residue system.

Residues can be used to determine the factors of composite numbers. Assume N is composite and equals $a \cdot b$ where a and b are prime. If either a or b is small, then the easiest way to detect these is simply to test the primes from 2 upwards. If on the other hand a and b are similar and approximately equal to $\sqrt{N}$, then this procedure will take some time. However, noting that:

$$N = a \cdot b = \frac{1}{2} \cdot (a+b)^2 - \frac{1}{2} \cdot (a-b)^2 = A^2 - B^2$$

then we can locate a and b simply by finding two integers with one greater than $\sqrt{N}$ such that the difference of their squares equals N. If, though, the factors are neither small nor comparable to $\sqrt{N}$, then neither of these approaches appeals. However, the latter can be generalized to overcome this problem. Assume numbers can be found such that:

$$< U^2 >_N = V^2$$

From the definition of residues, $(U + V, N)$ and $(U - V, N)$ must therefore be factors of N. A number of efficient factoring algorithms are based on this approach.

The Euclidean algorithm and the Chinese remainder theorem

It obviously requires considerable computational effort to locate the greatest common divisor of two arbitrary numbers if the definition given is followed. A reasonable question, then, is whether there isn't some easier way. The answer is yes and it is via a procedure known as the **Euclidean Algorithm**. Named after the famous Greek geometer Euclid, this is a very useful algorithm indeed and will reappear several times in the following sections of this book.

Two numbers A and B for which the GCD is to be determined can be linked via:

$$A = B \cdot q_0 + r_0$$

where $r_0 < B$. If r_0 is zero, q_0 must be the GCD. If not, then another step is needed. The GCD is the largest number which divides both A and B. As r_0 is less than B, we can relate the two via:

$$B = r_0 \cdot q_1 + r_1$$

where $r_1 < r_0$. If r_1 equals zero, by substituting into the first equation r_1 is found to be the GCD. If not, we repeat the operation to give a sequence of equations:

$$r_0 = r_1 \cdot q_2 + r_2 \qquad\qquad \text{where } r_2 < r_1$$

$$r_1 = r_2 \cdot q_3 + r_3 \qquad\qquad \text{where } r_3 < r_2$$

$$\cdot$$
$$\cdot$$
$$\cdot$$

$$r_{n-2} = r_{n-1} \cdot q_n + 0$$

The sequence $(r_0, r_1, r_2, \cdots, r_{n-1})$ is strictly decreasing and will always terminate. Thus r_{n-1} divides r_{n-2} which divides the previous remainder and so on. Therefore, it divides A and B and so is the GCD.

This algorithm is more simply described as:

$$r_0 = < A >_B$$

$$r_1 = < B >_{r_0}$$

$$r_2 = < r_0 >_{r_1}$$

$$\cdot$$
$$\cdot$$
$$\cdot$$

$$r_n = < r_{n-2} >_{r_{n-1}} = 0$$

The modulus r_{n-1} is therefore the greatest common divisor.

There is another way of expressing the Euclidean algorithm. The initial equation becomes:

$$\frac{A}{B} = q_0 + \frac{r_0}{B}$$

Similarly, the second equation can be expressed as:

$$\frac{B}{r_0} = q_1 + \frac{r_1}{r_0}$$

and so on. Combining these, the result is the **continuing fraction**:

$$\frac{A}{B} = q_0 + \cfrac{1}{q_1 + \cfrac{1}{q_2 + \cfrac{\cdot}{\cdot \; + \cfrac{1}{q_n}}}}$$

This suggests a series of approximations to the original equations be created of the form:

$$\frac{A_0}{B_0} = q_0 \qquad = \frac{q_0}{1}$$

$$\frac{A_1}{B_1} = q_0 + \frac{1}{q_1} \qquad = \frac{q_0 \cdot q_1 + 1}{1 \cdot q_1 + 0}$$

$$\frac{A_2}{B_2} = q_0 + \frac{1}{q_1 + \dfrac{1}{q_2}} \qquad = \frac{q_2 \cdot (q_0 \cdot q_1 + 1) + q_0}{q_2 \cdot (1 \cdot q_1 + 0) + 1}$$

$$\vdots$$

$$\frac{A_n}{B_n} = \frac{A}{B}$$

These approximations are termed **convergents**. Observing the expansions, we see that the two terms in the ratios can be expressed as:

$$A_k = q_k \cdot A_{k-1} + A_{k-2}$$
$$B_k = q_k \cdot B_{k-1} + B_{k-2}$$

where $2 \le k \le n$. Further, it can be shown:

$$B_k \cdot A_{k+1} - A_k \cdot B_{k+1} = (-1)^k \qquad\qquad k \ge 0$$

which implies:

$$(A_k, B_k) = 1$$

Another useful result is achieved by expressing the algorithm in slightly different form. The above equations can be written as:

$$r_0 = A - B \cdot q_0$$
$$r_1 = B - r_0 \cdot q_1$$
$$r_2 = r_0 - r_1 \cdot q_2$$
$$\vdots$$
$$r_n = r_{n-2} - r_{n-1} \cdot q_n$$

Substitute the first equation into the second. This gives:

$$r_1 = (1 + q_0 q_1) \cdot B - q_1 \cdot A$$

Substitute this into the third equation and continue the process. This will eventually give:

$$r_{n-1} = \alpha \cdot A + \beta \cdot B$$

where α and β are integers. But r_{n-1} is the GCD, thus the greatest common divisor of two numbers is a linear sum of those numbers. Another result which follows from the above is that, for any integer k:

$$(A, B) = (A, B - k \cdot A)$$

The Euclidean algorithm is closely linked to a very important class of equations known as the **Diophantine equations**. Like the algorithm, these have been known since antiquity. Given integers a, b and c, these are equations of the form:

$$a \cdot x + b \cdot y = c$$

and the problem is to find solutions for x and y which are also integers. Because of this condition, there can be many such equations that do not have a solution.

Consider approaches to solving these equations. Firstly, we will show all Diophantine equations can be reduced to a particular form. Let:

$$(a, b) = d$$

Then the equation becomes:

$$d \cdot A \cdot x + d \cdot B \cdot y = c$$

where A and B must be mutually prime. This equation has no solution unless d is a divisor of c. If it is, the Diophantine equation reduces to:

$$A \cdot x + B \cdot y = C$$

where

$$(A, B) = 1$$

$$C = \frac{c}{d}$$

To prove this, it is useful to examine the line of thought followed by the ancient Greeks. We rearrange the equation to:

$$a \cdot x = -b \cdot y + c$$

i.e.

$$x = -\frac{b}{a} \cdot y + \frac{c}{a}$$

This is the equation of a line in a plane. The grid points in Figure 3.1 show the positions of the integers in this plane. The infinite set of solutions of this equation are given by those

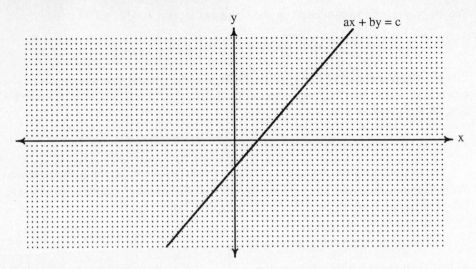

Figure 3.1   Geometric interpretation of the solutions of Diophantine equations

points on the grid intercepted by the line. Its slope in this plane is the ratio of two integers, namely $(-\frac{b}{a})$. Thus in:

$$x = -\frac{b}{a} \cdot y$$

for every (integral) spacing 'a' along the y axis of the grid, there will be an (integral) spacing 'b' along the x axis. These spacings can be reduced yet remain integral if these numbers have a common divisor. In fact, a spacing $(\frac{a}{d})$ along the y axis will be matched by an integral spacing $(\frac{b}{d})$ along the x axis, where d is the GCD of a and b. The question, therefore, is: under what conditions does $(\frac{c}{a})$ permit a solution? This offset can be described by $(\frac{c}{wd})$ where w is $(\frac{a}{d})$ and d is the GCD of a and b. If this is integral, then w and d are divisors of c and there are clearly solutions. However, that is not the general case. Intercepts occur and so solutions exist if this offset is such that when multiplied by w it becomes integral. If not, there can never be solutions. However, in order for $w \cdot c$ to be integral, d must be a divisor of c.

There is an alternative expression for this last Diophantine equation. It is:

$$A \cdot x' + B \cdot y' = 1$$

where $x' = \dfrac{x}{C}$ and $y' = \dfrac{y}{C}$

As there is no loss of generality with this, it is the form of the Diophantine equation commonly quoted. Now note the recurrence relations between the convergents in the Euclidean algorithm. Since A is A_n and B is B_n, we obtain by substitution that:

$$B_{n-1} \cdot A \ - \ A_{n-1} \cdot B = (-1)^{n-1}$$

i.e.
$$-B_{n-1} \cdot A + A_{n-1} \cdot B = (-1)^n$$
or
$$(-1)^{n+1} \cdot B_{n-1} \cdot A + (-1)^n \cdot A_{n-1} \cdot B = 1$$

As this has exactly the same form as the Diophantine equation, we deduce that:

$$x' = (-1)^{n+1} \cdot A_{n-1} \qquad \text{and} \qquad y' = (-1)^n \cdot B_{n-1}$$

is a solution to that equation. Substituting, we easily derive x and y.

This can only be one of an infinite number of such solutions. Returning to the original Diophantine equation, let a solution found by this method be described as x_0 and y_0. Then:

$$a \cdot x + b \cdot y = c = a \cdot x_0 + b \cdot y_0$$

Thus:

i.e.
$$a \cdot (x - x_0) + b \cdot (y - y_0) = 0$$
$$a \cdot (x - x_0) = (-b) \cdot (y - y_0) = b \cdot (y_0 - y)$$

A common factor on both sides must be (a,b) which is d. Dividing:

$$\left(\frac{a}{d}\right) \cdot (x - x_0) = \left(\frac{b}{d}\right) \cdot (y_0 - y)$$

As $(\frac{a}{d})$ and $(\frac{b}{d})$ are mutually prime, this implies $(\frac{a}{d})|(y_0 - y)$ and $(\frac{b}{d})|(x - x_0)$. Hence the other solutions are:

$$x = x_0 + k \cdot \left(\frac{b}{d}\right)$$

$$y = y_0 - k \cdot \left(\frac{a}{d}\right)$$

where k is an integer, $-\infty \leq k \leq \infty$. It is common just to quote x_0 and y_0 as the solution.

Some very useful congruence equations are related to Diophantine equations. Consider:

$$a \cdot x + M \cdot y = c$$

Taking the residue modulo M gives:

$$< a \cdot x >_M = c$$

We know the conditions for a solution to the Diophantine equation. Consequently, a solution exists to this congruence provided d divides c where:

$$(a, M) = d$$

There are d congruent solutions to this equation and they are:

$$\left\{ g, g + \frac{M}{d}, g + \frac{2M}{d}, \cdots, g + \frac{(d-1)M}{d} \right\}$$

where g is the solution to:

$$< \frac{a}{d} \cdot g >_{\frac{M}{d}} = \frac{c}{d}$$

Because of this, another way of expressing a Diophantine equation is:

$$< a \cdot x >_M = c \qquad\qquad\qquad (a, M) = 1$$

Now assume the Diophantine equation can be expressed as:

$$a \cdot x + b \cdot y = c = \alpha \cdot N + \beta$$

Taking the residue modulo N gives:

$$< a \cdot x + b \cdot y >_N = \beta$$

Solutions always exist if (a,b) is unity. In the more general case, solutions exist provided (a,b) divides c. If β is zero, that demands (a,b) divides N.

The Chinese remainder theorem, so-called because the first references to it seem to have been in ancient Chinese literature, is a very important theorem which relates to this class of equations. A statement of the theorem does not suggest its significance. It is:

THEOREM   The set of congruences:

$$< x >_{m_i} = r_i \qquad\qquad\qquad i = 1, 2, \cdots, n$$

have a unique solution modulo M where M is Πm_i provided that:

$$(m_i , m_j) = 1 \qquad\qquad\qquad i \neq j$$

The proof of this theorem is particularly important to later work and, indeed, parts of it are used more frequently than the theorem itself. To begin, construct a set of numbers:

$$M_1 = \frac{M}{m_1}$$

$$M_2 = \frac{M}{m_2}$$

.

.

.

$$M_n = \frac{M}{m_n}$$

If the m_i are mutually prime to each other, then (M_i , m_i) must be mutually prime. In this case, we can construct relations like:

$$< N_i \cdot M_i >_{m_i} = 1$$

where the numbers N_i are integers. This is true because it is the Diophantine equation considered earlier. Now construct a term S where:

$$S = r_1 \cdot N_1 \cdot M_1 + r_2 \cdot N_2 \cdot M_2 + \cdots + r_n \cdot N_n \cdot M_n$$

Given the form of the terms M_i, then:

$$< S >_{m_i} = < r_i \cdot N_i \cdot M_i >_{m_i}$$
$$= < r_i \cdot < N_i \cdot M_i >_{m_i} >_{m_i}$$
$$= r_i \qquad\qquad i = 1, \cdots, n$$

Thus S must be a solution to the set of congruences. Further, this solution is unique with respect to the modulus M.

Number sequences and primitive roots

The study of sequences of numbers and how they are generated is central to much of the application of number theory to digital signal processing. Our interest centers on finite sequences, such as $0, 1, 2, \cdots, M-1$, the complete residue system modulo M.

Of enormous assistance in the study of sequences is **Euler's totient function** $\Phi(N)$. This function specifically applies to the sequence:

$$1, 2, 3, 4, 5, \cdots, N-1$$

and it is the number of elements in this sequence that are relatively prime to N. That is:

$$\Phi(N) = \sum_{(x, N)=1} 1$$

Hence:

$$\Phi(1) = 1$$
$$\Phi(2) = 1$$
$$\Phi(3) = 2$$
$$\Phi(4) = 2$$

and so on.

An exact formula can be derived for the totient. Assume N is some prime p. From the definition of primes, clearly:

$$\Phi(p) = p - 1$$

N could also be the power of a prime. That is, p^a. Now here any value of x which is a multiple of p will have a GCD of p and so cannot be counted. Every other value will be relatively prime though and so:

$$\Phi(N) = p^a - p^{a-1}$$

$$\Phi(N) = N \cdot (1 - \frac{1}{p})$$

Finally, N can be composite. That is, N can be $p_1^{e_1} \cdot p_2^{e_2} \cdots p_N^{e_N}$. In this case:

$$\Phi(N) = N \cdot (1 - \frac{1}{p_1}) \cdot (1 - \frac{1}{p_2}) \cdot \cdots \cdot (1 - \frac{1}{p_n})$$

Proof of this relies on a most important property of the totient, namely that:

$$\Phi(A \cdot B) = \Phi(A) \cdot \Phi(B) \qquad \text{for } (A, B) = 1$$

This property can be shown using the Chinese remainder theorem. Let:

$$< x >_A = r_i \qquad < x >_B = s_i \qquad x = 0, 1, 2, \cdots$$

Let $\{ a_i \}$ be the set of $\Phi(A)$ numbers drawn from $\{ r_i \}$ that are relatively prime to A and $\{ b_i \}$ be the set of $\Phi(B)$ numbers drawn from $\{ s_i \}$ that are relatively prime to B. Now form $\Phi(AB)$ numbers:

$$c_i = a_i \cdot B + b_i \cdot A$$

This set $\{ c_i \}$ must be drawn from the set of numbers $\{ t_i \}$ where:

$$< x >_{AB} = t_i \qquad x = 0, 1, 2, \cdots$$

Let $\{ d_i \}$ be the set of $\Phi(AB)$ numbers drawn from $\{ t_i \}$ that are mutually prime to AB. Are the two sets $\{ c_i \}$ and $\{ d_i \}$ the same? Consider these c_i. Since A and B are relatively prime and the a_i are relatively prime to A, then the $a_i \cdot B$ must be relatively prime to A. Similarly, the $b_i \cdot A$ must be relatively prime to B and therefore these c_i must be relatively prime to AB. The theorem guarantees that they are the only such numbers, hence the result.

These exact formulae require a knowledge of the divisors of N and for large N these may not be known. The totient's value fluctuates quite strongly with N, hence an approximate formula can be significantly in error. Nevertheless, an asymptotic value is:

$$\Phi(N) \approx \frac{6N}{\pi^2}$$

It must be stressed this is only asymptotic. Derived by approximating to the average of $\Phi(N)$ taken over many values of N, it can be quite inaccurate for a given value of N.

One very useful application of Euler's totient is in classifying elements in a sequence. Consider the relation:

$$(x, N) = d_k \qquad\qquad 1 \le x \le N$$

Note that N has been added to this sequence. Given the definition of a greatest common divisor, the d_k can only be divisors of N. The totient can therefore classify elements into one of k classes depending on the value of d_k. How many belong to each class? For d_k equal to 1, it must be $\Phi(N)$. For other cases, since we can express the GCD as:

$$\left(\frac{x}{d_k} , \frac{N}{d_k} \right) = 1$$

the number of members must be $\Phi\!\left(\dfrac{N}{d_k} \right)$. All elements of the sequence will be classified and we deduce:

$$\sum_{k=1}^{d(N)} \Phi\!\left(\frac{N}{d_k} \right) = N$$

From the definition of divisors, this also means:

$$\sum_{k=1}^{d(N)} \Phi(d_k) = N$$

We now examine some number sequences, beginning with $< x >_N$. If x ranges over all the integers, then this operation maps that infinite set into a periodic sequence where each period consists of values drawn from the complete residue system:

$$\{ 0, 1, 2, \cdots, N{-}1 \}$$

An alternative to this sequence is $< a \cdot x >_N$. This is a Diophantine equation and it only has solutions if a and N are relatively prime. Again, the infinite set of integers are mapped into a periodic sequence which can only be drawn from this complete residue system. However, consider the order of the elements in this periodic sequence. If it is the same as the first case, then for a given x, say x_0, it would be the case that:

$$< a \cdot x_0 >_N = < x_0 >_N$$

which implies:

$$< a \cdot x_0 - x_0 >_N = 0$$

This can only be true if x_0 is zero or a is 1. Thus $< a \cdot x >_N$ is a **permutation** of $< x >_N$. From the properties of residues, there can only be $\Phi(N)$ such permuted sequences.

A much more interesting category of number sequences are those formed by $< a^n >_N$ where $a > 0$ and $n = 0, 1, 2, 3, \cdots$. This maps the infinite sequence given by the exponent n into a periodic sequence. However, there are two important differences from the previous type of sequences. First, the value 0 never appears; the first element of the sequence is always 1. Second, the period is not N and, in fact, varies according to the value of a. Note that mathematicians do not refer to periods but to the **order**. This is denoted $\text{ord}_N n$.

Consider the complete residue system $< x >_N$. Since $< a \cdot x >_N$ with a and N mutually prime is a permutation of this system, both have $\Phi(N)$ residues relatively prime to N and those residues are, of course, the same. Let values of the variable x generating those residues be $\{ x_1, x_2, \cdots, x_{\Phi(N)} \}$. Then those residues are also generated by $\{ ax_1, ax_2, \cdots, ax_{\Phi(N)} \}$. Proceeding, let:

$$ R = < x_1 >_N \cdot < x_2 >_N \cdot \quad \cdots \quad \cdot < x_{\Phi(N)} >_N $$

which, from the properties of residues, equals:

$$ R = < x_1 \cdot x_2 \cdot \quad \cdots \quad \cdot x_{\Phi(N)} >_N $$

Provided (a, N) is 1, R must also equal:

$$ R = < a \cdot x_1 >_N \cdot < a \cdot x_2 >_N \cdot \quad \cdots \quad \cdot < a \cdot x_{\Phi(N)} >_N $$

$$ = < a^{\Phi(N)} \cdot x_1 \cdot x_2 \cdot \quad \cdots \quad \cdot x_{\Phi(N)} >_N $$

$$ = << a^{\Phi(N)} >_N \cdot x_1 \cdot x_2 \cdot \quad \cdots \quad \cdot x_{\Phi(N)} >_N $$

which implies that:

$$ < a^{\Phi(N)} >_N = 1 \qquad\qquad\qquad (a, N) = 1 $$

This is **Euler's theorem**. A special case is when N is prime in which case:

$$ < a^{p-1} >_p = 1 $$

This is known as **Fermat's theorem**.

These two theorems occur repeatedly in number theory and are among the most useful tools available, especially in digital signal processing applications. For example, consider:

$$ < a \cdot x >_N = c \qquad\qquad\qquad (a, N) = 1 $$

Multiplying both sides by the non-zero term ($a^{\Phi(N)-1}$), then from Euler's theorem:

$$ < x >_N = c \cdot a^{\Phi(N)-1} $$

must be the solution. Another simple but very useful result applies to Mersenne numbers.

Assume p is an odd prime divisor of such a number. By definition $< 2^N >_p$ is 1, but from Fermat's theorem, so is $< 2^{p-1} >_p$. Then N must divide (p – 1) and if N is odd, we deduce p must satisfy the congruence:

$$< p >_{2N} = 1$$

This is very useful in locating factors. A similar result applies to Fermat numbers.

A very important application of Fermat's theorem is locating primes. It is not true that for an arbitrary number a, $< a^N >_N$ equal to 1 means N must be prime. However, if N is suspected of being prime and a number of a_i are tested and give 1, there is a high probability N must be prime. There is a problem in that a class of composites termed Carmichael numbers do satisfy the congruence for all a_i but these are relatively sparse.

The period of $< a^n >_N$, regardless of a and N, begins with 1 and so the most obvious step in the important task of finding that period is locating when the sequence is again 1. Euler's and Fermat's theorems have a role to play here. However, all they are stating is that a new period must begin after $\Phi(N)$ steps and so the maximum period must be $\Phi(N)$. Let the period of some element a be Q and, for generality, assume this does not equal $\Phi(N)$. The sequence is given by:

$$(1, a, a^2, \cdots, a^{Q-1}, 1, a^{Q+1}, \cdots, a^{2Q-1}, 1, a^{2Q+1}, \cdots, Q^{\Phi(N)-1})$$

Thus we deduce the important result Q must either be the totient or divide it.

Any value a which has a period equal to the maximum length $\Phi(N)$ is termed a **primitive root** of the modulus N. Digital signal processing has a particular interest in these. Hence the questions arise of how to find the primitive roots of some N, how many of them are there and what are their properties.

Locating primitive roots is, at first glance, difficult. However, the period must be a divisor of $\Phi(N)$. Let these divisors be { d_i }. Then all that is necessary to do is to calculate $< a^{d_i} >$ for each of the d_i. If any produce 1, they cannot be primitive roots. For N very large there is still a problem with this approach and, unfortunately, there is no easy solution to this. It can be shown, though, that the only moduli which support primitive roots are $1, 2, 4, p^k$ and $2p^k$ where p is an odd prime and $k \geq 1$.

Let g be a primitive root of some modulus N. The periodic sequence generated is:

$$\{ 1, g, g^2, \cdots, g^{\Phi(N)-1} \}$$

Consider these elements. It is known no element except the first can be 1 but of the others, if two should be the same, this would mean:

$$< g^i >_N = < g^j >_N \qquad \text{for } i < j < \Phi(N)$$

i.e. $$< g^{i-j} >_N = 1$$

However, g is a primitive root and so the period must be the totient. Therefore, the sequence members must be unique and so mutually incongruent.

It was shown earlier that the residue system $< a \cdot b >_N$ is a permutation of the residue system $< a >_N$ provided (b, N) is 1. Consider the residue system $< a^n >_N$ where n is 0, 1, $\cdots$, Q and Q is the period. Then we deduce $< (a^\alpha)^n >_N$ is just a permutation of this residue system provided (α, Q) is 1. Thus if g should be a primitive root of such a sequence, the other primitive roots can be very simply found by raising g to a power relatively prime to $\Phi(N)$.

This earlier work also answers another question on these multiplicative sequences. It was shown that for the sequence $1, 2, \cdots, N$, the number of elements sharing the same GCD of d_k with N was $\Phi(d_k)$ and that d_k had to be a divisor of N. Relating that to sequences of the form $< a^{\alpha n} >_N$, if d_k is a divisor of $\Phi(N)$, there are $\Phi(\Phi(d_k))$ elements in the sequence with period d_k. Hence there can only be $\Phi(\Phi(N))$ primitive roots.

Consider an element a with period Q. An interesting question is: what is the period M of a^α where α is some integer? We can immediately deduce, given $< a^{\alpha M} >_N$ is 1, that αM divides Q. From the general properties of the greatest common divisor:

$$(\alpha, Q) = d = r\alpha + sQ \qquad\qquad r, s = \text{integers}$$

Therefore:

$$dM = r\alpha M + sQM$$

If αM divides Q, then Q must divide dM. Expressed another way, M divides $\frac{Q}{d}$. But $\frac{Q}{d}$ divides M as:

$$< a^{\alpha\left(\frac{Q}{d}\right)} >_N \; = \; < a^{Q\left(\frac{\alpha}{d}\right)} >_N \; = \; <\left(a^Q\right)^{\left(\frac{\alpha}{d}\right)} >_N \; = 1$$

The only conclusion from these two results is that M must be $\frac{Q}{d}$. That is, if the element a^1 has a period Q, then a^α where α is some positive integer has a period $\frac{Q}{(\alpha,Q)}$.

Example

Consider the sequence produced by $< x >_N$ where N is 3^3 or 27. A primitive root for this sequence has a period:

$$\Phi(27) = 27(1 - \frac{1}{3}) = 18$$

As:

$$\Phi(18) = \Phi(2) \cdot \Phi(9) = 1 \cdot 9 \cdot (1 - \frac{1}{3}) = 6$$

there are six primitive roots. These roots must be mutually prime to 27, hence they can only be drawn from 2,4,5,7,8,10,11,13,14,16,17,19, 20, 22, 23, 25, and 26. To test which of these are roots, we need to raise them to a power equal to each of the divisors of the period and see if the result is 1 modulo 27. That is, we need to test for powers equal to 1, 2, 3, 6, and 9. Then:

$< 2^1 >_{27} = 2$	$< 2^2 >_{27} = 4$	$< 2^3 >_{27} = 8$	$< 2^6 >_{27} = 10$	$< 2^9 >_{27} = 26$
$< 4^1 >_{27} = 4$	$< 4^2 >_{27} = 16$	$< 4^3 >_{27} = 10$	$< 4^6 >_{27} = 19$	$< 4^9 >_{27} = 1$
$< 5^1 >_{27} = 5$	$< 5^2 >_{27} = 25$	$< 5^3 >_{27} = 17$	$< 5^6 >_{27} = 19$	$< 5^9 >_{27} = 26$
$< 7^1 >_{27} = 7$	$< 7^2 >_{27} = 22$	$< 7^3 >_{27} = 19$	$< 7^6 >_{27} = 10$	$< 7^9 >_{27} = 1$
$< 8^1 >_{27} = 8$	$< 8^2 >_{27} = 10$	$< 8^3 >_{27} = 26$	$< 8^6 >_{27} = 1$	
$< 10^1 >_{27} = 10$	$< 10^2 >_{27} = 19$	$< 10^3 >_{27} = 1$		
$< 11^1 >_{27} = 11$	$< 11^2 >_{27} = 13$	$< 11^3 >_{27} = 8$	$< 11^6 >_{27} = 10$	$< 11^9 >_{27} = 26$
$< 13^1 >_{27} = 13$	$< 13^2 >_{27} = 7$	$< 13^3 >_{27} = 10$	$< 13^6 >_{27} = 19$	$< 13^9 >_{27} = 1$
$< 14^1 >_{27} = 14$	$< 14^2 >_{27} = 7$	$< 14^3 >_{27} = 17$	$< 14^6 >_{27} = 19$	$< 14^9 >_{27} = 26$
$< 16^1 >_{27} = 16$	$< 16^2 >_{27} = 13$	$< 16^3 >_{27} = 19$	$< 16^6 >_{27} = 10$	$< 16^9 >_{27} = 1$
$< 17^1 >_{27} = 17$	$< 17^2 >_{27} = 19$	$< 17^3 >_{27} = 26$	$< 17^6 >_{27} = 1$	
$< 19^1 >_{27} = 19$	$< 19^2 >_{27} = 10$	$< 19^3 >_{27} = 1$		
$< 20^1 >_{27} = 20$	$< 20^2 >_{27} = 22$	$< 20^3 >_{27} = 8$	$< 20^6 >_{27} = 10$	$< 20^9 >_{27} = 26$
$< 22^1 >_{27} = 22$	$< 22^2 >_{27} = 25$	$< 22^3 >_{27} = 10$	$< 22^6 >_{27} = 19$	$< 22^9 >_{27} = 1$
$< 23^1 >_{27} = 23$	$< 23^2 >_{27} = 16$	$< 23^3 >_{27} = 17$	$< 23^6 >_{27} = 19$	$< 23^9 >_{27} = 26$
$< 25^1 >_{27} = 25$	$< 25^2 >_{27} = 4$	$< 25^3 >_{27} = 19$	$< 25^6 >_{27} = 10$	$< 25^9 >_{27} = 1$
$< 26^1 >_{27} = 26$	$< 26^2 >_{27} = 1$			

Thus the six primitive roots are 2, 5, 11, 14, 20, and 23. We can confirm this quite easily. The numbers less than 18 and relatively prime to it are 1, 5, 7, 11, 13, and 17. We know 2 is a primitive root, thus the complete set of roots must be:

$$< 2^1 >_{27} = 2 \qquad < 2^5 >_{27} = 5 \qquad < 2^7 >_{27} = 20$$
$$< 2^{11} >_{27} = 23 \qquad < 2^{13} >_{27} = 11 \qquad < 2^{17} >_{27} = 14$$

Finally, note the sequence generated by a primitive root. Choosing 2, we obtain:

$$< 2^0 >_{27} = 1 \qquad < 2^1 >_{27} = 2 \qquad < 2^2 >_{27} = 4 \qquad < 2^3 >_{27} = 8 \qquad < 2^4 >_{27} = 16$$
$$< 2^5 >_{27} = 5 \qquad < 2^6 >_{27} = 10 \qquad < 2^7 >_{27} = 20 \qquad < 2^8 >_{27} = 13 \qquad < 2^9 >_{27} = 26$$
$$< 2^{10} >_{27} = 25 \qquad < 2^{11} >_{27} = 23 \qquad < 2^{12} >_{27} = 19 \qquad < 2^{13} >_{27} = 11 \qquad < 2^{14} >_{27} = 22$$
$$< 2^{15} >_{27} = 17 \qquad < 2^{16} >_{27} = 7 \qquad < 2^{17} >_{27} = 14 \qquad < 2^{18} >_{27} = 1$$

Thus the roots generate { 1,2,4,5,7,8,10,11,13,14,16,17,19,20,22,23,25,26 }. That is, all elements of the complete residue system modulo 27 bar { 0,3,6,9,12,15,18,21,24 } which are multiples of the complete residue system modulo 3^2 or 9. This is to be expected. We showed numbers generated by a root must be mutually incongruent. These elements would violate that.

Quadratic residues

Finding a solution to the congruence $< x^2 >_p = a$ is to find square roots with respect to congruences. A solution is termed a **quadratic residue**, commonly denoted aRp. Similarly, any of the residues of p that are not quadratic residues are termed **quadratic non-residues** and denoted bNp. This is a further means of classifying the complete residue system modulo p { 0, 1, $\cdots$, p–1 }.

This result is quite general. To see this, consider the general quadratic congruence:

$$< \alpha y^2 + \beta y + \gamma >_M = 0$$

If M is some odd prime p and if 4α and p are relatively prime, then this implies:

$$< 4\alpha (\alpha y^2 + \beta y + \gamma) >_p = 0$$

which is:

$$< (2\alpha y + \beta)^2 - (\beta^2 - 4\alpha\gamma) >_p = 0$$

i.e.
$$< (2\alpha y + \beta)^2 >_p = < (\beta^2 - 4\alpha\gamma) >_p$$

and this has the form given above.

The usual quadratic equation has two solutions, say x_0 and $-x_0$. Negative values, though, are not defined in a complete residue system. However, if $(p - x_0)$ is substituted, it is found this is a solution and this offers an interpretation of unary negation. Unless otherwise stated then, $-x$ will be interpreted as being $(p - x)$ and, equally, if x is quoted as a solution it will be assumed the second solution of the equation is $(p - x)$.

Euler's criterion states that a number a is a quadratic residue for $0 < a < p$ where p is an odd prime if and only if:

$$< a^{\frac{(p-1)}{2}} >_p = 1$$

It is a quadratic non-residue if and only if:

$$< a^{\frac{(p-1)}{2}} >_p = -1$$

To see this, from Fermat's theorem and the properties of residues:

$$< a^{\frac{(p-1)}{2}} >_p \cdot < a^{\frac{(p-1)}{2}} >_p = 1$$

No restriction was placed on a. Consequently, any residue raised to the power $(\frac{p-1}{2})$ can only be 1 or -1 modulo p. The criterion is proved, therefore, by showing only quadratic residues produce 1. This can be done informally as follows. Let g be a primitive root of p. Then from the earlier work on primitive roots, $< g^n >_p$ generates the complete residue system of p, where $0 \le n \le p-2$. Make the transformation:

$$n = 2k \qquad 0 \le k \le \left(\frac{p-1}{2} \right) - 1 \text{ for n even}$$

$$= 2k - 1 \qquad 1 \le k \le \left(\frac{p-1}{2} \right) \qquad \text{for n odd}$$

Consider the terms g^{2k} and the effect of raising them to $(\frac{p-1}{2})$. From Fermat's theorem:

$$< (g^{2k})^{\frac{(p-1)}{2}} >_p \ = \ < g^{k(p-1)} >_p \ = \ < 1^k >_p \ = \ 1$$

As the solution to the congruence equation is g^k, these terms are quadratic residues. Equally, the odd terms are quadratic non-residues and so these terms raised to $(\frac{p-1}{2})$ must equal -1.

This outline suggests three other important properties of quadratic residues. First, that there must be $(\frac{p-1}{2})$ of them and so equally there must be the same number of non-residues. (Zero satisfies the condition of a quadratic residue but it is a trivial solution and so rarely considered as such.) Second, that in a sequence of a primitive root of some prime, the quadratic and non-quadratic residues must alternate. Third, given two quadratic residues, r_1 and r_2, of p and two non-residues, n_1 and n_2, then:

$$
\begin{aligned}
&r_1 \cdot r_2 &&\text{is a quadratic residue of p}\\
&r_1 \cdot n_1 &&\text{is a quadratic non-residue of p}\\
&n_2 \cdot r_2 &&\text{is a quadratic non-residue of p}\\
&n_1 \cdot n_2 &&\text{is a quadratic residue of p}
\end{aligned}
$$

A very useful tool in the study of quadratic residues is the **Legendre symbol** (a/p). This is a notation, not a function, and is defined as:

$$
\begin{aligned}
(a/p) &= &&1 &&\text{when a is a quadratic residue}\\
&= &&-1 &&\text{when a is a quadratic non-residue}\\
&= &&0 &&\text{when a equals zero}
\end{aligned}
$$

From the preceding work, this symbol could also be defined as:

$$(a/p) \ = \ < a^{\frac{(p-1)}{2}} >_p$$

Although a notation, it can be used as an operator to, say, transform a complete residue system into an alternating sequence of 1's and -1's. However, it is more interesting to consider its other properties. From the definition of a quadratic residue, 1, which is guaranteed to be a member of every sequence, must be a quadratic residue. Therefore:

$$(1/p) \ = \ 1$$

From the general properties of residues, it must also be true that:

$$(a/p) \cdot (b/p) \ = \ (ab/p)$$

Assume then, that a is b. This relation, together with the earlier result on the product of quadratic residues and non-residues, means:

$$(b^2/p) \ = \ 1$$

Incorporating this into the first relation:

$$(ab^2/p) = (a/p)$$

which can be very useful in locating quadratic residues. Another interesting result is:

$$(-1/p) = (-1)^{\frac{(p-1)}{2}}$$

These results require p to be an odd prime. Thus −1 must be a residue for primes of the type (4k + 1) and equally, a quadratic non-residue for primes of the type (4k + 3). Similarly, 2 can only be a quadratic residue for primes of the type (8k + 1) or (8k − 1).

Consider the set of residues { 1, 2, 3, $\cdots$, p − 1 }. Their product is just $< (p-1)! >_p$. Next, consider two sequences. The first is { k, 2k, 3k $\cdots$, (p − 1)k }. It was shown earlier that this sequence is a just a permutation of this residue system provided k is relatively prime to p. Thus:

$$< (p-1)! >_p = < k \cdot 2k \cdot 3k \cdot \ \cdots \ \cdot (p-1)k >_p$$

The second sequence is that generated by $< g^n >_p$ for $0 \le n \le p-2$ where g is a primitive root of p. This, too, is a permutation of the residue system and so the product of these terms is $< (p-1)! >_p$. By themselves, these results are trivial but they are useful steps in proving an important result.

It was shown earlier from Fermat's theorem that a unique solution exists for $< x >_p$ in $< \alpha \cdot x >_p = a$ provided $(\alpha, p) = 1$, and that solution is $a \cdot \alpha^{p-2}$. Interpreted another way, for every α in a residue system, there exists another unique element such that when multiplied modulo p the result is some third element a. Thus residues can be grouped into pairs in $< (p-1)! >_p$ and the product of those pairs is one of the elements of the residue system. Hence:

$$< (p-1)! >_p = a^{\frac{(p-1)}{2}}$$

However, assume that a is a quadratic residue. If that is the case, then one of the residues γ must be such that γ^2 equals a. This product pair is included in this result. Another pair, though, must be the solutions to the quadratic congruence and that product is:

$$< \gamma(p-\gamma) >_p = < \gamma p - \gamma^2 >_p = < -\gamma^2 >_p = < -a >_p$$

Therefore, in this case:

$$< (p-1)! >_p = < -a^{\frac{(p-1)}{2}} >_p$$

The general result may consequently be expressed as:

$$< (p-1)! >_p = < -(a/p) \cdot a^{\frac{(p-1)}{2}} >_p$$

Since 1 is always a quadratic residue, a special case of this is:

$$< (p-1)! >_p = < -1 >_p$$

which is **Wilson's theorem**. Incidentally, this general form is another proof of:

$$(a/p) = < a^{\frac{(p-1)}{2}} >_p$$

One of the most important results associated with the Legendre symbol is the **Law of Quadratic Reciprocity** which states that for two odd primes p and q:

$$(p/q)\cdot(q/p) = (-1)^{\frac{(p-1)(q-1)}{4}}$$

This, together with the other properties of the Legendre symbol, makes finding quadratic residues a very simple task. For example, consider the problem of finding whether or not 987 is a quadratic residue of the odd prime 2053. That is, of finding (987/2053). Now 987 is the product 3·7·47, hence:

$$(987/2053) = (3/2053)\cdot(7/2053)\cdot(47/2053)$$

Consider the first of these. From the law of reciprocity:

$$(3/2053)\cdot(2053/3) = (-1)^{\frac{2\cdot2052}{4}} = (-1)^{1026} = 1$$

Now:

$$(2053/3) = (684 + 1/3) = (1/3) = 1$$

as 1 is always a quadratic residue. Thus:

$$(3/2053)\cdot 1 = 1$$

The terms (7/2053) and (47/2053) can be treated in the same way so that:

$$(987/2053) = 1\cdot 1\cdot 1 = 1$$

hence 987 is a quadratic residue of 2053. A useful result for these calculations is that:

$$(2/p) = (-1)^\alpha$$

where α can be either $(\frac{p-1}{4})$ or $(\frac{p^2-1}{8})$.

An example of number theory applied to FFTs

Two related problems posed in Part 2 are ideal candidates for the application of number theory. They are discovering the location of the terms j, –j, 1, and –1 in short-length DFTs and also in the twiddle factors of the FFT. Solving them provides a good illustration of many of the techniques and results of number theory.

Both problems concern finding when $W_N^{<kn>}$ equals a purely real or purely imaginary value. For the twiddle factors, $0 \leq n \leq M-1$ and $0 \leq k \leq L-1$ where N is M·L, there are exactly $(M+L-1)$ terms equal to unity and the question is whether some other terms match -1, j or $-j$. For the short-length DFTs, where $0 \leq n \leq N-1$ and $0 \leq k \leq N-1$, the largest index is approximately N^2. While there are $(2N-1)$ matches to 1 in the first column and row, there can be other matches.

As W is periodic, it can be denoted $W^{<kn>_N}$. The question of whether W matches $1,-1$, j, or $-j$ is therefore the question of whether $< kn >_N$ matches $0, \frac{N}{4}, \frac{N}{2}$, or $\frac{3N}{4}$. Matching to $\frac{N}{4}$ is only possible if 2 is a divisor of N and, similarly, matching to $\frac{N}{4}$ is only possible if 4 is a divisor of N. Thus the N of greatest interest are the even numbers of the form $2^{\alpha} Q$.

Consider the problem of locating matches to these particular values of the indices when W^{nk} is part of a DFT. Here, the problem of interest can be expressed as finding matches to $\frac{N}{4}, \frac{N}{2}, \frac{3N}{4}$ and N within the set of sequences given by $\{ < kn >_N \}, (n, k = 1, 2, \cdots, N)$. Now if N is prime, there can only be matches to N. Similarly, for N composite, there can also only be matches to N provided 2 or 4 are divisors.

To begin, consider the sequence for the multiplier k equal to 1, namely $1, 2, 3, \cdots, N$. There is exactly one match to N and if N does have divisors of 2 and 4, one match to $\frac{N}{2}, \frac{N}{4}$ and $\frac{3N}{4}$. How many of the N sequences are like this? The answer is clearly $\Phi(N)$. Now k equal to 2 cannot be a member of this set as N is even and it produces the sequence $2, 4, 6, \cdots, 2N$. This has two matches to N and, provided N has 4 as a divisor, there are also two matches to $\frac{N}{2}$. If N is divisible by 8, there are two matches to $\frac{N}{4}$ and $\frac{3N}{4}$. Again, how many of the remaining sequences are just permutations of this one? It must be $\Phi(\frac{N}{2})$. So we may continue.

We can generalize. Let $\{ d_j \}$ be the set of divisors of N. This set must include 1 and N and if N is of the form $2^{\alpha} \cdot Q$, then it will also include terms of the type $\{ 2, 4, \cdots, \frac{Q}{4}, \frac{Q}{2}, Q \}$. We know the totient can be used to classify a sequence of elements into different classes. Each of the terms k in the sequence $\{ 1, 2, 3, \cdots, N-1 \}$ where $(k, N) = d_j$ belongs to a class with $\Phi(\frac{N}{d_j})$ members. Consider sequences of the form $\{ < kn >_N \}$, $(n, k = 1, 2, \cdots, N)$. These sequences can be divided into such classes plus one extra class given by $\{ < kn >_N \}, (n = 1, 2, \cdots, N \; k = N)$. Since all elements match N in this last class, it presents no problems. Of the other classes, each has d_j matches to N. If d_j is divisible by 2, then it also has d_j matches to $\frac{N}{2}$ and if divisible by 4, then d_j matches to $\frac{N}{4}$ and $\frac{3N}{4}$ as well. To summarize:

(a) If N is prime, there are matches to N only and there are $(2N-1)$ of these.
(b) If N is composite but has no factors of the form 2^{α}, then there are also only matches to N and, again, there are $(2N-1)$ of these.
(c) If N is composite and does have divisors of the form 2^{α} and if there are d(N) divisors in all, then there are:

$$\sum_{j=1}^{d(N)} d_j \cdot \Phi\left(\frac{N}{d_j}\right)$$

matches to N (i.e. matches to 1)

$$\sum_{j=1}^{d(N)} d_j \cdot P_2(d_j) \cdot \Phi\left(\frac{N}{d_j}\right) \quad \text{matches to} \quad \frac{N}{4} \text{(i.e. matches to } -1)$$

$$\sum_{j=1}^{d(N)} d_j \cdot P_4(d_j) \cdot \Phi\left(\frac{N}{d_j}\right) \quad \text{matches to} \quad \frac{N}{4} \text{ and } \frac{3N}{4} \text{(i.e. matches to } -j \text{ and } j)$$

where:

$$P_2(d_j) = 1 \quad \text{if } 2|d_j \text{ and } 0 \text{ otherwise}$$
$$P_4(d_j) = 1 \quad \text{if } 4|d_j \text{ and } 0 \text{ otherwise}$$

These results show there is little point in searching for these redundant operations. For software, the overheads in locating them would far exceed the savings made and for hardware there are too few to make much difference in cost or in performance. An exception, though, is when N is of the form 2^γ as there are $(\gamma + 1)$ divisors, namely $\{1, 2, 4, \cdots, \frac{N}{4}, \frac{N}{2}, N\}$. Of these, 1 is not divisible by 2 or 4 and 2 is not divisible by 4. The totient for a term like 2^β is $2^{\beta-1}$. Thus:

$$\text{number of matches to N} = N + \sum_{j=0}^{\gamma-1} 2^j \cdot \left(2^{\gamma-j-1}\right) = (\gamma + 2) \cdot \frac{N}{2}$$

where N is included separately here as the equation cannot be used for the divisor N but it is known there are N matches. Also:

$$\text{number of matches to } \frac{N}{2} = \sum_{j=0}^{\gamma-1} 2^j \cdot \left(2^{\gamma-j-1}\right) = \gamma \cdot \frac{N}{2}$$

and similarly, there are $(\gamma - 1) \cdot \frac{N}{2}$ matches to $\frac{N}{4}$ and $\frac{3N}{4}$. Each of these is proportional to γ which is $\log_2 N$. Substituting values gives Table 2.2. As mentioned there, while significant savings result they are not nearly as large as with an FFT.

The second problem is locating $1, -1, j$, and $-j$ within the terms of a twiddling matrix. This problem is a study of the sequences:

$$\{ < kn >_N \} \qquad n = 0, 1, \cdots, M-1 \qquad k = 0, 1, \cdots, L-1$$

where N is $M \cdot L$. Given that the terms for n and k zero match 1 and that no other exponent can do so, the problem here can be regarded as a study of:

$$\{ kn \} \qquad n = 1, 2, \cdots, M-1 \qquad k = 1, 2, \cdots, L-1$$

where the object is to find the number of matches to $\frac{N}{4}$, $\frac{N}{2}$, and $\frac{3N}{4}$. Again, this is only possible if N has 2 or 4 as a divisor. As the largest index of this expression is $(M-1) \cdot (L-1)$, which is $(N - M - L + 1)$, then a match is only possible if:

$$(N - M - L + 1) \geq \frac{N}{4}, \frac{N}{2} \text{ or } \frac{3N}{4}$$

Assuming M and L are comparable, this implies N must exceed approximately 8 for a match to $\frac{N}{4}$, 16 for a match to $\frac{N}{2}$ and 64 for a match to $\frac{N}{4}$.

This problem is similar to the first but is complicated by the different ranges of the indices. Consequently, it is not a study of permuted sequences. Now the object is to find those matches where:

$$kn = \frac{N}{4} \text{ or } \frac{N}{2} \text{ or } \frac{3N}{4}$$

Such matches require that 2 or 4 is a divisor of N and in addition, that n and k are divisors of $\frac{N}{4}$, $\frac{N}{2}$ or $\frac{3N}{4}$. Take $\frac{N}{4}$ as an example as the other two cases are similar. Let { d_j } be the set of divisors of $\frac{N}{4}$. Then:

$$kn = d_j \cdot \left(\frac{\frac{N}{4}}{d_j} \right)$$

describes the circumstances for a match but this will only occur if $d_j < L - 1$ and if the result of dividing $\frac{N}{4}$ by d_j is less than M–1. There can be at most $(d(\frac{N}{4}) - 1)$ matches as 1 must be a divisor but can never satisfy this condition.

The situation here is much the same as for the last problem. If N is prime, there can be no matches. For N composite, there are also no matches unless N is divisible by 2 or 4. The only case which seems to be worth investigating is again N equal to 2^γ. However, this has at most $(\gamma - 2)$ matches to $\frac{N}{4}$ and $(\gamma - 3)$ to $\frac{3N}{4}$. For N equal to 1024, quite a large value, these are only 8 and 7 respectively and that is hardly worth detecting even for hardware implementations.

THEORY

Finite mathematical structures

We have studied residues and so modulo addition and multiplication in some detail. The reason is that these operations act upon a finite set of elements and are therefore ideal candidates for constructing finite mathematical structures.

A group requires a non-empty set of elements and an associative binary operation with a right identity and right inverse. The set of elements { 0, 1, 2, $\cdots$, N –1 } together with modulo addition satisfies these requirements. The identity is 0 and the inverse to some element a is (N – a). Further, this operation is commutative, thus this is an abelian group. Modulo multiplication and the set of elements { 1, 2, 3, $\cdots$, N –1 } can also satisfy these conditions and, again, as multiplication is commutative, form an abelian group with 1 as the identity. However, from $< a \cdot x >_N = 1$, an inverse can only exist in this group for every element a if N is prime. For obvious reasons, these two groups are usually called the **additive** and **multiplicative** groups modulo N respectively.

A **cyclic group** is a group where the elements are given by { i, g, g², $\cdots$, g^{N-1} } where g^i is interpreted as being the operation concerned repeated i times upon the element g. Both of the above groups are cyclic. The multiplicative, which is cyclic because if N is prime a primitive root is guaranteed, is by far the more important .

A ring consists of a set of elements R and two operations. The operations modulo addition and multiplication with the set of elements { 0, 1, 2, $\cdots$, N –1 } form a ring. Further, it is a commutative ring with identity. Now this ring is an **integral domain** if a cancellation property holds. That is, if three elements a, b, and c are related by:

$$< ab >_N = < ac >_N$$

where $< a >_N$ is not zero, then a cancellation law implies:

$$< b >_N = < c >_N$$

This property is sometimes expressed as follows. If a is not zero and if $< ab >_N = 0$, then this implies b is zero. In general, this is not the case for a ring. For example, if p is a prime divisor of N, then:

$$< p \cdot (\frac{N}{p}) >_N = 0$$

It is true if N is prime, though, and so the ring can be an integral domain.

A division ring is simply a ring where the elements { 1, 2, 3, $\cdots$, N –1 } and multiplication forms a group. A field is a division ring where these elements form a commutative group with multiplication. The only possible finite number fields are those for which N is a prime p. These are termed **Galois fields** after the French mathematician who first investigated them and are denoted GF(p). Galois fields are extremely important in a number of areas including digital signal processing. One consequence of this is that it is quite common for a primitive root to be defined as one of $\Phi(p - 1)$ elements of order (p – 1) which generate the multiplicative group of this field.

Rader's theorem and its implications

Rader's theorem is a very simple result of considerable importance to transform techniques for digital signal processing. Now the set of elements { 1, 2, 3, $\cdots$, p – 1 } where p is some odd prime, together with multiplication modulo p, forms a multiplicative group. A primitive root must exist, thus this set is equivalent to { g^0, g^1, g², $\cdots$, g^{p-2} }. The exponents of the terms in this set also form a group but with respect to addition modulo (p – 1). Then we may state the following:

THEOREM   The additive group formed from modulo addition with the complete residue system modulo (p – 1), p an odd prime, is isomorphic to the multiplicative group formed from the non-zero residues and multiplication modulo p.

Rader's theorem is a direct application of this result. It is:

THEOREM   For N prime, the discrete Fourier transform is equivalent to cyclic convolution.

To show this, we begin with the DFT equations:

$$X(k) = \sum_{n=0}^{N-1} x(n) \cdot W^{kn} \qquad\qquad k = 0, 1, \cdots, N-1$$

where W is $\exp(-\frac{j2\pi}{N})$. Then:

$$X(0) = \sum_{n=0}^{N-1} x(n)$$

$$X'(k) = X(k) - x(0) = \sum_{n=1}^{N-1} x(n) \cdot W^{kn} \qquad\qquad \begin{matrix} k = 1, \cdots, N-1 \end{matrix}$$

The key calculation is the set of equations X'(k). With W periodic, the exponent kn in these belongs to the residue set of $< kn >_N$ and this describes a multiplicative group. Now if N is some odd prime p, this group is isomorphic to the additive group modulo (p − 1). The isomorphism can be described as:

$$n \rightarrow g^{p-1-u} \equiv g^{-u} \qquad\qquad k \rightarrow g^v$$

where g is a primitive root of p. Substituting:

$$X'\left(g^v\right) = \sum_{u=0}^{p-2} x\left(g^{-u}\right) \cdot W_N^{g^{v-u}} \qquad\qquad v = 0, 1, \cdots, p-2$$

and this is the equation for the cyclic convolution of a sequence $x(g^{-u})$ with a sequence $W_N^{g^u}$ to give a sequence $X'(g^v)$. Choosing n as g^u gives cyclic correlation.

Example

Consider the DFT of 7 points. The equations are:

$$X(k) = \sum_{n=0}^{6} x(n) \cdot W^{kn} \qquad\qquad k = 0, 1, 2, 3, 4, 5, 6$$

where W is $\exp(-\frac{j2\pi}{7})$. The key calculation is:

$$X'(k) = X(k) - x(0) = \sum_{n=1}^{6} x(n) \cdot W^{kn} \qquad\qquad k = 1, 2, 3, 4, 5, 6$$

A primitive root of 7 is 3 and:

$$< 3^0 >_7 = 1 \quad < 3^1 >_7 = 3 \quad < 3^2 >_7 = 2 \quad < 3^3 >_7 = 6 \quad < 3^4 >_7 = 4 \quad < 3^5 >_7 = 5$$

Then the mappings are:

$$k = 3^v \qquad\qquad\qquad n = 3^{-u}$$

Substituting gives a calculation identical to the cyclic convolution of the sequence:

$$\{ x(1), x(5), x(4), x(6), x(2), x(3) \} \quad \text{with} \quad \{ W_7^1, W_7^3, W_7^2, W_7^6, W_7^4, W_7^5 \}$$

to produce the sequence:

$$\{ X'(1), X'(3), X'(2), X'(6), X'(4), X'(5) \}$$

That is:

$$
\begin{bmatrix} X'(1) \\ X'(3) \\ X'(2) \\ X'(6) \\ X'(4) \\ X'(5) \end{bmatrix}
=
\begin{bmatrix}
W_7^1 & W_7^5 & W_7^4 & W_7^6 & W_7^2 & W_7^3 \\
W_7^3 & W_7^1 & W_7^5 & W_7^4 & W_7^6 & W_7^2 \\
W_7^2 & W_7^3 & W_7^1 & W_7^5 & W_7^4 & W_7^6 \\
W_7^6 & W_7^2 & W_7^3 & W_7^1 & W_7^5 & W_7^4 \\
W_7^4 & W_7^6 & W_7^2 & W_7^3 & W_7^1 & W_7^5 \\
W_7^5 & W_7^4 & W_7^6 & W_7^2 & W_7^3 & W_7^1
\end{bmatrix}
\begin{bmatrix} x(1) \\ x(5) \\ x(4) \\ x(6) \\ x(2) \\ x(3) \end{bmatrix}
$$

Up to this point, development of the FFT was constrained to be a technique for combining prime length DFTs into longer length DFTs. This theorem shows those prime length DFTs can be expressed as ($p-1$)-point cyclic convolutions. Now as N increases, so the distance between primes increases. For an arbitrarily large prime p therefore, it is almost certain that ($p-1$) will be highly composite. A technique for efficiently computing highly composite cyclic convolutions is known: fast Fourier transforms. Thus an efficient discrete Fourier transform technique exists for any N.

From the practical viewpoint, the theorem shows that only very short-length prime order DFTs need retain any interest for us. Say, no more than those for N equal to 2, 3, 5, 7, 11, 13, 17, 19, and possibly 23.

While Rader's theorem is most important, it does have some practical limitations and, in many respects, is of far more value to hardwired implementations than software. The reason is the essential nature of the theorem. Expressed another way, it is saying that a prime length DFT can always have its terms permuted in such a way as to give a cyclic convolution. The key issue here is permutation. With hardwired implementations, that is invariably simple, but in software it can be a significant overhead. To avoid it, programs need to explicitly list all equations and that makes them lengthy.

The key requirement for Rader's theorem is that N has a primitive root. Primitive roots exist for 2, 4, p^α, and $2p^\alpha$ where p is an odd prime and α is a non-zero integer and therefore the theorem can be used to develop a decomposition procedure. Assume N is p^α. A primitive root of N generates a sequence of order $\Phi(N)$, which here is $p^\alpha(1 - \frac{1}{p})$. The $\frac{N}{p}$ other terms must be those that are not relatively prime to N and these are easily determined. The DFT equation, then, can be separated into the two sets:

$$X(pk) = \sum_{n=0}^{N-1} x(n) \cdot W^{pkn} \qquad\qquad k = 0, 1, \cdots, p^{\alpha-1}-1$$

$$X(k) = \sum_{(n,N)\neq1} x(n) \cdot W^{kn} + \sum_{(n,N)=1} x(n) \cdot W^{kn} \qquad (k, N) = 1$$

The second term of the second set of equations meets Rader's criterion and so is equivalent to a $p^{\alpha}(1 - \frac{1}{p})$-point cyclic convolution. For the first set of equations, let:

$$n = p^{\alpha-1}q + r \qquad 0 \leq q \leq p-1 \qquad 0 \leq r \leq p^{\alpha-1} -1$$

Substituting:

$$X(pk) \quad = \sum_{r=0}^{\frac{N}{p}-1} \left(\sum_{q=0}^{p-1} x(p^{\alpha-1}q + r) \right) \cdot W^{pkr} \qquad k = 0, 1, \cdots, p^{\alpha-1}-1$$

$$= \sum_{r=0}^{\frac{N}{p}-1} y(r) \cdot W^{pkr}$$

This describes a ($p^{\alpha-1} - 1$)-point DFT. So too, does the first term of the second set of equations. Hence this decomposition has resulted in:

(a) two ($p^{\alpha-1} - 1$)-point DFTs;
(b) a cyclic convolution of length $\Phi(N)$, which is $p^{\alpha-1} \cdot (p - 1)$ points;
(c) a set of $p^{\alpha-1} \cdot (p - 1)$ real additions.

   This decomposition has practical value if it is easier to implement than other methods or if it has fewer operations. The first is clearly not satisfied so consider the second. Now a $p^{\alpha-1} \cdot (p - 1)$-point cyclic convolution can be performed via DFTs and, if done, that will involve $2(p - 1) p^{\alpha-1}$-point DFTs plus many other terms. Ignore these for the moment and take the number of operations as being:

$$2p \ p^{\alpha-1}\text{-point DFTs}$$
$$p^{\alpha-1} \cdot (p - 1) \ \text{real additions}$$

A recurrence formula can be constructed from this and from that Table 3.1 constructed. Compare this to a p^{α} algorithm. From previous work, this requires $(\frac{p}{p-1}) \cdot (p^{\alpha-1}-1)$ p-point DFTs and no additions other than those for these DFTs.

   Both approaches require many extra terms. The base-p algorithm requires a substantial number of twiddles. The Rader decomposition requires some twiddles plus many 2-point and ($p - 1$)-point DFTs and numerous permutations. The implication of this simple analysis, though, is that the Rader decomposition is not going to offer any significant reduction in operations.

Table 3.1

Algorithm	P-point DFTs	Additions
p^2	$2p$	$p \cdot (p-1)$
p^3	$4p^2$	$p^2 \cdot (p-1) + 2p \cdot (p-1)$
.		
.		
.		
p^α	$2^{\alpha-1} \cdot p^{\alpha-1}$	$p^\alpha \cdot (\dfrac{(p-1)}{(p-2)}) \cdot (1 - (\dfrac{2}{p})\alpha - 1$

The complexity of this procedure, the number of operations and the considerable number of permutations of terms all ensure this decomposition is of little practical interest. What it does show though, is that there are alternative methods of constructing algorithms to those discussed earlier. Also, of course, Rader's theorem by itself is most important. Although this particular decomposition must be discarded, in Part 4 we shall show a generalization of Rader's theorem does lead to efficient algorithms for p^α and these are attractive alternatives to mixed radix algorithms.

Number theoretic transforms

The concept of transforms on finite structures is very appealing to digital signal processing for practical reasons. The close link with number theory accounts for the name of these transforms; **number theoretic transforms** or NTTs. Although a wide variety of finite structures could be considered, again for implementation reasons interest has focused almost exclusively on rings and fields employing modulo addition and multiplication. Specifically, on rings based on arbitrary moduli and Galois fields of the type GF(p). The significant difference between the two is simply that in the field $a \cdot b = 0$ implies either a is zero or b is zero whereas in the ring both a and b can be finite.

Although only fields were considered in the development of the general conditions for a transform to exist in Part 1, most of that theory applies to rings as well. Those conditions require an element Ω to exist in the structure such that the transforming matrix $[\Omega^{kn}]$ has rows $\{ i, \omega^0, \omega^1, \cdots, \omega^{N-1} \}$ where i is the multiplicative identity and $\omega^N = i$. The rows are all elements of a multiplicative group and are permutations of each other.

With the work on number theory covered in this Part, the conditions for the existence of transforms can be more closely examined. Since elements of a ring or field exist with these properties, there is some certainty transforms exist. Furthermore, as we know there can be several such elements, we can anticipate that several number theoretic transforms may exist. Unfortunately, some of these expectations must be blunted by the rather rigid conditions the mathematics place on the transforms.

An issue not examined in Part 1 was the conditions which the inverse transform places on the problem. We must now tackle this question and, as well, investigate the specific conditions which modulo arithmetic places on the transform. To commence, a general expression for the transform pair is:

$$X(k) = \left\langle \sum_{n=0}^{N-1} x(n) \cdot \omega^{kn} \right\rangle_M \qquad\qquad k = 0, 1, \cdots, N-1$$

$$x(n) = \left\langle \sum_{k=0}^{N-1} x(k) \cdot \upsilon^{nk} \right\rangle_M \qquad\qquad n = 0, 1, \cdots, N-1$$

We have specified the transform as being of size N while the modulus of the field or ring is of size M. These are Diophantine equations, hence from earlier work on these and Euler's theorem, we deduce that:

$$(\omega, M) = (\upsilon, M) = 1$$

and that N must divide $\Phi(M)$ are conditions for the existence of the transform. Substituting the inverse equation into the forward gives:

$$x(k) = \left\langle \sum_{n=0}^{N-1} \left\langle \sum_{i=0}^{N-1} X(i) \cdot \upsilon^{ni} \right\rangle_M \omega^{kn} \right\rangle_M$$

$$= \left\langle \sum_{n=0}^{N-1} \left\langle \sum_{i=0}^{N-1} X(i) \cdot \upsilon^{ni} \cdot \omega^{kn} \right\rangle_M \right\rangle_M$$

$$= \left\langle \sum_{i=0}^{N-1} X(i) \left\langle \sum_{n=0}^{N-1} \upsilon^{ni} \cdot \omega^{kn} \right\rangle_M \right\rangle_M$$

The transform only exists, therefore, if:

$$\left\langle \sum_{n=0}^{N-1} \upsilon^{ni} \cdot \omega^{kn} \right\rangle_M = 1 \text{ for } i = k$$

$$= 0 \text{ otherwise}$$

Now consider the existence of such a transform in a field. Every element α of a Galois field GF(p) has a unique multiplicative inverse. That is, a second element β exists such that $\alpha \cdot \beta$ is 1. The element β is a power of α and usually denoted α^{-1}. It needs to be remembered, though, that the exponent is actually ($\Phi(p) - 1$). In addition, since the modulus is prime, any ω will be relatively prime to M. Then it would seem the condition for the transform to exist is just that $\upsilon^{kn} = \omega^{-kn}$. However, as:

$$\sum_{n=0}^{N-1} 1 = N$$

we must require that:

$$\upsilon^{kn} = N^{-1} \cdot \omega^{-kn}$$

where N^{-1} is the multiplicative inverse of N and guaranteed to exist.

To summarize, the transform exists in the Galois field GF(p) provided:

1. there exists a root ω with period N;
2. N divides $\Phi(p)$; that is, N divides (p – 1).

If these conditions are satisfied, the transform pair is:

$$X(k) = \left\langle \sum_{n=0}^{N-1} x(n) \cdot \omega^{kn} \right\rangle_p \qquad k = 0, 1, \cdots, N{-}1$$

$$x(n) = \left\langle N^{-1} \sum_{k=0}^{N-1} X(k) \cdot \omega^{-nk} \right\rangle_p \qquad n = 0, 1, \cdots, N{-}1$$

From the general properties of fields, there must be $\Phi(N)$ choices for ω and hence that many transforms.

While a ring allows M to be composite, there is the distinct difficulty that in the general case a ring has no multiplicative inverse. Consequently, to find a transform like the above, we must completely review each stage of the development of the field transform, identify where multiplicative inverses are employed and find additional conditions to ensure their existence in the ring.

We require an element ω of the ring with period N. From Euler's theorem, N must be relatively prime to M and divide $\Phi(M)$. Given any arbitrary r, this implies:

$$< \omega^r \cdot \omega^{N-r} >_M = 1$$

so that ω has an inverse in all circumstances. Clearly, N must be an element of the ring but is this true of N^{-1}? It can be and the condition for existence arises from the congruence:

$$< N \cdot x >_M = 1$$

which has a solution provided N is relatively prime to M. That solution, of course, is N^{-1}.

Assuming these conditions hold, all that remains to prove the transform exists is to show:

$$\left\langle \sum_{n=0}^{N-1} \omega^{jn} \right\rangle_M = N \text{ for } j = 0$$
$$= 0 \text{ for } j = 1, 2, \cdots, N{-}1$$

The first case clearly holds, hence we only need to be concerned with the second. Now consider:

$$< \omega^j - 1 >_M \cdot < 1 + \omega^j + \omega^{2j} + \cdots + \omega^{(N-1)j} >_M = < \omega^{Nj} - 1 >_M = 0$$

This product is zero because ω is a root with period N. Also because of that, $< \omega^j - 1 >_M$ is non-zero. However, since this is a ring, it does not follow the second term is necessarily zero. If it were integral domain, this would be the case, but a finite integral domain is a field not a ring. Now this product is in the form of a congruence equation and so from earlier work on the solutions to such equations the second term will always equal zero provided M and also $(\omega^j - 1)$ are relatively prime. This is quite demanding as M is composite and it is quite likely for some given ω that $(\omega^j - 1)$ will divide M.

In summary, a transform exists on a ring R_M in the general case provided:

1. an element ω exists relatively prime to M with period N;
2. N is relatively prime to M and a divisor of $\Phi(M)$;
3. $(\omega^j - 1)$ is relatively prime to M for $1 \le j \le N-1$.

A little aside. Previously we considered the problem of: given some ω, what is its period? Here, it is more a case of: given some N, which elements have that as a period while also satisfying for all j, $1 \le j \le N-1$, $(\omega^j - 1, M)$ is 1? Since N is the required period, there is no point in considering elements ω other than those for which $< \omega^N >_M$ is 1. However, this alone does not guarantee the period of ω is N. Now if j is a divisor of N:

$$< \omega^N >_M \ = \ < \omega^{\left(\frac{N}{j}\right)j} >_M \ = \ < (\omega^{\left(\frac{N}{j}\right)j}) >_M$$

We previously discovered it is only necessary to test ω for these divisors but this equation can be used to further reduce the workload. Earlier, it was shown that if ω has a period N, then ω^α has a period $\frac{N}{d}$ where d is (N, α). Here then, $\omega^{\frac{N}{j}}$ will have a period N if $\frac{N}{j}$ is relatively prime to N. Given N is entirely arbitrary, this is only possible in the general case if $\frac{N}{j}$ is itself prime. Hence the problem reduces to one of testing whether $(\omega^j - 1)$ is relatively prime to M for those j such that $\frac{N}{j}$ is prime.

The simplest ring is where M is of the form p^e where e is some integer. Since ω must be relatively prime to M, it must therefore be relatively prime to p as p is a divisor of M. Also, N must be a divisor of $(p-1)$ as N must divide $\Phi(M)$ and $\Phi(M)$ is $\frac{M}{p} \cdot (p-1)$. Consequently, ω must be $g^{\left(\frac{M}{p}\right)}$ where g is a primitive root of p to satisfy the final condition.

The more general case of a ring is where M is given by:

$$M = p_1^{e_1} \cdot p_2^{e_2} \cdot \ \cdots \ \cdot p_r^{e_r}$$

From the Chinese remainder theorem, a set of congruences taken with respect to factors $M_1, M_2, \cdots, M_r$ will have a unique solution modulo $M_1 \cdot M_2 \cdot \cdots \cdot M_r$. In this case, those congruences are chosen to be the terms $p_r^{e_i}$. The problem of a transform in such a system has just been analyzed. The Chinese remainder theorem is therefore indicating that N must now be a divisor of each of $(p_1-1), (p_2-1), \cdots, (p_r-1)$ where $p_1, p_2, \cdots p_r$ are the prime divisors of M. Thus N must divide the GCD of $(p_1-1) \cdot (p_2-1) \cdot \cdots \cdot (p_r-1)$ and that clearly limits the size of N. Choosing ω is also difficult and requires use of the Chinese remainder theorem.

To be more specific then, a transform exists in a finite ring provided:

1. an element ω exists with period N where both ω and N are relatively prime to M;
2. N is a divisor of the GCD of $(p_1-1)\cdot(p_2-1)\cdot\ \cdots\ \cdot(p_r-1)$ where the p_i are the prime factors of the modulus M;
3. (ω^j-1) is relatively prime to M for $1 \le j \le N-1$.

Efficient algorithms for determining the DFT, fast Fourier transforms, were found in the infinite field. The question arises of whether an equivalent algorithm exists here. The answer is a conditional yes. One of the reasons FFTs are successful is that any arbitrary N in the infinite field is likely to be highly composite. The process of computing short prime-length DFTs and then combining them leads to a significant reduction in computational effort. Here, however, many restrictions are placed on N. Ensuring these are met and, in addition, ensuring that N remains highly composite, is a daunting task. Finding an N that retains a simple structure equivalent to, say, the base-B FFT is almost impossible. A fast NTT, therefore, in general means a fast mixed radix algorithm.

Conceptually at least, there is no difficulty in developing a fast algorithm. In fact, most of the results of Part 2 are quite easily adapted. For hardwired implementations, these are worth considering but not for software. The reason is that, unlike the FFT, such algorithms tend to be very dependent on the specific parameters of the NTT. There is no general algorithm in the sense of a base-2 FFT and, consequently, while general programs can be written they must include so many exceptions for the different values of N that they lose efficiency and are quite difficult to program.

DEVELOPMENTS

The prime factors algorithm

The prime factors algorithm (PFA) is a very elegant application of number theory to the development of an FFT. Alternatively known as the Good-Thomas algorithm, it is an algorithm with quite a long history. Like much of the older FFT work, it was overlooked in the initial burst of FFT-related developments in the late 1960s to early 1970s but, particularly following investigations into number theory, its true worth was recognized.

The PFA is a mixed radix FFT. What distinguishes it from other such techniques is that number theory, especially the Chinese remainder theorem, is applied to the indices. That leads to a more efficient algorithm. It can be developed in similar form to those of Part 2, which means via the development of a computational section. Here, we shall consider a DIT version of the algorithm.

To begin, since W is periodic, we may express the DFT equations as:

$$X(k) = \sum_{n=0}^{N-1} x(n) \cdot W^{(kn)_N} \qquad\qquad k = 0, 1, \cdots, N-1$$

Calculation of the indices may therefore be via congruence equations. To develop a computational section, assume that $N = M \cdot L$, but here we shall require that $(M, L) = 1$.

With this condition, we can apply the Chinese remainder theorem or, more particularly, the proof of the theorem given earlier.

Define two mappings:

$$< k >_M = k_1$$
$$< k >_L = k_2$$

From the Chinese remainder theorem, a unique solution exists to these two congruences modulo N. We already have that, namely k. From the proof of the theorem we know we can construct two sets of terms:

$$< s_1 >_M = 1 \qquad \text{and} \qquad < s_1 >_L = 0$$

and:

$$< s_2 >_M = 0 \qquad \text{and} \qquad < s_2 >_L = 1$$

and from these, that a unique solution is given by:

$$k = < s_1 \cdot k_1 + s_2 \cdot k_2 >_N$$

We can repeat this procedure with n but here choose:

$$< r_1 \cdot n >_M = n_1 \qquad\qquad \text{where } < L >_M = r_1$$

and:

$$< r_2 \cdot n >_L = n_2 \qquad\qquad \text{where } < M >_L = r_2$$

where this also follows from the proof of the remainder theorem. Construct:

$$n' = < n_1 \cdot L + n_2 \cdot M >_N$$

This is designated n' as it is NOT n. Consequently, this equation describes a mapping of input values in much the same way as scrambling does for the FFT. The difference is that scrambling involves digit reversal whereas this involves computation of a congruence.

As with the FFT, these two sets of mappings can now be substituted and so convert the DFT equations into a set of two-dimensional equations. That is, into:

$$X(k_1 k_2) = \sum_{n_1=0}^{M-1} \sum_{n_2=0}^{L-1} x(n_1, n_2) \cdot W_N^{<s_1 k_1 + s_2 k_2>_N <n_1 L + n_2 M>_N}$$

Consider the exponent. It may be written:

$$E = << s_1 \cdot k_1 + s_2 \cdot k_2 >_N \cdot n_1 \cdot L + < s_1 \cdot k_1 + s_2 \cdot k_2 >_N \cdot n_2 \cdot M >_N$$

$$= E_1 + E_2$$

Thus:

$$W_N^E = W_N^{E_1} \cdot W_N^{E_2}$$

However, N has factors M and L and M is a factor of E_2 while L is a factor of E_1. Hence:

$$W_N^E = W_M^{e_1} \cdot W_L^{e_2}$$

where
$$e_1 = \ll s_1 \cdot k_1 + s_2 \cdot k_2 >_N \cdot n_1 >_M = n_1 \cdot k_1$$
$$e_2 = \ll s_1 \cdot k_1 + s_2 \cdot k_2 >_N \cdot n_2 >_L = n_2 \cdot k_2$$

Therefore:

$$X(k_1 k_2) = \sum_{n_1=0}^{M-1} \sum_{n_2=0}^{L-1} x(n_1, n_2) \cdot W_M^{k_1 n_1} W_L^{k_2 n_2}$$

$$= \sum_{n_1=0}^{M-1} W_M^{k_1 n_1} \left(\sum_{n_2=0}^{L-1} x(n_1, n_2) W_L^{k_2 n_2} \right)$$

These equations are almost exactly the same as the computational section derived in Part 2 with the small but critical difference that there are no twiddling terms.

The absence of twiddling makes the PFA a more efficient algorithm than the mixed radix algorithms of Part 2. The cost is the requirement that the factors must be relatively prime to one another and the quite complex scrambling operation. Ensuring the divisors of N are relatively prime is difficult. However, given that in general:

$$N = p_1^{\alpha_1} \cdot p_2^{\alpha_2} \cdots p_r^{\alpha_r}$$

this suggests using base-p algorithms for each of the terms $p_i^{\alpha_i}$ and then the PFA to derive the final result. That eliminates many of the problems of the mixed radix algorithm while introducing economies in operations.

Since only the scrambling is different, the general properties of the PFA computational section are the same as the DFT section discussed in Part 2. However, as there is no twiddling, there are ($N - M - L + 1$) fewer complex multiplications. Countering this is the cost of the scrambling operation and for software that can be quite a burden. For hardware implementations though, it is insignificant.

Example

The PFA will be compared against the earlier DFT computational section. Choose:

$$M = 15 = 3 \cdot 5$$

and take the input sequence as being:

$$a \; b \; c \; d \; e \; f \; g \; h \; i \; j \; k \; l \; m \; n \; o$$

For both algorithms, map this sequence into a two-dimensional array. For the FFT section, that mapping is defined by:

$$n = 3 \cdot n_1 + n_2 \qquad 0 \leq n_1 \leq 4 \qquad 0 \leq n_2 \leq 2$$

and so the array becomes:

n_2 \ n_1	0	1	2	3	4
0	a	d	g	j	m
1	b	e	h	k	n
2	c	f	i	l	o

Now follows actions upon the array. They are:

(a) DFTs on the rows of this array;
(b) twiddling of the array;
(c) DFTs of the columns of the array;

and this gives a final matrix:

k_1 \ k_2	0	1	2	3	4
0	A	B	C	D	E
1	F	G	H	I	J
2	K	L	M	N	O

There is a mapping to read out the data from this array into a one-dimensional output sequence and it is:

$$k = 5 \cdot k_1 + k_2 \qquad 0 \leq k_1 \leq 2 \qquad 0 \leq k_2 \leq 4$$

From this, we obtain the transformed sequence as:

$$A \; B \; C \; D \; E \; F \; G \; H \; I \; J \; K \; L \; M \; N \; O$$

The initial mapping here defines the scrambling.

The PFA is similar but the input mapping is quite different. We require the two terms:

$$r_1 = \; < 5 >_3 \; = 2$$
$$r_2 = \; < 3 >_5 \; = 3$$

and so:

$$n_1 = \; < 2n >_3$$
$$n_2 = \; < 3n >_5$$

which leads to the input mapping:

$$n' = \; < 5n_1 + 3n_2 >_{15}$$

Consider the following table:

n	0	1	2	3	4	5	6	7	8	9	10	11	12	13	14
n_1	0	2	1	0	2	1	0	2	1	0	2	1	0	2	1
n_2	0	3	1	4	2	0	3	1	4	2	0	3	1	4	2
n'	0	4	8	12	1	5	9	13	2	6	10	14	3	7	11

In this table, n gives the position of the elements in the input sequence. The two coordinates n_1 and n_2 indicate where an element is to be placed in the array and that element is the one pointed to in the input sequence by n'. Thus the array is:

n_1 \ n_2	0	1	2	3	4
0	a	d	g	j	m
1	f	i	l	o	c
2	k	n	b	e	h

This is now processed by performing:

(a) DFTs on the columns;
(b) DFTs on the rows;

and then a one-dimensional sequence can be constructed according to the output mapping. The array at this stage is:

k_1 \ k_2	0	1	2	3	4
0	A	G	M	D	J
1	K	B	H	N	E
2	F	L	C	I	O

The output mapping is:

$$k = < s_1 \cdot k_1 + s_2 \cdot k_2 >_{15}$$

where

$$k_1 = < k >_3$$

$$k_2 = < k >_5$$

$$< s_1 >_3 = 1 \quad \text{and} \quad < s_1 >_5 = 0$$

$$< s_2 >_3 = 0 \quad \text{and} \quad < s_2 >_5 = 1$$

We interpret this as meaning that a given value of k in the output is derived from a location (k_1, k_2). Before applying this though, we need to find values for s_1 and s_2. For s_1, we need a number α that is a multiple of 5 but ($\alpha - 1$) is divisible by 3. Testing the multiples of 5, the smallest suitable value is:

$$s_1 = 10$$

Similarly, for s_2 we need a number β that is a multiple of 3 but ($\beta - 1$) is divisible by 5. The smallest value is:

$$s_2 = 6$$

Thus the output sequence required is derived from:

$$k = < 10 \cdot k_1 + 6 \cdot k_2 >_{15}$$

As can be seen, the mapping operations in the PFA are not especially difficult but they are computationally expensive on a general-purpose computer unless look-up tables or some similar approach is used. However, they are no more so than digit reversal counters of normal mixed radix algorithms and so the PFA is certainly worthy of close attention.

Some issues in the development of NTTs

As mentioned in the Introduction, NTTs have a limited attraction for software implementations. The reason is simply that they use modulo operations and the nature of most programmable systems does not permit these to be readily or efficiently executed. For hardware, the attraction of NTTs is the extreme simplicity of that hardware compared to an equivalent FFT realization. Apart from lower costs, which is always appealing, it also means very high speed designs are much more easily implemented

The appeal of NTTs is that they have no round-off or overflow or quantization effects at all. This has a number of specific implications in different areas but note what it means in general terms. In an FFT realization, all of these problems occur and in a very general sense they depend on N, the order of the transform. While for short-length transforms they cannot be considered a serious difficulty, they can be of concern for long-length transforms. Here, NTTs would appear to have a distinct practical advantage. Our analysis, therefore, will tend to be biased towards how well the different NTTs examined can be applied to long-length transforms. In addition, since software implementation is less likely, discussion will be framed in terms of hardware implementations.

There are two key practical considerations in creating an NTT. They are the choice of the modulus M and the selection of the transform length N. N needs to be as close to M as possible for efficiency reasons. The value of M in turn must allow easy implementation of the modulo operations required. Further, it must be sufficiently large to cover the dynamic range of the input and output. Since the largest number employed in the system is ($M - 1$), then naturally, it is desirable to have some flexibility over the choice of M and, in general, to keep it as small as possible.

The obvious choice for M is some value based on a power of 2. Digital electronic systems are based on binary operations with counters being inherently modulo and units like multipliers or adders easily made modulo. Any other choice would need to be embedded into a binary system in some way and that could lead to inefficiency and complexity. Nevertheless, in view of some of the difficulties which we shall encounter, it may not be wise to totally dismiss other choices.

There are only three choices for a modulus based on 2. They are:

$$M = 2^Q - 1 \qquad \text{or } 2^Q \qquad \text{or } 2^Q + 1$$

It is a requirement of an NTT that the transform length N divides the totient. For 2^Q the totient is 2^{Q-1}. Therefore, N can be 1 or 2^α, where $1 \le \alpha \le Q-1$. However, N must be relatively prime to M and that means the only possible choice is 1. Clearly, that is of no value and so only two moduli can be considered. Since ($2^Q - 1$) is a Mersenne number, NTTs based on this moduli are naturally termed **Mersenne number theoretic transforms** or just **Mersenne number transforms** (MNTs). Those based on the modulus ($2^Q + 1$) are termed **Fermat number theoretic transforms** or **Fermat number transforms** (FNTs).

It is important to mention that there is a class of NTTs called **pseudo Mersenne** and **pseudo Fermat** transforms. These names are slightly misleading as they imply some form of variation of Mersenne or Fermat numbers. Pseudo Fermat transforms do indeed use a variation of Fermat numbers but pseudo Mersenne have the name for other reasons. An NTT is a congruence relation of the form:

$$< f(x) >_M = y$$

where M is any composite. Given another number μ, then from the general properties of residues, this relation can be converted to:

$$< < f(x) >_{\mu M} >_M = y$$

The final reduction is modulo M but this only needs to be done once so it isn't of any real concern what M should be. Most modulo operations are modulo $\mu \cdot M$ and this can be selected to give simple binary arithmetic by being either Mersenne or (pseudo) Fermat.

These pseudo transforms lose some efficiency over MNTs and FNTs but, provided μ is small, this needn't be very great. What is gained is a greater range of sequence lengths and a quite simple implementation. That makes this approach quite appealing. However, there is one additional constraint in that in practice there are limits on the choice of M. As a result, while there is a greater choice of sequence lengths, it is still not a large number.

Mersenne number theoretic transforms

Arithmetic modulo ($2^Q - 1$) is one's complement arithmetic. A review is given in Appendix 1.

A Mersenne number M is of the form ($2^Q - 1$) where Q is any integer. Some values of Q are prime and some are composite. If Q is composite, then M must be composite as:

$$M = 2^Q - 1 = 2^{mn} - 1$$
$$= (2^n - 1) \cdot (2^{n(m-1)} + 2^{n(m-2)} + \cdots + 1)$$

If Q is a prime, there is the possibility M is a **Mersenne prime**. This, however, is not always true. Appendix 2 lists some Mersenne numbers and, as can be seen, the primes 11, 23, and 29 do not generate Mersenne primes. There are therefore three different cases of Mersenne number transforms to consider: Q prime and M prime, Q prime but M composite, and Q and M composite.

Of the 30 or so known Mersenne primes, only 22 are generated by primes less than ten thousand. Of these, only those generated by 2, 3, 5, 7, 13, 17, 19, 31, 61, 89, 107, and 127 are of practical interest. Clearly, this particular class of transforms is very small.

Assume that M is a Mersenne prime. There are three conditions to be satisfied for an NTT to exist. Both ω and N must be relatively prime to M but as M is prime then, provided both are smaller than M, that is always the case. Next, there must exist an element ω with a sequence length N and, third, N must divide $\Phi(M)$. Now:

$$\Phi(M) = (2^p - 2) = 2 \cdot (2^{p-1} - 1)$$

Fermat's theorem states that $< a^p >_p$ is congruent to 1 for a < p where p is prime. From the definition of residues, that means p must divide ($a^{p-1} - 1$). This suggests a choice for the parameters is $\omega = 2$ and N = p. This is a convenient choice but not the only one. For example, the expression for $\Phi(M)$ shows 2p must be a divisor. If ω equals 2 though, the values ω^i, $0 < i < N-1$, cannot be distinct and so N cannot be 2p. If ω equals -2, they are distinct and so a sequence length of 2p is possible. This is composite and hence this root can have a fast transform. A problem is that -2 is not a simple number.

As this Mersenne number transform is within a field, a unique N^{-1} must exist. It can be found via Fermat's theorem but there is an easier approach. If M is prime, then:

$$< M - \Phi(M) >_M = < M - M + 1 >_M = 1$$

Therefore:

$$< \alpha M - \Phi(M) >_M = 1$$

If α is a divisor of $\Phi(M)$, then:

$$< \alpha M - \Phi(M) >_M = < \alpha \cdot \left(M - \frac{\Phi(M)}{\alpha} \right) >_M = 1$$

and the multiplicative inverse of α is ($M - \frac{\Phi(M)}{\alpha}$). In this case, N is a divisor of the totient and so N^{-1}, which is either p^{-1} or ($2p$)$^{-1}$, is given by ($2^p - 1 - 2 \cdot (\frac{2^{P-1}-1}{p})$) or ($2^p - 1 - (\frac{2^{P-1}-1}{p})$).

We move on to the class of Mersenne transforms based on a composite modulus. It was shown earlier that any Mersenne number ($2^N - 1$) has divisors ($2^{d_i} - 1$) where the d_i are the divisors of N. Consider two applications of this. If N is even it must have 2 as a divisor, thus the Mersenne number must have a divisor of 3. As the condition on sequence length is that it must divide each of the ($p_i - 1$) where p_i are the prime divisors, then the sequence length of these Mersenne numbers can at most be 2. Similarly, every third Mersenne number can give a sequence length of at most 6.

The implication is clear. Mersenne transforms based on composite moduli generated from composite powers are of little value. An exception can be made for Q equal to 9 as it generates a sequence length of 6. For the others, though, sequence lengths are just too short to have any practical application.

Now consider composite Mersenne numbers generated by a prime power p. From Appendix 2, there are three which seem likely to have practical application, namely 11, 23, and 29. Each of these, even though composite, appears to behave like a Mersenne prime. That is to say, each has a sequence length of p and 2p. This is to be expected. It was shown earlier that if U is a prime divisor of a Mersenne number ($2^p - 1$) where p is an odd prime, then $< U >_{2p}$ must be 1. Consequently:

$$U = 2k \cdot p + 1$$

where k is any integer. The sequence length must divide the prime divisors of the number and with this result, p and 2p can be values for N. Further, as $< 2^{U-1} >_U$ is congruent to 1 so an appropriate value of ω is 2. It is easily shown -2 is also suitable but this only has a sequence length of 2p. Thus these transforms are almost the same as the class when the power is a prime.

From these results, we may now state that the Mersenne Number Transform pair is:

$$X(k) = \left\langle \sum_{n=0}^{N-1} x(n) \cdot 2^{kn} \right\rangle_M \qquad k = 0, 1, \cdots, N{-}1$$

$$x(n) = \left\langle N^{-1} \cdot \sum_{k=0}^{N-1} X(k) \cdot 2^{-nk} \right\rangle_M \qquad n = 0, 1, \cdots, N{-}1$$

where $\qquad M = 2^p - 1$ with p an odd prime

$\qquad\qquad N = p$

$$N^{-1} = (2^p - 1 - 2 \cdot \frac{(2^{p-1} - 1)}{p})$$

and where 2^{-nk} is interpreted as $2^{(p-1)nk}$. Alternatively, if $(-2)^{kn}$ replaces 2^{kn} in these equations, then N can be 2p and that will make N^{-1} equal to $(2^p - 1 - \frac{(2^{p-1} - 1)}{p})$.

It must be stressed that this is not the only possibility but the transform pair with maximum practical application. The equation for the inverse applies to the case of M composite as well as M prime.

Looking at the forward transform equation, transforming p points requires $p \cdot (p-1)$ real additions and $(p-1)^2$ multiplications by a power of 2. That simply means shifting in digital electronics and shifting is far simpler than multiplication (but not necessarily faster unless barrel shifters are used). That promises simple and inexpensive hardware implementations. Even in general-purpose systems, shifting can be quite fast compared to multiplication and so this approach still appeals for some software implementations.

The number of additions in an MNT is the same as in an equivalent complex DFT. Therefore, for long sequence lengths an MNT will always have a substantially greater number of additions than an FFT implementation. This could only be reduced via a fast algorithm but, as the only factoring of an MNT is when the sequence length is 2p, that is not a significant reduction. This can be a major disadvantage of this approach.

Unfortunately, there are two other severe practical disadvantages of the MNT as expressed here. The modulus is $(2^p - 1)$ where p must be a prime, which means the word size must be p bits. The sequence length, though, is only p or 2p. The standard computer/microprocessor word size is 32 bits. This only permits ten MNTs to be implemented, of which the longest is based on the prime $(2^{31} - 1)$. A sequence length of only 31 or 62 hardly seems adequate.

Clearly, the limited choice of sequence lengths and the very short sequence lengths are major problems. They counter one of the major reasons for considering NTTs and consequently must exclude MNTs from many applications. The exceptions are where quantization error cannot be tolerated under any circumstance, where a short transform length is acceptable or where, for some reason, multipliers cannot be used. In practice, that just means very high speed processing such as in radar systems, possibly very specialized processing chips for tasks such as interpolation or some forms of speech processing, or some very low-frequency signal processing tasks.

What may be termed practical Mersenne number theoretic transforms are those where the power Q is prime. It makes little real difference whether M is composite or not. In fact, the only difference is that when the modulus is prime, there exists a primitive root and so there is the potential for a sequence length of $(M - 1)$. There also exists the potential for roots of other orders, all higher than those quoted earlier. For composite M, however, the maximum sequence length is 2p.

Little can be done about these problems. Finding a different value of ω defeats the main practical advantage of this approach. A root raised to a power is also a root (but not necessarily of the same order). However, there is no practical advantage to be gained by creating roots based on powers of 2 or –2 here. There is, though, one possible advantage in considering a different value of ω. For some VLSI realizations or by using look-up tables for multiplication or similar techniques, the longer sequence length and the possibility of a composite value leading to a fast algorithm may compensate for the additional difficulties in multiplication.

Example

Consider the Mersenne number theoretic transform for:

$$M = 2^p - 1 = 2^7 - 1 = 127$$

For the MNT based on 2, the sequence length N is p, or 7. Also:

$$N^{-1} = (2^p - 1 - 2 \cdot \frac{(2^{p-1} - 1)}{p}) = 109$$

Thus the transform pair is:

$$X(k) = \left\langle \sum_{n=0}^{N-1} x(n) \cdot 2^{kn} \right\rangle_M \qquad k = 0, 1, \cdots, N-1$$

$$= \left\langle \sum_{n=0}^{6} x(n) \cdot 2^{kn} \right\rangle_{127} \qquad k = 0, 1, \cdots, 6$$

$$x(n) = \left\langle N^{-1} \cdot \sum_{k=0}^{N-1} X(k) \cdot 2^{-nk} \right\rangle_M \qquad n = 0, 1, \cdots, N-1$$

$$= \left\langle 109 \cdot \sum_{k=0}^{6} X(k) \cdot 2^{-nk} \right\rangle_{127} \qquad n = 0, 1, \cdots, 6$$

A useful simplification in these calculations is:

$$\left\langle 2^Q \right\rangle_{2^p-1} = 2^{\langle q \rangle_p}$$

and:

$$\left\langle 2^{-Q} \right\rangle_{2^p-1} = 2^{\langle (p-1)Q \rangle_p}$$

We shall use this transform pair to convolve the two sequences:

x(n) :	1	2	3	4	3	2	1
h(n) :	1	2	3	4	5	6	7

First, x(n) will be transformed. The equation is:

$$X(k) = < 1 + 2^{k+1} + 3 \cdot 2^{2k} + 2^{3k+2} + 3 \cdot 2^{4k} + 2^{5k+1} + 2^{6k} >_{127}$$

The calculations are quite straightforward. A typical value is:

$$
\begin{aligned}
X(5) &= < 1 + 2^6 + 3 \cdot 2^{10} + 2^{17} + 3 \cdot 2^{20} + 2^{26} + 2^{30} >_{127} \\
&= < 1 + 2^6 + 3 \cdot 2^3 + 2^3 + 3 \cdot 2^6 + 2^5 + 2^2 >_{127} \\
&= 71
\end{aligned}
$$

The complete set of transformed values is:

$$X(k) : 16 \quad 98 \quad 113 \quad 87 \quad 61 \quad 71 \quad 69$$

The transform for $h(n)$ is:

$$H(k) = < 1 + 2^{k+1} + 3 \cdot 2^{2k} + 2^{3k+2} + 5 \cdot 2^{4k} + 3 \cdot 2^{5k+1} + 7 \cdot 2^{6k} >_{127}$$

The complete set of transformed values is:

$$H(k) : 28 \quad 7 \quad 87 \quad 1 \quad 119 \quad 33 \quad 113$$

The convolution is:

$$Y(k) = < X(k) \cdot H(k) >_M \qquad\qquad k = 0, 1, \cdots, N-1$$

Thus:

$$
\begin{aligned}
Y(0) &= < 28 \cdot 16 >_{127} &= 67 \\
Y(1) &= < 7 \cdot 98 >_{127} &= 51 \\
Y(2) &= < 87 \cdot 113 >_{127} &= 52 \\
Y(3) &= < 1 \cdot 87 >_{127} &= 87 \\
Y(4) &= < 119 \cdot 61 >_{127} &= 20 \\
Y(5) &= < 33 \cdot 71 >_{127} &= 57 \\
Y(6) &= < 113 \cdot 69 >_{127} &= 50
\end{aligned}
$$

The inverse transform is:

$$y(n) = \left\langle 109 \cdot \sum_{k=0}^{6} Y(k) \cdot 2^{-nk} \right\rangle_{127} \qquad\qquad n = 0, 1, \cdots, 6$$

$$= < 109 \cdot (67 + 51 \cdot 2^{6n} + 52 \cdot 2^{12n} + 87 \cdot 2^{18n} + 20 \cdot 2^{24n} + 57 \cdot 2^{30n} + 50 \cdot 2^{36n}) >_{127}$$

A typical calculation is:

$$
\begin{aligned}
y(5) &= < 109 \cdot (67 + 51 \cdot 2^{30} + 52 \cdot 2^{60} + 87 \cdot 2^{90} + 20 \cdot 2^{120} + 57 \cdot 2^{150} + 50 \cdot 2^{180}) >_{127} \\
&= < 109 \cdot (67 + 51 \cdot 2^2 + 52 \cdot 2^4 + 87 \cdot 2^6 + 20 \cdot 2 + 57 \cdot 2^3 + 50 \cdot 2^5) >_{127}
\end{aligned}
$$

$$= \; < 109 \cdot (4) >_{127}$$
$$= 55$$

The result of the convolution is therefore:

$$y(n) : 73 \quad 75 \quad 70 \quad 58 \quad 53 \quad 55 \quad 64$$

An MNT of sequence length 14 is possible here by using a root of –2. In this case, –2 is 125 and N^{-1} is 118. Using these would present some practical difficulties.

Since M was chosen to be prime here, primitive roots exist. In fact, $\Phi(\Phi(127)) = 36$ roots. An MNT based on these will have a sequence length of M. Finding a primitive root can be done by testing the candidates modulo M raised in turn to a power equal to each of the divisors of $\Phi(M)$. Since:

$$\Phi(127) = 126 = 2 \cdot 3 \cdot 3 \cdot 7$$

the divisors are 1,2,3,6,7,9,14,18,21,42,63, and 126. For practical reasons, a primitive root should be a small number whose powers are easily implemented in binary. The smallest number to consider for a root is 3. Testing:

$< 3^1 >_{127}$	=	3	$< 3^2 >_{127}$	=	9	$< 3^3 >_{127}$	= 127
$< 3^6 >_{127}$	=	94	$< 3^7 >_{127}$	=	28	$< 3^9 >_{127}$	= 125
$< 3^{14} >_{127}$	=	22	$< 3^{18} >_{127}$	=	4	$< 3^{21} >_{127}$	= 108
$< 3^{42} >_{127}$	=	107	$< 3^{63} >_{127}$	=	126		

Hence 3 is a primitive root. So too are 6, 7, 12, 14, 23, 29, 39, 43, 45, 46, 48, 53, 55, 56, 57, 58, 65, 67, 78, 83, 85, 86, 91, 92, 93, 96, 97, 101, 106, 109, 110, 112, 114, 116, and 118. None give powers easily implemented in binary.

Pseudo Mersenne number theoretic transforms

As mentioned earlier, pseudo Mersenne transforms are number theoretic transforms which are Mersenne for most operations but conclude with a reduction modulo some other value. That is, they are an implementation of:

$$<< f(x) >_{\mu M} >_M = y$$

where $\mu \cdot M$ is some Mersenne number and μ, for practical reasons, needs to be a small value.

The prime concern with this NTT is proving it exists modulo M. Using operations internally that are modulo a Mersenne number is simply a practical convenience. The choices for the modulus M, though, are not arbitrary. A composite Mersenne number of the form $(2^{nm} - 1)$ can be described as:

$$M_{nm} = M_n \cdot (2^{n(m-1)} + 2^{n(m-2)} + \cdots + 1)$$

Clearly, $\mu \cdot M$ being Mersenne limits μ or M to be one of these two terms or their factors. There is no point in making M a Mersenne number which means that μ should be either the lowest Mersenne prime factor of M_{nm} or a factor of some low order Mersenne number. In general, the former is preferred. Hence if μ is M_n, the modulus is fixed to:

$$M = (2^{n(m-1)} + 2^{n(m-2)} + \cdots + 1)$$

i.e.
$$M = \frac{(2^{nm} - 1)}{(2^n - 1)}$$

and this provides a limit.

We showed in the last section that if Q is even, then $(2^Q - 1)$ is composite and has 3 as a divisor. Consequently, another limit is that the maximum sequence length is 2 if mn is even. That is unacceptable, hence mn must be odd.

Return to the question of whether one or more NTTs exist given these constraints on M. They do and we can prove this by considering the general case for any NTT of this type. Let ω be an element of the complete residue system formed modulo M where:

$$M = \frac{(\omega^N - 1)}{(\omega^{N_o} - 1)}$$

We require that $N = p^q$ and $N_o = p^{q-1}$ (where p is an odd prime, q is a positive integer) and that:

$$< \omega >_p \neq 1$$

If these conditions are satisfied, then an NTT of sequence length N exists.

We must show three things to prove this. First, we need to show ω has an order N in this system. That is, does $< \omega^N >_M$ equal 1? Now:

$$< \omega^N >_{M'} = 1 \qquad\qquad \text{where } M' = \omega^N - 1$$

But $(\omega^N - 1)$ is a factor of the modulus M and so:

$$< \omega^N >_M = 1$$

Next, we need to prove the existence of an inverse to N. From the earlier work on congruences, we know the inverse exists if (N, M) is 1. Since N is p^q and p is an odd prime, this reduces to showing N is relatively prime to p. Now from Fermat's theorem:

$$< \omega^p >_p = < \omega >_p = < \omega^{p + \alpha p} >_p$$

That is, ω raised to any integral power of p modulo p is ω. Expanding M:

$$M = \omega^{(p-1)\frac{N}{P}} + \omega^{(p-2)\frac{N}{P}} + \cdots + \omega^{\frac{N}{P}} + 1$$

Since (p – 1) is even and there are (p – 1) terms involving ω raised to an integral power of p, then:

$$< M >_p = 1$$

This, though, means M and p must be relatively prime. Consequently, the inverse exists.

The final condition is that the terms ($\omega^j - 1$), where $1 \leq j \leq N-1$, must be mutually prime to M. As discussed in THEORY, that reduces to testing those values of j such that $\frac{N}{j}$ is prime. As N is p^q, the only such term is p^{q-1}. Thus we must show:

$$(\omega^{\frac{N}{P}} - 1, M) = 1$$

The above expansion for M may be expressed as:

$$M = (\omega^{\frac{N}{P}} - 1) \cdot (\omega^{(p-2)\frac{N}{P}} + 2\omega^{(p-3)\frac{N}{P}} + \cdots + p - 1) + p$$

Consequently, $(\omega^{\frac{N}{P}} - 1)$ is relatively prime to M if it is relatively prime to p. We have already established $< M >_p$ is 1 and $< \omega^{\frac{N}{P}} >_p$ is $< \omega >_p$. Then provided $< \omega >_p$ is not congruent to 1, the two terms are mutually prime and the NTT exists.

As these results are quite general, several pseudo NTTs can exist. Our particular interest is in pseudo MNTs, however, and that means limiting ω to 2, –2, or some power of these.

Two pseudo MNTs have been closely investigated. In one, N is chosen to be p^2 thus:

$$M = \frac{(2^N - 1)}{(2^P - 1)}$$

Operations in this system require an N-bit word size but the final reduction means the eventual word size is approximately (N – p). The problem with this choice is the extremely small number of candidate NTTs. If we limit the word size to a maximum of 64, there are only three possible transforms:

$$p = 3 \quad \text{which gives } N = 9$$
$$p = 5 \quad \text{which gives } N = 25$$
$$p = 7 \quad \text{which gives } N = 49$$

These results are unimpressive. Consider Mersenne transforms based on the primes 5 and 7. Both support primitive roots and using the MNT methods of the last section with any of these roots, sequence lengths of 30 and 126 respectively can be obtained. Further, while it is true M for each of these terms is quite composite, for example:

$$2^{49} - 1 = 127 \cdot 4, 432, 676, 798, 593$$

and so permits a fast algorithm to be developed, this is also true of the MNT proper.

The other choice is no better but it is more flexible. Choose ω to be 2^q and N as p. Thus:

$$M = \frac{(2^{qp} - 1)}{(2^p - 1)}$$

Internal operations are therefore with a word size of $q \cdot p$ bits and the final reduction leads to a word size of approximately $q \cdot (p-1)$. The problem is to find q and p such that their product is odd. Even if found, p can at best be approximately equal to the square root of that prime and that gives quite a short sequence length.

Fermat number theoretic transforms

Fermat numbers are of the form ($2^q + 1$) and so require (q + 1) bits. However, given this modulus, numbers within this system are the set $\{ 0, 1, 2, \cdots, 2^q \}$ which means only one number ever has a most significant bit of 1. Arithmetic modulo a Fermat number and efficient methods of representing these numbers are discussed in Appendix 1.

Fermat numbers are commonly denoted F_m. The first five, taken from Appendix 3, are:

$$
\begin{aligned}
F_0 &= & 3 \\
F_1 &= & 5 \\
F_2 &= & 17 \\
F_3 &= & 257 \\
F_4 &= & 65,537
\end{aligned}
$$

are all prime. They are, in fact, the only known Fermat primes and there is some evidence that they may be the only such primes. For implementation, these require word sizes of 2, 3, 5, 9, and 17 bits respectively. Therefore, these numbers plus:

$$F_5 = 4,294,967,297$$

and:

$$F_6 = 18,446,744,073,709,551,617$$

which require word sizes of 33 and 65 bits respectively, are the only Fermat numbers of interest to most digital signal processing.

Most Fermat number transforms of practical interest are in fields and so constructing an NTT reduces to choosing a sequence length N and finding a root ω. The maximum sequence length is the totient, which is ($F_m - 1$) or 2^q where q is 2^m. As F_m is prime, N must divide this. Consequently N must be of the form 2^r where r is 2^s and $s \le m$.

As discussed earlier in this Part, a divisor of a composite Fermat number F_m is a number of the form ($k \cdot 2^{m+2} + 1$). If a composite Fermat number transform is desired, then from the general theory of NTTs, N must divide 2^{m+2}. As this is substantially smaller than F_m a prime, there is clearly a problem with long sequence length FMTs.

Finding a suitable ω is relatively simple but as with Mersenne transforms the interest is in roots which are related to 2 or powers of 2. Now:

$$< F_m - 1 >_{F_m} = -1 = < 2^q >_{F_m} \text{ where } q = 2^m$$

and from the work on quadratic residues:

$$< 2^{2q} >_{F_m} = 1$$

Given, too, that N must be of the form 2^r we can derive a series of roots and corresponding values of N. However, the maximum value of N is 2^{m+1} for ω equal to 2. For these values, the transform pair is:

$$X(k) = \left\langle \sum_{n=0}^{2^{m+1}-1} x(n) \cdot 2^{kn} \right\rangle_M \qquad\qquad k = 0, 1, \cdots, 2^{m+1}-1$$

$$x(n) = \left\langle \left(-2^{q-m-1}\right) \cdot \sum_{k=0}^{2^{m+1}-1} X(k) \cdot 2^{-nk} \right\rangle_M \qquad\qquad n = 0, 1, \cdots, 2^{m+1}-1$$

where
$$\begin{aligned} M &= 2^q + 1 \\ q &= 2^m \\ 2^{-nk} &= 2^{M-1-nk} \end{aligned}$$

When F_m is composite, the transform is in a ring and there is no guarantee of an inverse for N. To prove the transform exists in this case then, it is necessary to prove both the inverse does exist and that the terms ($\omega^j - 1$) are relatively prime to F_m. This is easily done, hence this transform pair also applies to F_m composite. It is also true for F_m composite that a family of transforms exists based on a root of 2^α, where α is 2^γ, with sequence length $2^{m-1-\gamma}$.

The sequence length N for this transform and the others based on powers of 2 is also a power of 2 and so highly composite. Therefore, a fast algorithm can be found directly analogous to a base-2 FFT. Its development is quite straightforward. Indeed, it is just a base-2 algorithm with 2 replacing W and all terms subject to modulo F_m reduction. Thus it has $\log_2 N$ real additions and $(\frac{N}{2}) \cdot \log_2 N$ shifts.

Fermat number transforms are simple, there is a wide choice of sequence lengths, although all are based on a power of 2, and there are fast algorithms. These are powerful advantages over Mersenne number transforms. However, in order, the first seven Fermat numbers create transforms with maximum sequence lengths of only 2, 4, 8, 16, 32, 64, and 128. Fermat number transforms, consequently, are little better than Mersenne for executing long-length transforms.

To overcome this problem for the Fermat primes requires a root other than 2. That would mean general multiplications modulo F_m which is unattractive in most circumstances. The obvious choice, though, is 3 or a power of 3 as 3 is a primitive root of any prime Fermat number and so has a sequence length of ($F_m - 1$). Multiplication by 3 is still relatively simple and as N can only be a power of 2, such FMTs would also have a fast algorithm. That certainly makes them more appealing than any equivalent Mersenne transform.

For the composite Fermat numbers, a simple step can be taken to double the sequence length to the maximum possible of 2^{m+2}. In a complete residue system modulo F_m, a number exists equivalent to $\sqrt{2}$ as a solution exists to the quadratic congruence:

$$< x^2 >_{F_m} = 2$$

If 2 has a sequence length of 2^{m+1}, then $\sqrt{2}$ must have a length 2^{m+2}. It can be shown:

$$\sqrt{2} = 2^{\frac{q}{4}} \cdot (2^{\frac{q}{2}} - 1) \qquad \text{where } q = 2^m$$

While relatively simple expressions in binary, these numbers are still large. Any odd power of this root has the same sequence length, thus there is some flexibility in choosing more useful values for implementation. Note that it is necessary to have general multiplications with these roots for half the operations and shifts for the other half. Those general multiplications, though, can be reduced to one multiplication by $\sqrt{2}$ plus shifts.

Example

This example is based on:

$$F_2 = 2^4 + 1 = 17$$

The maximum sequence length is therefore:

$$N = 2^{2+1} = 8$$

The transform pair is:

$$X(k) = \left\langle \sum_{n=0}^{2^{m+1}-1} x(n) \cdot 2^{kn} \right\rangle_M \qquad k = 0, 1, \cdots, 2^{m+1}-1$$

$$= \left\langle \sum_{n=0}^{7} x(n) \cdot 2^{kn} \right\rangle_{17} \qquad k = 0, 1, \cdots, 7$$

$$x(n) = \left\langle \left(-2^{q-m-1}\right) \cdot \sum_{k=0}^{2^{m+1}-1} X(k) \cdot 2^{-nk} \right\rangle_M \qquad n = 0, 1, \cdots, 2^{m+1}-1$$

$$= \left\langle 15 \cdot \sum_{k=0}^{7} X(k) \cdot 2^{-nk} \right\rangle_{17} \qquad n = 0, 1, \cdots, 7$$

This transform will be used to compute the convolution of the two sequences:

x(n):	0	1	2	1	2	1	2	0
h(n):	0	1	2	1	0	2	2	1

First, we transform x(n). Thus:

$$X(k) = \ <\ 2^k + 2^{2k+1} + 2^{3k} + 2^{4k+1} + 2^{5k} + 2^{6k+1}\ >_{17}$$

A typical calculation is:

$$
\begin{aligned}
X(5) &= \ <\ 2^5 + 2^{11} + 2^{15} + 2^{21} + 2^{25} + 2^{31}\ >_{17} \\
&= \ <\ -2 + 2^3 - 2^3 - 2 + 2 - 2^3\ >_{17} \\
&= \ <\ -10\ >_{17} \\
&= 7
\end{aligned}
$$

A useful simplification here is:

$$< 2^{aM+b} >_{2^M+1} = (-1)^a \cdot 2^b$$

The set of transformed values, then, is:

$$X(k): \quad 9 \quad 6 \quad 2 \quad 0 \quad 3 \quad 7 \quad 11 \quad 13$$

Next, $h(n)$ is transformed:

$$H(k) = \ <\ 2^k + 2^{2k+1} + 2^{3k} + 2^{5k+1} + 2^{6k+1} + 2^{7k}\ >_{17}$$

The transformed values are:

$$H(k): \quad 9 \quad 15 \quad 0 \quad 9 \quad 16 \quad 2 \quad 9 \quad 8$$

The convolution is therefore:

$$
\begin{aligned}
Y(0) &= \ <\ 9 \cdot 9\ >_{17} &= 13 \\
Y(1) &= \ <\ 15 \cdot 6\ >_{17} &= 5 \\
Y(2) &= \ <\ 0 \cdot 2\ >_{17} &= 0 \\
Y(3) &= \ <\ 9 \cdot 0\ >_{17} &= 0 \\
Y(4) &= \ <\ 16 \cdot 3\ >_{17} &= 14 \\
Y(5) &= \ <\ 2 \cdot 7\ >_{17} &= 14 \\
Y(6) &= \ <\ 9 \cdot 11\ >_{17} &= 14 \\
Y(7) &= \ <\ 8 \cdot 13\ >_{17} &= 2
\end{aligned}
$$

The inverse transform is:

$$y(n) = \ <\ 15 \cdot (13 + 5 \cdot 2^{-n} + 14 \cdot 2^{-4n} + 14 \cdot 2^{-5n} + 14 \cdot 2^{-6n} + 2 \cdot 2^{-7n}\ >_{17}$$

Obtaining the transform is made easier using:

$$< 2^{-(aM+b)} >_{2^M+1} = (-1)^{a+1} \cdot 2^{M-b}$$

The resulting convolution is:

$$y(n): \quad 12 \quad 10 \quad 8 \quad 12 \quad 11 \quad 8 \quad 8 \quad 12$$

Pseudo Fermat number theoretic transforms

The name pseudo Fermat number transforms is a little misleading. A general class of NTTs exists using a modulus:

$$M = \frac{(\omega^Q + 1)}{(\omega^{\frac{Q}{P}} + 1)} \qquad Q = p^r, \ r > 0, \ p \text{ an odd prime}$$

As with pseudo Mersenne transforms, the object is to perform most operations modulo a number ($\omega^Q + 1$) with a final reduction modulo M. The most significant subclass of this transform is where ω is 2 in which case most operations are modulo ($2^Q + 1$). This is similar in form to a Fermat number, hence the name.

Provided $< \omega >_p$ is not congruent to -1, an NTT exists for this modulus with ω being a root of order 2Q. Proving this requires showing ω indeed has an order N equal to 2Q, that N has an inverse and that M is relatively prime to the terms ($\omega^j - 1$) for $0 \le j \le N-1$. The proof is very similar to the earlier work for pseudo Mersenne transforms.

The first requirement is easily shown. Since $< \omega^Q >_M$ is -1 for a modulus ($\omega^Q + 1$), then $< \omega^{2Q} >_M$ (which is $< \omega^N >_M$) is 1 and so ω is a root of order 2Q for this modulus. But the root must also be of order 2Q for M as ($\omega^Q + 1$) is a factor of M.

Proving an inverse to N exists is a little difficult given the involved modulus but, as before, it only involves proving (M, N) is 1. Now N is 2Q, and so even, thus this condition is only true if M is odd. Furthermore, since Q is p^r and p is an odd prime, it is also necessary that (p, M) is unity.

Expanding M, we obtain:

$$M = \omega^{(p-1)\frac{Q}{P}} - \omega^{(p-2)\frac{Q}{P}} + \omega^{(p-3)\frac{Q}{P}} - \omega^{(p-4)\frac{Q}{P}} + \cdots + \omega^{\frac{2Q}{P}} - \omega^{\frac{Q}{P}} + 1$$

Applying Fermat's theorem, noting that (p – 1) is even and that there are (p – 1) terms involving integral powers of p, we deduce:

$$< M >_p = 1$$

which in turn establishes M and p as relatively prime. Continuing with this expression:

$$M = \omega^{(p-2)\frac{Q}{P}} \cdot \left(\omega^{\frac{Q}{P}} - 1 \right) + \omega^{(p-4)\frac{Q}{P}} \cdot \left(\omega^{\frac{Q}{P}} - 1 \right) + \cdots + \omega^{\frac{Q}{P}} \cdot \left(\omega^{\frac{Q}{P}} - 1 \right) + 1$$

$$= \omega^{\frac{Q}{P}} \cdot \left(\omega^{\frac{Q}{P}} - 1 \right) \bullet \left(\omega^{(p-3)\frac{Q}{P}} + \omega^{(p-5)\frac{Q}{P}} + \cdots + 1 \right) + 1$$

The third term of the first part of this expression is a polynomial of even powers of ω. Hence regardless of whether ω is even or odd, it is always even and so M must be odd.

The final requirement reduces to testing those values of j for which $\frac{N}{j}$ is prime. Since N is $2p^r$, j can only be p^r or $2p^{r-1}$. Thus we need to show:

$$(\omega^Q - 1, M) = 1 \qquad \text{and} \qquad (\omega^{\frac{2Q}{P}} - 1, M) = 1$$

Now:

$$\omega^Q - 1 = \omega^Q + 1 - 2 = M \cdot \left(\omega^{\frac{Q}{P}} + 1 \right) - 2$$

This implies $(\omega^Q - 1, M)$ is 1 provided $(2, M)$ is 1. Since M is odd, this is always the case. For the second, note:

$$\omega^{\frac{2Q}{P}} - 1 = \left(\omega^{\frac{Q}{P}} - 1 \right) \cdot \left(\omega^{\frac{Q}{P}} + 1 \right)$$

As $(\omega^{\frac{Q}{P}} - 1)$ is a factor of $(\omega^Q - 1)$ and this is relatively prime to M, the problem reduces to proving $(\omega^{\frac{Q}{P}} + 1, M)$ is 1. Now we can express $< M >_p$ as:

$$< M >_p = < \left(\omega^{\frac{Q}{P}} + 1 \right) \cdot \left(\omega^{(p-2)\frac{Q}{P}} - 2\omega^{(p-3)\frac{Q}{P}} + 3\omega^{(p-3)\frac{Q}{P}} - \cdots - p + 1 \right) + p >_p$$

Then $(\omega^{\frac{Q}{P}} + 1)$ is relatively prime to M provided p is not a divisor of $(\omega^{\frac{Q}{P}} + 1)$. Fermat's theorem shows this is not possible.

Since this is a very general result, a wide family of pseudo transforms exist. The ones of interest are again those where ω is 2. In this case, the modulus M is:

$$M = \frac{(2^Q + 1)}{(2^{\frac{Q}{P}} + 1)}$$

where Q is p^r and N is $2p^r$. This still leaves many options. For example, the modulus is $\frac{(2^P + 1)}{3}$ when r is 1 and N is 2p. Or, if r is 2, then N is $2p^2$ and the modulus then becomes $\frac{(2^{\frac{N}{2}} + 1)}{(2^P + 1)}$. Rather than use 2 as a root, we could also use a power of 2.

While in general terms pseudo Fermat transforms are more flexible than pseudo Mersenne transforms, there is still a problem with the relationship between word size and sequence length N. That word size needs to be approximately $(\frac{N}{2} + 1)$ bits.

These results can be extended to Q even. As $(2^{\frac{N}{2}} - 1)$ has factors $(2^{\frac{N}{2}} - 1)$ and $(2^{\frac{N}{2}} + 1)$ when N is even, that is, a Mersenne and pseudo Fermat number, then:

$$< 2^N > 2^{\frac{N}{2}} + 1 = 1$$

That is, 2 is a root of order N for the pseudo Fermat number $(2^{\frac{N}{2}} + 1)$. Reviewing the proof just given for Q odd, it can be seen that if Q is even, then an NTT exists of order 2Q for the root of 2 provided M is odd. This gives quite a large number of transforms.

Complex number theoretic transforms

There are two approaches to complex number theoretic transforms. One is to simply apply the previous theory to each of the parts of a complex calculation. The other is to develop theory based on complex (that is, Gaussian) integers. In general terms, this is better treated by the methods of the next Part.

Assume two complex sequences are to be convolved. Then:

$$y_R(k) + jy_I(k) = \sum_{n=0}^{N-1} (h_R(k-n) + jh_I(k-n)) \cdot (x_R(n) + jx_I(n)) \qquad k = 0, 1, \cdots, N-1$$

This can be executed via four separate real convolutions, namely:

$$y_R(k) = \sum_{n=0}^{N-1} h_R(k-n)x_R(n) - \sum_{n=0}^{N-1} h_I(k-n)x_I(n) \qquad k = 0, 1, \cdots, N-1$$

$$y_I(k) = \sum_{n=0}^{N-1} h_I(k-n)x_R(n) + \sum_{n=0}^{N-1} h_R(k-n)x_I(n) \qquad k = 0, 1, \cdots, N-1$$

In the ring of integers with modulus M equal to ($2^Q + 1$):

$$< 2^Q >_M = -1$$

and so if Q is even j is equivalent to $2^{\frac{Q}{2}}$. Define two intermediate equations:

$$u(k) = \left\langle \sum_{n=0}^{N-1} \left(h_R(k-n) + 2^{\frac{Q}{2}} \cdot h_I(k-n) \right) \cdot \left(x_R(n) + 2^{\frac{Q}{2}} \cdot x_I(n) \right) \right\rangle_M$$

$$v(k) = \left\langle \sum_{n=0}^{N-1} \left(h_R(k-n) - 2^{\frac{Q}{2}} \cdot h_I(k-n) \right) \cdot \left(x_R(n) - 2^{\frac{Q}{2}} \cdot x_I(n) \right) \right\rangle_M$$

so that the convolution becomes:

$$y_R(k) = < \frac{1}{2}(u(k) + v(k)) >_M$$

$$y_I(k) = < \frac{1}{2}(-2^{\frac{Q}{2}} \cdot (u(k) - v(k))) >_M$$

There are now only two real convolutions to perform to execute this complex convolution; a significant saving. There is the problem, of course, that a Fermat or pseudo Fermat transform must be used and that will limit the choice of the sequence length N.

We shall look briefly at NTTs based on Gaussian integers. Since an NTT is based on a finite mathematical structure, those integers will be of the form (a + jb) where a and b are both drawn from a Galois field GF(p). Thus the structure must have p^2 elements and -1 must be a quadratic non-residue.

The basis of our previous work has been to find a root in the finite structure and then use that property to construct a transform. In order to repeat that here, we need to study the properties of powers of complex numbers in finite structures and derive some equivalent of Euler's theorem or Fermat's theorem for complex structures. Consider $(a + jb)^p$. This is a binomial polynomial and can be expanded to:

$$(a + jb)^p = a^p + \cdots + c_i \cdot a^{p-i} \cdot j^i \cdot b^i + \cdots + j^p \cdot b^p$$

where $c_i = \dfrac{p!}{i! \cdot (p-i)!}$

The coefficients c_i must be integers and therefore the denominator must divide the numerator. However, p is prime, thus the denominator must divide $(p-1)!$ rather than p. The coefficients c_i must consequently have the form $\beta_i p$ and so:

$$(a + jb)^p = a^p + j \cdot \beta_1 \cdot p \cdot a^{p-1} \cdot b + \cdots + j^i \cdot \beta_i \cdot p \cdot a^{p-i} \cdot b^i + \cdots + j^p \cdot b^p$$

Taking the residue modulo p of both sides of this equation gives:

$$< (a + jb)^p >_p = < a^p + j^p \cdot b^p >_p$$

and applying Fermat's theorem:

$$< (a + jb)^p >_p = a + j^p \cdot b$$

A complex structure, though, has p^2 elements, so repeating the process we finally derive:

$$< (a + jb)^{p^2} >_p = < a + j^{p^2} \cdot b >_p$$

A problem in this expression is the term j^{p^2}. If $< p^2 >_4$ is 1, then j^{p^2} must be j, thus:

$$< (a + jb)^{p^2} >_p = a + j \cdot b$$

This is a generalized form of Fermat's theorem for Gaussian integers.

So far, two conditions have been set for p in this structure, namely that -1 must be a quadratic residue in GF(p) and that $< p^2 >_4$ is 1. Values of p satisfying this have been discussed earlier with respect to quadratic residues. That is, p has to be of the form:

$$
\begin{aligned}
p &= 4k + 3 \\
&= 3, 7, 11, 19, 23, 31, 43, 47, 59, 67, 71, \cdots
\end{aligned}
$$

Mersenne numbers are some of the primes that satisfy this condition. Hence a class of complex Mersenne NTTs exist.

The next question to consider is the order of roots in a complex structure. From the work on real structures and in view of the above results, the order N of any complex root must divide $(p^2 - 1)$. For a Mersenne prime $(2^q - 1)$ then, N must divide $2^{q+1} \cdot (2^{q-1} - 1)$.

The simplest approach to take in creating complex transforms is to extend the results obtained with real Mersenne transforms. Assume again the Mersenne prime p is given by

$(2^q - 1)$. The order N of any root in the real transform must divide the totient and that is $2(2^{q-1} - 1)$. Therefore, any root supported by the real transform is also supported by the complex. Hence we know immediately that a root of $(2 + j0)$ has a sequence length q and a root $(-2 + j0)$ has a sequence length 2q.

A complex structure allows additional roots to be created along these lines. There are eight possible roots based on 2, namely $2, 2(1+j), 2j, 2(-1+j), -2, -2(1+j), -2j$ and finally, $2(1-j)$. Substituting and testing, we find the roots $2j$ and $-2j$ both have sequence lengths of 4q and that the remainder (other than 2 and –2) have a sequence length of 8q.

These results are most attractive. With these sequence lengths, a fast algorithm with three or four sections can be developed for the Mersenne transforms. There are still no multiplications and with a sequence length of 8q, N is now approaching a reasonable value. For example, q equal to 31 allows sequence lengths of 31, 124 or 248. Further, being Mersenne number transforms, the arithmetic is particularly simple. In terms of practical application, these NTTs present only minor problems. Complex data does not occur very often, but that is circumvented by simultaneously transforming two data sequences as described in Part 1.

Although these results are quite pleasing, it is possible to do better. Since N must divide the maximum sequence length $2^{q+1}(2^{q-1} - 1)$, we could have a sequence length of 2^{q+1}. This appeals as it is a power of 2 and so an FFT-like base-2 algorithm can be developed. Let $(a + j \cdot b)$ be a root with this sequence length. We wish to find specific values for a and b. Since N is even, then we find:

$$\langle (a+j\cdot b)^{\frac{N}{2}} \rangle_M = \langle (a+j\cdot b)\cdot(a+j\cdot b)^{\frac{N}{2}-1} \rangle_M$$

$$= \langle (a+j\cdot b)\cdot(a+j^{\frac{N}{2}-1}\cdot b) \rangle_M$$

$$= \langle (a+j\cdot b)\cdot(a+j\cdot j^{\frac{N}{2}-2}\cdot b) \rangle_M$$

$$= \langle (a+j\cdot b)\cdot(a-j\cdot b) \rangle_M$$

$$= \langle (a^2+b^2) \rangle_M$$

By definition:

$$\langle (a+j\cdot b)^N \rangle_M = 1$$

thus:

$$\langle (a+j\cdot b)^{\frac{N}{2}} \rangle_M = -1 = \langle (a^2+b^2) \rangle_M$$

Let:

$$U = \langle a^2 \rangle_M \qquad\qquad V = \langle -b^2 \rangle_M$$

then

$$\langle U+1 \rangle_M = V$$

Clearly, U is a quadratic residue. From previous work, V must be a quadratic non-residue. Thus we need two numbers U and V from $\{\ 0, 1,\ \cdots, 2^q - 1\ \}$ where V is (U + 1) and U is a square. Expressed another way, two consecutive numbers are needed from this sequence such that the first is a quadratic residue and the second is not. Testing, it is found the smallest values that meet this requirement are 2 and 3. Hence the problem is solved by solving the Diophantine equations:

$$< a^2 >_M\ =\ 2$$

$$< -b^2 >_M =\ 3\quad \text{i.e.}\ < b^2 >_M\ =\ -3$$

We can solve these equations in the following manner. First, provided $\frac{(M^2 - 1)}{8}$ is even then 2 is a quadratic residue of the odd prime M. Now:

$$\frac{(M^2 - 1)}{8}\ =\ 2^{q-2} \cdot (2^{q-1} - 1)$$

and this is even for q greater than 3. In the complete residue system $0, 1,\ \cdots, M-1$ where M is an odd prime, 2 is a quadratic residue. From Euler's criterion:

$$< 2^{\frac{(M-1)}{2}} >_M\ =\ 1$$

where $\dfrac{(M - 1)}{2}$ is $2^{q-1} - 1$. Multiplying by 2:

$$< 2 \cdot 2^{\frac{(M-1)}{2}} >_M\ =\ < 2^R >_M = 2$$

where $R\ =\ \dfrac{(M - 1)}{2} + 1\ =\ 2^{q-1} - 1 + 1\ =\ 2^{q-1}$

Thus R must be even and therefore:

$$< 2^{\frac{R}{2}} \cdot 2^{\frac{R}{2}} >_M = 2$$

Consequently, a solution to the congruence is that:

$$< a >_M\ =\ \pm 2^{\frac{R}{2}}$$

Following a similar procedure for the other equation, we finally obtain the result that the roots (a + jb) with order 2^{q+1} are of the form:

$$< \pm 2^\alpha \pm j\,(-3\,)^\alpha >_M \qquad\qquad \text{where } \alpha\ =\ 2^{\,q-2}$$

The presence of 3 means these roots no longer involve just shifts but must employ general multiplications. Note that these roots raised to an odd power have the same sequence length and so one of these may prove to be more suitable for implementation.

So far, we have considered the complex field for Mersenne primes. The question arises of whether transforms exist based on (complex) rings of composite Mersenne numbers. These must satisfy the usual requirements for rings in that the sequence length N must divide the greatest common divisor of the terms ($p_i - 1$) where these are the prime divisors of the modulus M given by ($2^q - 1$). In addition, if roots of order 2q, 4q or 8q are sought, which would be the principal interest in these structures, then 2 must be relatively prime to q. That is, q must be odd and have only odd divisors.

There are other conditions. The simplest approach to finding roots is to transfer known roots from the real transform to the complex. The roots of interest in the real ring are those based on a power of 2. Assume then, a root G of order N is of the form 2^r. Thus:

$$< G^N >_M = < 2^{rN} >_M = 1$$

Translated to the complex ring, we deduce that $< rN >_q$ must be zero and that the indices $< ri >_q$ in the transform must be unique. We seek roots based on the square, fourth, and eighth roots of this G as these may have sequence lengths of 2N, 4N, and 8N respectively. A root with a sequence length of 8N based on an eighth root of 2^m in this ring will be of the form $(1 + j)^r$. Therefore:

$$< (1 + j)^{r8N} >_M = 1$$

The earlier work on the permutation of sequences and the first equation here implies that r must be odd. Since q must also be odd and $< rN >_q$ zero, then N, too, must be odd. Hence NTTs with a sequence length 8N do exist in complex Mersenne rings provided in the root 2^r of the real ring r is odd, the sequence length N is odd and q is odd and only has odd factors in the modulus ($2^q - 1$).

For the most part, this result is of limited use. Even the composites which do satisfy this condition frequently have very short sequence lengths. For example, many have a sequence length of only 3. Going to a complex ring to achieve an eightfold increase in sequence length does not seem much of a reward when in real terms N becomes only 24. The exception is those composites generated by primes. Then the sequence length N is p where M is ($2^p - 1$). Thus the useful transforms for a sequence length N up to 64 are 11, 23, 29, 37, 41, 43, 47, 53, and 59. In the complex ring, these generate quite significant sequence lengths.

Example

As an example, consider:

$$M = 2^9 - 1 = 7 \cdot 73$$

Thus the terms ($p_i - 1$) are 6 and 72. The GCD is 6 and as N can only be an odd divisor of this, it can only be 1 or 3. For N equal to 3, we need a root 2^r where r is odd, 2^{3r} is 1 modulo M and the 2^{ir} terms, $0 < i < 2$, must be unique. 2^3 satisfies this. As a second example, consider

$$M = 2^{53} - 1 = 6,361 \cdot 69,431 \cdot 20,394,401$$

The GCD of the factors $(p_i - 1)$ is 530, hence N must divide 265. This is $5 \cdot 53$ and it is easily shown that no root exists where the 2^{ir} are unique for this value. However, the root of 2 for N equal to 53 does satisfy all conditions.

As was shown earlier, some (real) pseudo Mersenne transforms have quite reasonable sequence lengths. Converting these to a complex ring can give an excellent result. Conditions for the existence of transforms are essentially as above. It can be shown that if an integer ring supports a transform of length N with a root 2^r where r and the modulus M are odd, then the complex ring supports a transform with root 2j and also a transform with root $(1+j)$. If the sequence length N is of the form:

$$N = 2^Q \cdot \eta$$

where n is odd, then the root 2j has a sequence length:

$$
\begin{array}{ll}
4\eta & \text{if } Q = 0 \\
2\eta & \text{if } Q = 1 \\
\dfrac{\eta}{2} & \text{if } Q = 2 \\
\eta & \text{if } Q > 2
\end{array}
$$

The root $(1+j)$ has a sequence length twice that of the root 2j.

For example, the transform based on the modulus $\frac{(2^{27}-1)}{511}$ with a root 2 in a real ring has a sequence length of 27 and satisfies the conditions. Hence it has a root $(1+j)$ in a complex ring with a sequence length of 216. Other moduli with a word size less than 64 bits of interest are $\frac{(2^{25}-1)}{31}$, $\frac{(2^{35}-1)}{37}$, and $\frac{(2^{49}-1)}{127}$, where each has a root 2 and sequence lengths of 25, 35, and 49 respectively.

Multi-dimensional number theoretic transforms

Two major problems have been encountered with number theoretic transforms so far. The sequence length is linearly related to the word size and there is a limited choice of sequence lengths. A way to avoid both of these exists but at a cost of some complexity. It is to employ multi-dimensional techniques.

The principles are simple. Transforms with the cyclic convolution property are linear and so the transform of multi-dimensional data is accomplished by the repeated application of the one-dimensional transform along each data axis. A problem with NTTs is the relatively small value of N for the one-dimensional transform. Mapping data into an n-dimensional structure means that an N-point transform can be applied to N^n data points. The cost is an input and output mapping procedure but the gain is a sequence length which is both long and proportional to a power of the word size.

Let x(n) and h(n) be two N-point data sequences to be cyclically convolved. Thus:

$$y(n) = \sum_{m=0}^{N-1} h(m) \cdot x(n-m) \qquad\qquad n = 0, 1, \cdots, N-1$$

Assume N has two factors M and L and define a mapping:

$$n = r \cdot L + s$$

where $0 \leq r \leq M-1$ and $0 \leq s \leq L-1$. Using this mapping, convert x(n) and h(n) into arrays. That is:

$$\begin{aligned} x(n) &= x(r \cdot L + s) \rightarrow x(r, s) \\ h(n) &= h(r \cdot L + s) \rightarrow h(r, s) \end{aligned}$$

The convolution can now be transformed to reflect this change in the data structures. For that, we need another mapping:

$$m = u \cdot L + v$$

where $0 \leq u \leq M-1$ and $0 \leq v \leq L-1$. Thus the convolution becomes:

$$y(rL + s) = \sum_{v=0}^{L-1} \sum_{u=0}^{M-1} h(uL + v) \cdot x((r-u)L - (s-v))$$

This equation needs to be carefully interpreted. In particular, consider the inner set of convolutions. While the overall convolution is cyclic and so is the outer set, these inner convolutions are not as the indices address terms outside their range. This is to be expected. It was shown in Part 1 that splitting a large convolution results in a set of aperiodic convolutions and that is exactly the situation here.

Extending this algorithm to higher dimensions gives similar results. Operations on one axis are cyclic but on all others aperiodic. As cyclic transforms can only be used to implement cyclic convolutions, mapping this convolution to two (or more) dimensions so that a cyclic transform can be used demands that extended structures be created.

One possibility is to extend the r axis so that there are (2M – 1) terms in all. Thus the h(u,v) array is mapped into a new array H(u,v) where the extension consists of all zeroes and the x(u,v) array is mapped into a new array X(u,v) where the extension consists of the repeated terms needed in the convolution. This is not the only choice. Overlap and save or overlap and add can both be used here.

With these extended matrices, the convolution can be performed by transforming each of the structures, pointwise multiplying them and then inverse transforming the result. This will naturally give another extended array but extracting the relevant part is simple.

This approach offers nothing for FFT algorithms. There is no difference in the number of operations from transforming a one-dimensional convolution but there is the extra burden of the mappings. Its only appeal is when there are severe memory constraints. For NTTs, it is also true there are no fewer operations and there is the burden of the mappings. However, NTTs are severely constrained by the nexus of a linear relationship between word size and sequence length. This approach breaks that nexus. Thus the potential number of sequence lengths for NTTs can be greatly increased but not the word size. Achieving that means a transform becomes available with wide application but with no multiplications at all and a reasonable number of additions. Thus it is practical considerations which make this approach important, not theoretical.

This approach is rather wasteful of memory and that is a problem. Number theory offers a solution but at a cost of some additional constraints on sequence length. Consider again:

$$y(n) = \sum_{m=0}^{N-1} h(m) \cdot x(n-m) \qquad\qquad n = 0, 1, \cdots, N-1$$

The two indices here are interpreted modulo N. Now assume N is $M \cdot L$ and that (M, L) is 1. With these conditions, the Chinese remainder theorem may be applied and it indicates that two separate sets of convolutions, with indices modulo M and L, exist and have a unique solution modulo N. These require no extension and so no additional memory. However, they do require unique factorization of N and a more complex mapping procedure. Although this procedure is of some importance to this Part, the techniques are more appropriate to the next and so it will be covered there.

IMPLEMENTATION ISSUES

Multiple transform techniques

The link between word size and sequence length is a theoretical problem of some consequence but it need not be a practical problem. Assume a very long sequence length is required. Therefore, the word size must also be large, say over 127, even though the input data may be only 8 bit. Now in Part 1, methods of more efficiently utilizing transform techniques by declaring one set of data as the imaginary part of another were examined. This enabled two sets of data to be transformed at once with no substantial increase in the number of operations. A similar approach can be adopted here. If the transform word size is large, define a new input signal:

$$x(n) = x_0(n) + 2^M \cdot x_1(n) + 2^{2M} \cdot x_2(n) + \cdots + 2^{(N-1)M} \cdot x_{N-1}(n)$$

where $x_0(n)$, $x_1(n)$, $x_2(n)$, $\cdots$, $x_{N-1}(n)$ are successive sections of some input signal, or N different input signals. If 2^M exceeds the largest expected value of any of these separate inputs, then transforming $x(n)$ will give an output in which the individual transformed inputs are easily separated. Thus we can transform N data sets with the one long sequence length NTT. It requires no extra operations, just some encoding prior to transformation and some decoding following.

Word sizes in number theoretic transforms

The question of word size is of some importance in number theoretic transforms because of the close relationship between word size and sequence length. However, it is also an important practical parameter.

Let $x(n)$ be some digital signal to be cyclically convolved with a system described by $h(n)$. The result $y(n)$ is given by:

$$y(n) = \sum_{m=0}^{N-1} h(m) \cdot x(n-m) \qquad\qquad n = 0, 1, \cdots, N-1$$

From Schwartz' inequality:

$$| \, y(n) \, |_{Max} \leq N \cdot | \, h(n) \, |_{Max} \cdot | \, x(n) \, |_{Max}$$

Taking the upper bound:

$$| \, y(n) \, |_{Max} = N \cdot | \, h(n) \, |_{Max} \cdot | \, x(n) \, |_{Max}$$

The maximum word size W_o is just $Log_2 | \, y(n) |_{max}$, hence:

$$W_o = Log_2 N + Log_2 (\, | \, x(n) \, |_{Max}) + Log_2 (\, | \, h(n) \, |_{Max}) \qquad \text{bits}$$

Assuming $h(n)$ and $x(n)$ are similar, which is usually the case, then:

$$W_o = Log_2 N + 2 Log_2 (\, | \, x(n) \, |_{Max}) \text{ bits}$$

If the convolution is performed by an NTT with modulus M, then the maximum value of W_o is:

$$W_o = Log_2 M \text{ bits}$$

Let W_s be the maximum word size of the input to the system. Therefore:

$$Log_2 M = Log_2 N + 2 W_s$$

and so:

$$W_s = \frac{1}{2} \cdot Log_2 \left(\frac{M}{N} \right)$$

Hence the input signal is limited to a range $\left(0, \sqrt{\dfrac{M}{N}} \right)$.

Example

Earlier we studied an example of an MNT where:

$$M = 2^7 - 1 = 127$$

The sequence length N is 7 and the two sequences convolved were:

$x(n):$	1	2	3	4	3	2	1
$h(n):$	1	2	3	4	5	6	7

Checking with this formula, the maximum permitted range of the signals is (0,4). This shows we were rather fortunate in obtaining the correct result in that example.

We also considered a problem with a Fermat prime:

$$F_2 = 2^4 + 1$$

The sequence length N is 8. Here, the maximum range of the input signals is calculated as (0, 1). If, therefore, we had attempted to duplicate the first example and convolve, say:

x(n) :	0	1	2	3	4	3	2	1
h(n) :	0	1	2	3	4	5	6	7

we would expect significant errors. Indeed, convolution of these two sequences gives:

y(n) :	67	72	72	64	48	40	40	48

and all of these exceed the modulus.

This formula is rather stringent and can give a misleading view of the problem as the Mersenne example shows. However, it is a very general result not dependent on specific values of inputs. Using other mathematical techniques, it is possible to find better formulae but what these and the above formula show is that it can be necessary to scale the input data which can introduce quantization noise. Consequently, while the NTT itself may not produce finite register effects, the practical application of the transform may well do so. This needs to be carefully considered in any detailed systems design.

Residue number systems

The Chinese remainder theorem suggests a number system well suited to high speed arithmetic. However, while often spoken of with much favor, it has rarely been employed and with advances in monolithic multipliers plus falling costs, may never be.

Consider a set of numbers $\{ 0, 1, 2, \cdots, M-1 \}$. Let M have m relatively prime factors M_i. Then some given number k taken from this set can be represented by the set of residues $< k >_{M_i}$. Describe each by k_i. Thus $\{ k_1, k_2, \cdots, k_m \}$ is consequently a set of digits in this number system and, from the Chinese Remainder theorem, a unique set.

The attraction of this number system is in its arithmetic. Here:

(a) the digits of the sum or difference of two numbers A and B are given by:

$$c_i = < a_i \pm b_i >_{M_i}$$

(b) the digits of the product of two numbers A and B are given by:

$$c_i = < a_i \cdot b_i >_{M_i}$$

The latter is considerably simpler than the usual binary multiplication.

This **residue number system** (RNS) allows operations to be performed in parallel but in stark contrast to conventional binary multiplication, requires no carry bits and that permits great speed. Multiplications can be easily implemented via ROM look-up tables and, if the M_i are primes, very few are needed to get quite large numbers. For example:

$$
\begin{array}{lll}
2\times3\times5\times7\times11\times13\times17 & = \ 510,510 & \text{which is comparable to } 2^{16} \\
2\times3\times5\times7\times11\times13\times17\times19\times23 & = \ 2,230,928,700 & \text{which is comparable to } 2^{32}
\end{array}
$$

Since the digits cannot exceed the size of these primes, the largest table needed for look-up multiplication is only 23×23 or 529 5-bit words.

While these are powerful advantages, there are, unfortunately, some severe disadvantages. There is no division operation; that can only be implemented by converting to conventional binary. Another problem is that it is purely an integer number system. Implementation presents two other problems. First, the conversion of a binary input to RNS is not simple. While it can be done via look-up tables, they must be large tables. Second, an RNS system cannot represent as wide a range of numbers as an equivalent binary number as primes, apart from 2, are not represented efficiently in binary. For example, converting the primes 2, 3, 5, 7, 11, and 13 to binary requires 22 bits and can represent the numbers to 510, 510, but 2^{22} is 4, 194, 304.

In the case of arithmetic for NTTs, there is no need for division or any numbers other than integers. For that reason, hardware for NTTs is one area where RNS has been seriously investigated.

SUMMARY

Number theoretic transforms are transforms defined in finite rings and fields. They employ the set of integers { 0, 1, 2, $\cdots$, M–1 } as elements and modulo addition and modulo multiplication as operations. If M is a prime, then the structure is a Galois field, denoted GF(p). Otherwise, it is a ring. The transforms are N-point, where N is usually a value significantly different from M.

The general number theoretic transform pair is:

$$
X(k) = \left\langle \sum_{n=0}^{N-1} x(n) \cdot \omega^{kn} \right\rangle_M \qquad\qquad k = 0, 1, \cdots, N-1
$$

$$
x(n) = \left\langle N^{-1} \sum_{k=0}^{N-1} X(k) \cdot \omega^{-nk} \right\rangle_M \qquad\qquad n = 0, 1, \cdots, N-1
$$

where $\quad \omega \quad$ = is a root of order N i. e. $\ \langle \omega^n \rangle_M \ \neq 1$ for $0 < n < N$
$$\qquad\qquad\qquad\qquad\qquad\qquad\qquad\quad = 1 \text{ for } n = N$$
$\qquad N^{-1}$ = a number such that $\langle N^{-1} \cdot N \rangle_M = 1$

For the transform to exist in a field GF(p), N must divide $\Phi(p)$ which is $(p-1)$. For a given p, there are $\Phi(p)$ choices for ω and so that many different possible transforms.

For the transform to exist in a ring with composite modulus M, the requirements are:

1. there exists an ω with period N where both ω and N are relatively prime to M;
2. N divides $GCD((p_1-1)\cdot(p_2-1)\cdot \cdots \cdot(p_r-1))$

where $p_1, p_2, \cdots, p_r$ are the prime divisors of M.

The appeal of NTTs is largely for hardware implementation of long-length digital signal processors. They offer systems with simple electronics and no round-off, quantization or overflow problems. In order for this latter condition to hold, the modulus M cannot be any arbitrary value but related to a power of 2. In fact, it can only be either $(2^Q - 1)$, in which case it is termed a Mersenne number transform or MNT, or $(2^Q + 1)$. If Q in this latter case is 2^m, the transform is a Fermat number transform or FNT. Otherwise, it belongs to a class of pseudo Fermat transforms.

If, in addition, ω should also be of the form 2^α where $\alpha > 1$, then the transform pair requires no general multiplications at all. The multiplications in the transform are solely by numbers based on powers of 2 which in a binary system are shifts. That, too, is of great practical appeal and, consequently, virtually all investigations into MNTs and FNTs have been into this group of transforms.

There is no one Mersenne number transform pair even with the rather rigid requirement that the root must be a power of 2. Rather, there are families of transforms. Q can be prime or composite. For Q prime, the Mersenne number resulting can be prime or composite. For these three cases:

(a) If Q is composite and even, it has a maximum transform length of 3. If it is composite and odd, then it will probably have a sequence length of 7 and if not that, then 31, or some other low value. In fact, the longest sequence lengths for Q composite up to 64 are:

$$
\begin{aligned}
Q &= 25 \text{ gives a sequence length of } & 31 \\
Q &= 35 \text{ gives a sequence length of } & 31 \\
Q &= 49 \text{ gives a sequence length of } & 127 \\
Q &= 55 \text{ gives a sequence length of } & 23
\end{aligned}
$$

This group is therefore of no practical interest.

(b) There are only around 30 known primes which generate Mersenne primes and of these, only 2, 3, 5, 7, 13, 17, 19, 31, and 61 are less than 64. This set of nine plus possibly 89, 107, and 127 are the only possible candidates for practical application. The usual MNT of this type uses roots of 2 or -2 and this allows sequence lengths of p or 2p where p is the prime concerned. Note that -2 only has a sequence length of 2p and:

$$
p^{-1} \quad \text{is} \quad (2^p - 1 - 2 \cdot \frac{(2^{p-1} - 1)}{p})
$$

$$
(2p)^{-1} \text{ is} \quad (2^p - 1 - \frac{(2^{p-1} - 1)}{p})
$$

(c) All other primes generate composite Mersenne numbers. There are nine such primes up to 64, namely 11, 23, 29, 37, 41, 43, 47, 53, and 59. These MNTs are identical to the prime MNTs.

Although the last two cases give identical NTTs, the prime MNTs can employ other roots while the composite MNTs cannot. However, none are as simple as 2 or –2.

The advantage of MNTs is the very simple arithmetic for implementation. The disadvantages are the small number of possible transform lengths, the lack of a fast algorithm (although a sequence length of 2p is possible) and the fact the sequence length and word size needed in implementation are closely linked.

A general class of number transforms exists involving the execution of most operations of the transform modulo a number ($\omega^N - 1$) but the final reduction in the transform is modulo $\dfrac{(\omega^N - 1)}{(\omega^{N_o} - 1)}$. Here, N is p^q where p must be odd and q is a positive integer and N_o is p^{q-1}. When ω is 2, –2 or a power of these, the result is a pseudo Mersenne number. Again, several classes of transform exist but two are of most interest. They are where ω is 2 and N is p^2 and where ω is 2^m and N is n.

In both sets of transforms, the sequence length N is known. The problems are keeping N, the working word size, small and not too different from the output word size, which is approximately ($N - p$). At the same time, ($\omega^N - 1$) should be highly composite to allow a fast transform to be developed. For the first case, the only choices up to order 64 are for p equal to 3, 5, and 7, leading to sequence lengths of 9, 25, and 49 respectively. For the second, choices of m and n are (3, 3), (3, 9), (3, 11), and (3, 13), leading to sequence lengths of 3, 9, 11, and 13 respectively. Values beyond 64 are not much better.

Fermat number transforms use a Fermat number as modulus. That is, a number ($2^m + 1$) where m is 2^q. F_0, F_1, F_2, and F_3 are all prime numbers and in fact the only Fermat primes. For these prime values, the sequence length of the NTT must divide 2^m and so is of the form 2^r. As $< 2^m >_M$ is 1, there are a wide choice of potential FNTs based on powers of 2. However, a root of 2 gives the longest sequence length of 2^q. In the case of the composite Fermat moduli, much the same applies. However, here the maximum sequence length is 2^{m+2} for a root of 2.

The advantages of FNTs are that as the sequence length is a power of 2, a fast algorithm exists and it is very similar to the base-2 FFT. There are two disadvantages though. First, the arithmetic needed to execute modulo addition and multiplication modulo a Fermat prime is quite complex. Second, up to a word size of 65, there are only 7 possible transforms.

Two points of interest concerning Fermat transforms are that the equivalent of $\sqrt{2}$ exists and that 3 is a primitive root. It can be shown:

$$\sqrt{2} = 2^{\frac{q}{4}} \cdot (2^{\frac{q}{2}} - 1)$$

which unfortunately is not a simple number but if used as a root, it does double the sequence length. Note that this root introduces general multiplications. Any odd power of this is also a root, so that may be employed to find a simpler value. That 3 is a primitive root is interesting but in general, the resulting arithmetic is too complicated to use.

A class of pseudo Fermat transforms with moduli of the form $\dfrac{(\omega^A + 1)}{(\omega^B + 1)}$ exist and are pseudo Fermat as ω is not necessarily 2 nor is A or B necessarily 2^r. Most operations with these transforms are modulo ($\omega^Q + 1$).

The simplest category of these transforms is when A is even and the modulus as a whole is odd. The transform usually uses a root of 2 and the sequence length is 2A. B is chosen as the lowest possible power of 2. Up to 64, transforms exist for all even numbers except 2, 4, 8, 16, 32, and 64.

A more complex category of transforms exist for A equal to p^r where p is an odd prime and r > 0 and B equal to p^{r-1}. The transform has a length $2p^r$ provided ω is not 1 modulo p. One class of transforms sets r to 1, thus the modulus is $\frac{(2^p+1)}{3}$ and the sequence length is 2p. Choices of p up to 64 are 3, 5, 7, 11, 13, 17, 19, 23, 29, 31, 37, 41, 43, 47, 53, and 59. Another choice is r equal to 2, which gives a sequence length of $2p^2$. Choices here for a word size below 64 bits are 3, 5, and 7.

Complex NTTs are an important class and are based on Gaussian integers. The most useful are complex Mersenne transforms which share many of the same characteristics of real Mersenne transforms. There are two groups. One is based on complex fields, which means the Mersenne moduli must be prime. The other is based on complex rings with composite Mersenne numbers as moduli. In both cases, only roots of 2 or their powers are of interest and only those roots which have a transform in a real Mersenne ring or field. These can generate transforms of size 8N where N is the transform length in the real structure. This is a composite number, hence fast algorithms can be devised for these NTTs. As well, being complex, two sets of data can be transformed simultaneously.

If a root 2 of order N exists in a real Mersenne field, then in the complex field a root 2j of order 4N and a root (1 + j) of order 8N exist. A similar result applies to complex rings. However, now the sequence length must divide the GCD (which has terms ($p_i - 1$) where the p_i are the prime factors of the modulus) and it must be odd, the Mersenne number must be generated by an odd number and the power of the root must be odd. Under these conditions, the only complex rings of practical interest are those where the Mersenne number is generated by a prime.

Apart from NTTs, the Part has examined some other important topics. Two are simple results of number theory. They are Rader's theorem and the prime factors algorithm. Rader's theorem shows that for a prime length DFT, there exists a permutation of the indices which can convert the transform to a convolution. Therefore, the two are synonymous and any fast method for computing convolutions may also be employed for these DFTs. The theorem suggests a new form of decomposition for long-length transforms but it is complex. The prime factors algorithm shows that if a transform length for a DFT can be expressed as the product of relatively prime factors, then a mapping can be developed to express the transform as the combination of short-length transforms but without any twiddling. The mapping becomes two-stage rather than three and eliminates a very significant number of multiplications. The indices of a DFT may be regarded as a Diophantine equation and both these approaches are very simple exploitations of some of the properties of such equations.

Although several different NTTs were examined, a problem with most of them was a strict requirement between word size of the hardware implementing the transform and the sequence length. That presents some practical difficulties. Consequently, a most important topic examined was multi-dimensional transform techniques. All the approaches examined have application to any transform technique but, since they do nothing to reduce the number of operations and introduce the added burden of the multi-dimensional mapping, are of limited interest. In the case of NTTs though, they allow the word size/sequence length nexus to be broken and so are of great importance.

One approach examined was very simple but quite general. If N is the sequence length of some NTT, it allows an NTT to be constructed of size N^n. While there is some increase in the number of operations due to the one to multi-dimensional mapping, the major practical problem is the increase in memory requirements. Each dimensional axis (except one) must be expanded so that convolutions are all cyclic and, if n is large, that can mean considerable extra storage. The second approach overcomes this but it can only treat situations where the sequence length is $N_1 \cdot N_2 \cdot \cdots \cdot N_n$ and each of the N_i are relatively prime to one another. That too, is quite restrictive for practical applications.

DISCUSSION

Reading this Part, it is not at all difficult to see why NTTs have often been pushed to one side in discussions of techniques for digital convolution. There is a range of NTTs but each seems to be suitable for only a small group of sequence lengths. Further, there seems little to link these different methods and that creates an impression of a 'grab bag' of recipes to be used as required. In contrast, the last Part has the FFT computational section as a guiding approach and with its comforting solidity we were able, with very little effort, to construct any algorithm desired.

In all of this, there is an element of truth but much is just a matter of perspective. Granted an FFT can be constructed for any composite N (and from this chapter, for any prime value as well) and the result is well structured but the resulting algorithms can be difficult to implement. It only requires a limited exposure to digital electronics or to the problems of programming a digital signal processing chip to show these implementations can be quite significant undertakings. With the necessary compromises that must be made in the practical situation, then, the attraction of these FFT techniques can be more aesthetic than real. Note, too, that if the hardware for executing the transform is limited to a 64-bit word size, then there are only 64 possible base-2 FFTs, or 16 base-4 FFTs.

Our purpose, though, is not to denigrate FFT techniques but to put NTTs into perspective with them. We must begin by stressing that the nature of NTTs tends to orient them strongly to hardware rather than software implementations, and that of necessity makes them a specialized subject. Additionally, even though a range of techniques was examined, that should not be construed as a limitation but more realistically as an example of the flexibility of NTTs.

FFT algorithms are based on prime length DFTs and a combination procedure. That combination procedure can be the general but basic approaches discussed in Part 2, or the more restrictive but efficient prime factors approach discussed here. While a wide range of FFTs can be considered, practical reasons limit interest to base-2 or base-4.

Now consider NTTs. Here also, two combination procedures could be employed. Again, one was general and one was restricted to cases where there are relatively prime factors. Unlike the FFT algorithms though, no efficiency gains are made by using these procedures. However, neither do they introduce any multiplications. These combination procedures, when combined with the short-length NTTs discussed, enable an algorithm for almost any value of N to be created. In fact, given the range of techniques introduced here, we may omit 'almost' and state not only that an algorithm can be created for any N but that in many cases there are several possible algorithms.

Now let us look at the short-length NTTs more closely. In many respects, these correspond to the prime-length DFTs employed in FFT algorithms. The difference is that these cover both prime and composite sequence lengths. A quick tally of the algorithms discussed will show that there are slightly more than 64 based on a root of 2. Because these are based on 2, for all but a very small number, there are no multiplications needed, only shifts. We should stress that these are the algorithms based on 2. If other roots are chosen, other algorithms result but they do involve multiplications.

The set of algorithms based on one of the combination procedures and these short-length NTTs based on a root of 2 involve no, or possibly a few, general multiplications. They just involve shifts and additions. Consequently, overall they can be much more efficient than equivalent FFTs. Further, from the practical point of view, they can be much more cost effective. However, it is important to note the number of additions can be very large.

We have, of course, attempted to put NTTs in a rosy light but even so, the evidence shows they compare well against FFT algorithms. The one exception is in implementation. For FFTs, a general architecture can be employed for a very wide range of algorithms. For NTTs, the need to implement modulo arithmetic means a given sequence length must have a specific word size and so a tailored architecture. Given the hardware orientation of NTTs, that cannot be considered a serious difficulty. The electronic realization of NTTs is comparatively simple and there is ample scope for implementations ranging from very low speed to ultra high speed and to single chips.

Having decided NTTs do have value, let us look a little more critically at them. First, there is one small implementation problem. Engineering problems begin with a value for the sequence length and then raise the question of the most suitable transform for that value. If this Part is reviewed, it can be seen NTTs are not well oriented to such an approach and certainly are not as well oriented as FFTs. That can be partially overcome by tables but it is still discomforting. However, to aid in that decision making, we make another important observation of NTTs as a whole. Again in reviewing the Part, it will be seen that, of the possible NTTs to choose, the ones most likely to be of value are the Fermat number transforms and the complex. The latter in particular seem very well suited to many applications and offer significant practical advantages.

A pertinent question to ask to lead into Part 4 is what strengths have been uncovered in this Part and what seems to warrant further and more detailed theoretical examination. For the strengths, we have found that not all multiplications are the same. A range of algorithms have been studied where the multiplications are by powers of 2 and in digital systems, that means the multiplication can be performed as a shift, an easier operation. In terms of areas worthy of greater study, the obvious candidate is more flexible NTTs. However, as mentioned above, the range discovered is enough to ensure an NTT can be created for any N via a combination procedure and the short-length NTTs employed are very efficient. Rather, it is the combination procedure that is more important to study. In particular, are better or more flexible procedures possible and in what way? Another issue is the number of additions as a quick review over the Part will show the number of these in the algorithms is quite high and means to minimize them are of practical concern.

The material of this Part has been developed relatively recently. Most has only appeared in the literature since the early 1970s. Even so, it reflects a historical concern that is increasingly being overcome. During the 1970s, the concerns were to minimize multiplications as they were difficult and expensive to implement and to minimize the total number of circuits employed. Thus the interest in NTTs based on roots of 2. The current capabilities of VLSI technology permit a change in this attitude. While it certainly will not lead to more efficient realizations, it is now worthwhile considering NTTs based on roots other than 2. The attraction is simply the much longer sequence lengths possible. Why bother with NTTs with general multiplications though, when an FFT could be employed? Simply that the NTT remains a real arithmetic process and has the advantage of no errors. It is much more oriented to VLSI implementations than any FFT and so should be the first choice in this case. Equally of course, the FFT should be the first choice for any general-purpose programmable system.

FURTHER STUDY

References

One book has stood out for some decades as an outstanding introduction to number theory. It is the classic:

G.H. HARDY and E.M. WRIGHT, *An Introduction to the Theory of Numbers*, 4th edition, Oxford University Press, 1975

Although a very detailed book, it is also highly readable and requires little mathematical background. A useful adjunct to Hardy and Wright is:

M.R. SCHROEDER, *Number Theory in Science and Communications*, Springer-Verlag, 2nd edition, 1987

It is not nearly as comprehensive as Hardy and Wright but is much more informally written and devotes much effort to applications of number theory in fields such as cryptography and coding.

One of the first books on this subject is:

J.H. McCLELLAN and C.M. RADER, *Number Theory in Digital Signal Processing*, Prentice Hall, 1979

It begins with a highly readable and very tutorial section on number theory as it pertains to digital signal processing and then concludes with reprints of a number of key papers published in the field. The papers, of course, were addressed to an expert audience and so can be difficult to read, but the introductory section is a worthy reference for their study.

A more recent book covering both the topics discussed here plus a range of other transforms is:

D.F. ELLIOT and K.R. RAO, *Fast Transforms; Algorithms, Analyses, Applications*, Academic Press, 1982

It also discusses some aspects of transform theory with respect to FIR signal processors. Another book that warrants close attention is:

H.J. NUSSBAUMER, *Fast Fourier Transform and Convolution Algorithms*, Springer-Verlag, 1981

Essentially a research monograph, this is a tightly written but very comprehensive work. It is aimed at an expert audience but a more general reader can still gain by studying it.

For those interested in residue arithmetic, an early but comprehensive book to read is:

N.S. Szabo and R.I. Tanaka, *Residue Arithmetic and its Applications to Computer Technology*, McGraw-Hill, 1967

Subsequent work is well covered by the recent compendium of papers:

M.A. Soderstrand, W.A. Jenkins, G.A., Jullien, and F.J. Taylor (eds), *Residue Number System Arithmetic : Modern Applications in Digital Signal Processing*, I.E.E.E. Press, 1986

The outstanding journal in this field is the *I.E.E.E. Transactions on Acoustics, Speech and Signal Processing*. Numerous landmark papers have appeared in this journal and it remains the favored publication for much research in this field. Many are referenced in the two books quoted above and will not be re-listed here. Two of particular relevance to this Part that are worth quoting are:

E. Dubois and A.N. Venestanopoulos, The Generalized Discrete Fourier Transform in Rings of Algebraic Integers, *I.E.E.E. Transactions on Acoustics, Speech and Signal Processing*, V 28 N 2, April 1980, pp 169–75

J.B. Martens, Number Theoretic Transforms for the Calculation of Convolutions, *I.E.E.E. Transactions on Acoustics, Speech and Signal Processing*, V 31 N 4, August 1983, pp 969–78

The first of these has relevance to one or two of the Problems and develops some important results for complex NTTs. The second summarizes previous work and shows the directions of current research. A final pair of papers to note is the following:

R.L. Nevin, Application of the Rader-Brenner FFT Algorithm to Number Theoretic Transforms, *I.E.E.E. Transactions on Acoustics, Speech and Signal Processing*, V 25 N 1, April 1977, pp 196–8

B. Arembepola and P.J.W. Rayner, Discrete Transforms over Residue Class Polynomial Rings with Applications in Computing Multi-dimensional Convolutions, *I.E.E.E. Transactions on Acoustics, Speech and Signal Processing*, V 28 N 4, August 1980, pp 407–14

Exercises

1. What Mersenne and Fermat numbers have a primitive root?

2. Determine $< 15^{3132} >_7$.

3. Show that if a and b are odd, then $(a^2 - b^2)$ is divisible by 8.

4. Prove that n^5 and n have the same last digit for any n.

5. A means of shuffling cards is to cut the pack in half and then form a new pack by taking a card from the upper half, then the lower, then the upper, and so on. Will this procedure restore the pack to its original configuration in a finite number of shuffles?

6. Find the first twenty Gaussian and Eisenstein primes.

7. Show that for any Mersenne prime, possible sequence lengths are 2, 3, p, 2p, and 3p.

8. A company places an order for 10,000 integrated circuits worth 21 cents each and 5,000 worth 36 cents each with The International Dependable Electronics Supply Company (No Liability). Three weeks later, they get a single box full of devices, together with a note saying: " Sorry. Insufficient in stock but you can have all we have got" and a bill for $2,406.30. How many of each chip were supplied?

9. Show that for p a prime:

$$< (a+b)^p >_p = < a^p + b^p >_p$$

10. Prove that if g is a primitive root of p^k where p is an odd prime, then it is also a root of $2p^k$.

11. Prove the totient is always even for any N greater than 2.

12. Julia just loves Choc-o-Fruity Bombs. These, according to their manufacturer are a delicious dark chocolate center flavored with the essence of either strawberry, apple, orange, passionfruit, cherry, pineapple, and 'the mystery of the tropics', and coated in pink, green, orange, purple, red, yellow, and black toffee. Each packet, the manufacturer assures us, has numerous examples of each. When Julia gets a pack, she carefully groups them so she can eat a complete set at once. However, in the latest pack, she finds there are six pink, one green, three orange, two purple, and five yellow bombs left over. How many were there in the original packet?

13. Why cannot a Mersenne number transform exist for a root of 2 and a sequence length of 3p?

14. Why cannot a pseudo Fermat number transform exist of order 36?

15. Show every prime except 2 and 3 is of the form (6k + 1) or (6k – 1).

16. Derive fast Fermat number transforms corresponding to the decimation in time and decimation in frequency FFT algorithms.

17. Examine the roots of the first 20 Mersenne numbers. Is there a simple way to determine these? Is there any order to the roots with maximum sequence lengths?

18. A root of $\sqrt{2}$ is usually found to be a most inconvenient number in most mathematical structures. Can a power be used? If so, prove which and determine the most practical values for five structures.

19. Find all possible roots of F_5 and M_{31}. From these, determine all possible NTTs. In many

modern digital signal processing chips, shifters, and multipliers have the same execution time. Consequently, there is no substantial practical penalty for using either. On that basis, do your results suggest there may be a practical advantage in choosing a root other than 2 for the Mersenne number transforms? Further, do they suggest this might be a better choice than the Fermat transform?

20. Use Rader's theorem and the decomposition suggested by it to develop a 49-point FFT algorithm.

21. Develop the DIF form of the PFA.

22. Develop an NTT for a 31-point FFT. Do you consider this would be practical for a VLSI realization?

23. In a complex NTT, the root $< (\pm 2^\alpha \pm (-3)^\alpha)^r >_Q$ gives a sequence length of 2^{p+1}. Why must r be odd? For p equal to 31, list all values of this root and its powers with this sequence length and determine which is the most practically useful. On what basis are you making that decision?

24. Is a base-4 or base-8 FNT possible? If so, is there any advantage in using them over the base-2?

25. How would you determine $\sqrt{3}$ in some integer field or ring? Derive this for a representative set.

26. Prove there can be no more than three Friday the thirteenths in a year.

27. As mentioned in the text, if some number N satisfies $< a^N >_N = 1$, then it does not imply N is prime. However, if for a randomly chosen set of values $\{ a_1, a_2, a_3, \cdots, a_k \}$ this is true then there is a high probability N is prime. What should k be for that probability to exceed 0.99?

PROBLEMS

1. The first five problems have a common background. We firstly introduce that, then list the problems.

   The next Part introduces cyclotomic polynomials. It is not important at this stage to know their details except that they are factors of the polynomial $(z^N - 1)$. If the substitution 2^q is made for z, then this polynomial becomes $(2^{Nq} - 1)$, a Mersenne number and so the cyclotomic polynomials become the factors of such a number. Thus:

$$2^{2q} - 1 = (2^q - 1) \cdot (2^q + 1)$$
$$2^{3q} - 1 = (2^q - 1) \cdot (2^{2q} + 2^q + 1)$$
$$2^{4q} - 1 = (2^q - 1) \cdot (2^q + 1) \cdot (2^{2q} + 1)$$
$$2^{5q} - 1 = (2^q - 1) \cdot (2^{4q} + 2^{3q} + 2^{2q} + 2^q + 1)$$

$$2^{6q} - 1 = (2^q - 1) \cdot (2^q + 1) \cdot (2^{2q} - 2^q + 1) \cdot (2^{2q} + 2^q + 1)$$
$$2^{7q} - 1 = (2^q - 1) \cdot (2^{6q} + 2^{5q} + 2^{4q} + 2^{3q} + 2^{2q} + 2^q + 1)$$
$$2^{8q} - 1 = (2^q - 1) \cdot (2^q + 1) \cdot (2^{2q} + 1) \cdot (2^{4q} + 1)$$

and so on. For a complex field based on a Mersenne prime ($2^p - 1$), the number of elements is M_p^2. The totient in this case is $2^{p+1} \cdot (2^{p-1} - 1)$. This generates some slightly different forms to the above.

Our examination of pseudo transforms was rather limited. Using these results, we can develop quite a range of interesting transforms and that is our object here. The question here, then, is concerned with a modulus M given by:

$$M = (2^a + 1) \cdot (2^b + 1)$$

(a) Does this modulus have a root of 2 and if so, what is its order? What conditions apply to a and b for this root to exist?
(b) List suitable moduli for implementation on 32-bit hardware.

2. Following from problem 1, determine conditions for the existence of a number theoretic transform for:

$$M = \frac{(2^a + 1) \cdot (2^b + 1)}{(2^c + 1) \cdot (2^d + 1)}$$

List moduli suitable for hardware implementation up to 32 bits and comment on their practicality.

3. Determine conditions for the existence of number theoretic transforms with the modulus:

$$M = \frac{(2^{2a} + 2^a + 1)}{(2^{2b} + 2^b + 1)}$$

List moduli suitable for hardware implementation up to 32 bits and comment on their practicality.

4. Repeat problem 3 for the modulus:

$$M = \frac{(2^{2a} - 2^a + 1)}{(2^{2b} - 2^b + 1)}$$

Consider problem 2. Would it be an advantage using such a modulus in an Eisenstein field or ring?

5. Given the form of the factors of ($2^{Nq} - 1$), what conditions apply for 2 to have a root of order Nq modulo these factors?

6. How do you solve $< a \cdot x >_N = 0$ for N composite? Give an example.

7. Under what conditions do NTTs exist in:

   (a) the ring of Eisenstein integers?
   (b) the field of Eisenstein integers?

8. If g is a primitive root modulo N, then by definition, $< g^{\Phi(N)} >_N$ is 1. Assume a root of order r exists. Then again by definition, r divides the totient and so a second root given by $\frac{\Phi(N)}{r}$ must exist. Consider some number $p_i^{e_i}$ where p_i is some odd prime and the exponent is greater than zero. If g is a primitive root of this number, show how the Chinese remainder theorem can be used to create a root for:

$$N = p_i^{e_i} \cdot P_{2^{e_2}} \cdot \cdots \cdot P_r^{e_r}$$

9. The attraction of 2 as a root in an NTT is that terms like $< x(n) \cdot \omega^{kn} >_M$ can be implemented via a set of shifts upon $x(n)$ rather than a general multiplication. In a more general case though, it would be possible to implement the multiplications via a look-up table. Given current commercially available read-only memories, what is the largest sized NTT that could be handled in this fashion and how?

10. As mentioned in the Part, due to the large word size required for long-length NTTs, it is possible to consider transforming several signals at once by multiplying them by a suitable power of 2 and adding them. How does this compare in terms of operations and practical convenience to a complex NTT?

   For a VLSI realization, there is the distinct problem of pin count. Would there be any advantage in feeding back a signal from the output of the reverse transform unit in a digital signal processor repeatedly to overcome this and other problems? Discuss.

PART

Digital Convolution in Finite Polynomial Structures

INTRODUCTION

Polynomials are one of the most familiar mathematical concepts. Once a mastery of arithmetic has been achieved, the next step in a mathematical education is to introduce the broad concepts of abstraction. For generations, that progression has been founded on a study of polynomials. Polynomial theory, though, is not confined to elementary algebra. Number theory is one of numerous branches of mathematics that employs polynomials and, because of that, it was inevitable they would be examined for application to digital signal processing. Polynomial theory has proved very useful indeed, allowing both a uniform framework for tackling the general problem of digital convolution and a much deeper insight into this problem.

A question to address immediately is: what constitutes a polynomial? That seems obvious. The reader was undoubtedly instructed at an early age that expressions like:

$$P(z) = a_0 \cdot z^n + a_1 \cdot z^{n-1} + \cdots + a_{n-1} \cdot z + a_n$$

are polynomials. Further, that z represents some unknown number and so, in effect, the polynomial defines a formula for calculating a value. However, this is by no means the only interpretation. Assume that $P(z)$ is the z-transform of some sequence. Then here, each of the numbers $\{ a_0, a_1, \cdots, a_n \}$ is an element of a sequence, and so the different powers of z merely serve to identify position within the sequence. Multiplying by z^n where n is any positive or negative integer is just a shift operation, thus z can be viewed as a shift operator.

Assume the set of coefficients $\{ a_0, a_1, \cdots, a_n \}$ of a polynomial $P(z)$ are drawn from some finite ring or field. We can interpret z as being an 'unknown' in the usual algebraic sense of representing any of the elements of the same structure. Since finite rings and fields are closed structures, then the repeated operation z^i simply represents the repeated application of one of the structure operators upon a single element that will, as a result, generate one of the other elements. In this perspective, the polynomial fits the traditional algebraic concept but at the same time is quite comfortable with the idea of multiplication by z acting in a similar fashion to an operator.

If the coefficients are drawn from a finite ring or field, if z^i is interpreted as repeated application of an operator $\cdot$, if $\cdot$ and $+$ are interpreted as the ring or field operators and if z represents any element of the structure, then $P(z)$ also represents an element of the structure. A polynomial therefore represents a transformation or mapping acting upon a ring or field and with respect to the mappings discussed in Part 1, these are just mapping in terms of the coefficients or **fixed elements** of the structure. Such studies are generally termed the study of **polynomials defined on a ring or field**.

Studying polynomials defined on a ring or field is quite different from studying the set of coefficients $\{ a_0, a_1, \cdots, a_n \}$ drawn from families of polynomials. This, too, is a subject of obvious interest to digital signal processing. Now each of these sets of coefficients forms an element, an n-tuple, and operations can be defined to act upon these. Hence there is the possibility of forming **polynomial rings** and **fields**. By the same token of course, any mathematical structure formed from tuples can be interpreted as representing polynomials.

For these structures, the unknown element z reverts to being just a place operator. It makes no difference to these structures what z represents. It can be elements from the same field as the coefficients, or it can be elements from some entirely different structure. It can even be polynomials or more complex structures in some circumstances. It can also represent nothing at all. All of these are quite reasonable interpretations as interest is in the coefficients not in what is being manipulated.

This part will be examining both of these polynomial concepts. In particular, it will examine operations within finite polynomial rings and fields with a view to determining efficient means of executing those operations on those finite structures.

In TOOLS, we briefly look at polynomial congruences and related matters; briefly, because it is really just a case of extending the results of Part 3. We also consider polynomial rings and fields and formally define what we mean by these. That leads in to the polynomial Chinese remainder theorem. We also examine cyclotomic polynomials, a topic that proves of some importance in DEVELOPMENTS. To complete TOOLS, we will make a brief and very restricted examination of tensors.

In THEORY, three topics are studied beginning with Winograd's minimal complexity algorithm; a most important result that has influenced most developments in this area. This leads into a study of minimal multiplication algorithms and then polynomial transforms.

In DEVELOPMENTS, we shall closely investigate the application of Winograd's theorem and of the Chinese remainder theorem. The former offers powerful means of constructing very efficient short-length convolutions. That leads into very efficient means of computing prime length DFTs and some more general DFTs. The Chinese remainder theorem is an integral part of this work as it is a simple yet very effective tool for combining polynomial expressions. It suffers from some shortcomings as the size of the polynomials increase, thus we shall examine some alternatives for combining short-length convolutions into long-length.

In the course of DEVELOPMENTS, we shall apparently detour on occasions to study some DFT algorithms. However, it is only apparent. What these sections will show is the strong connection between convolution and DFT techniques, and how fruitful techniques developed in one area can be used to great advantage in the other.

There are no implementation issues to discuss in this Part. The only relevant issues have already been covered within the other Parts.

TOOLS

Polynomial congruences

An **nth degree polynomial** $P(z)$ over some field F is:

$$P(z) = \sum_{i=0}^{n} a_i \cdot z^i \qquad\qquad n > 0$$

where the fixed elements $\{ a_i \}$ are drawn from F. In much of this Part, F will be GF(p). Some of the elements can be zero but a_n cannot. If a_n is unity, then the polynomial is termed **monic**. The degree of the polynomial will be denoted Deg[$P(z)$]. Another term for degree is the **order** of the polynomial. We specifically avoid any conditions on z and simply leave it as 'unknown'. Where it needs to be specified, then for the most part it will be required to represent an element of F.

Assume two polynomials P(z) and Q(z) meet this definition. If there exists a third polynomial D(z) such that:

$$P(z) = Q(z) \cdot D(z)$$

then D(z) is termed a **divisor** of P(z). This is denoted D(z) | P(z). If P(z) can only be divided by polynomials of degree 0 (that is, numbers) or polynomials of degree n, then P(z) is termed **irreducible** over the field F. Sometimes they are referred to as **prime** for obvious reasons. Consequently, in general terms any polynomial can be described by:

$$P(z) = K \prod_{i=1}^{m} (p_i(z))^{e_i}$$

where   K   = a constant
          $p_i(z)$ = irreducible monic polynomials

$$Deg[P(z)] = \sum_{i=1}^{m} e_i \cdot Deg[p_i(z)]$$

The fundamental theorem of algebra states an nth degree polynomial has no more than n roots, thus $m \leq n$ in this equation.

The structure of the monic polynomials $p_i(z)$ is strongly dependent on the field F. To quote a classic example, $(z^2 + 1)$ is irreducible in the rational field, the field of real numbers and the integer field. However, it has factors $(z + j)$ and $(z - j)$ in the complex field. Further, in GF(2) which only has the two elements { 0, 1 }, the factors are $(z + 1)$ and $(z + 1)$. The nature of the field, then, figures rather largely in the study of polynomials and needs to be very carefully examined.

The **greatest common divisor** (GCD) of two polynomials is just the largest monic polynomial that divides both of them. Thus:

$$(P(z), Q(z)) = G(z)$$

If G(z) is 1 then the two polynomials have no common factors and are **mutually prime** or **coprime**. Two useful results adapted from number theory are that:

$$(x^n - 1, x^m - 1) = x^{(n, m)} - 1$$

and for some arbitrary polynomial R(z):

$$(P(z), Q(z)) = (P(z), Q(z) - R(z) \cdot P(z))$$

Following from the work in number theory and these results, any two polynomials can be related via:

$$P(z) = Q(z) \cdot D(z) + R(z)$$

where Deg[R(z)] < Deg[P(z)]. It can be shown this representation is unique. If the remainder polynomial $R(z)$ is zero, then $D(z)$ is a divisor. It naturally follows that we can denote this equation as:

$$< P(z) >_{D(z)} = R(z)$$

or more traditionally, that $R(z)$ is $P(z)$ modulo $D(z)$. Then two polynomials are **congruent modulo D(z)** if:

$$< A(z) >_{D(z)} = < B(z) >_{D(z)}$$

From the definition of residue polynomials, polynomial residue reduction is not influenced by scalar multiplications of $D(z)$. Consequently, it is usual to take $D(z)$ as being monic.

Clearly, the process of obtaining residue polynomials is more complex than obtaining residues modulo a number. However, it is far from being a formidable problem. Indeed, only three cases need to be considered:

1. By conventional long division, we can easily show that if $D(z)$ is a first order polynomial of the form $(z - b)$, then:

$$< P(z) >_{D(z)} = P(b)$$

2. By similar means, $P(z)$ can be reduced if:

$$D(z) = z^n + \sum_{i=0}^{n-1} d_i \cdot z^i$$

by replacing z^n and its powers everywhere in $P(z)$ with $\left(-\sum_{i=0}^{n-1} d_i \cdot z^i \right)$

3. From the definition of polynomial residues:

$$< F(z) >_{D(z)} = << F(z) >_{T(z)} >_{D(z)}$$

if $T(z)$ is $D(z) \cdot S(z)$ where $S(z)$ is any polynomial. This can be used to great effect if $T(z)$ can be made simpler than $D(z)$.

Example

Consider:

$$< z^8 - z^7 + z^5 - z^4 + z^3 - z + 1 >_{D(z)}$$

where $D(z)$ is $z^2 + z + 1$. We reduce by substituting $-(z + 1)$ for z^2. This can be done by creating a list such as:

$$z^2 = -z - 1$$
$$z^3 = z \cdot z^2 = z \cdot (-z - 1) = -z^2 - z = -(-z - 1) - z = 1$$
$$z^4 = z \cdot z^3 = z$$
$$z^5 = z \cdot z^4 = z \cdot z = -z - 1$$
$$z^7 = z^2 \cdot z^5 = (-z - 1) \cdot (-z - 1) = z^2 + 2 \cdot z + 1 = z$$
$$z^8 = z \cdot z^7 = z \cdot z = -z - 1$$

Substituting and collecting terms, then:

$$< z^8 - z^7 + z^5 - z^4 + z^3 - z + 1 >_{D(z)} = -5z$$

To verify, we note that:

$$(z^2 + z + 1) \cdot (z - 1) = z^3 - 1$$

and:

$$<< z^8 - z^7 + z^5 - z^4 + z^3 - z + 1 >_{z^3 -1} >_{D(z)} = < z^2 - z + z^2 - z + 1 - z + 1 >_{D(z)}$$
$$= < 2z^2 - 3z + 2 >_{z^2 +z+1}$$
$$= -5z$$

Consider the reduction:

$$< P(z) >_{z^N -1} = < a_m \cdot z^m + a_{m-1} \cdot z^{m-1} + \cdots + a_1 \cdot z + a_0 >_{z^N -1}$$
$$= A_{N-1} \cdot z^{N-1} + A_{N-2} \cdot z^{N-2} + \cdots + A_1 \cdot z + A_0$$

From the second reduction rule, any term in an mth order polynomial of the form $a_i \cdot z^i$ where $i > N$, reduces to $a_i \cdot z^j$ where j is $< i >_N$. Therefore, a coefficient A_i of the resulting reduction is given by:

$$A_i = a_i + a_{i+N} + a_{i+2N} + \cdots + a_{i+kN}$$

where $kN \le m$. Thus a periodicity has been introduced into the reduction.

It was shown in Part 1 that convolution can be expressed as the product of two polynomials. Consider the general polynomial congruence:

$$Y(z) = < H(z) \cdot X(z) >_{P(z)}$$

where $H(z)$ and $X(z)$ are two polynomials defined over the same field as $Y(z)$. If:

$$Deg[P(z)] > Deg[H(z)] + Deg[X(z)]$$

then this describes aperiodic convolution. If, though, all three polynomials have a degree of N and $P(z)$ is the polynomial $(z^N - 1)$, then it describes cyclic convolution.

The equations of the DFT can be expressed as the two congruence equations:

$$X(z) = \left\langle \sum_{n=0}^{N-1} x(n) \cdot z^n \right\rangle_{P(z)} \qquad \text{where } P(z) = z^N - 1$$

$$X(k) = \langle X(z) \rangle_{Q(z)} \qquad \text{where } Q(z) = z - W^k$$

The reduction in the first equation is redundant but useful to include for the reason that we may now express cyclic convolution by the two equations:

$$\begin{aligned}
Y(k) &= H(k) \cdot X(k) \\
&= \langle\langle H(z) \cdot X(z) \rangle_{P(z)} \rangle_{Q(z)}
\end{aligned}$$

$$y(n) = \left\langle N^{-1} \sum_{k=0}^{N-1} Y(k) \cdot z^n \right\rangle_{Q^*(z)} \qquad \text{where } Q^*(z) = z + W^n$$

Although this describes the DFT and so a transform within the infinite complex field, it is not difficult to consider other fields by replacing W with some root ω of order N. Thus polynomial congruences and transform techniques as a whole can be regarded as simply different perspectives of the one problem.

Polynomial rings and fields

Polynomial rings and fields were introduced in Part 1. Here we shall investigate them in more detail. A most important result foreshadowed in Part 3 is:

THEOREM   Every finite field has p^k elements where p is a prime and $k \geq 1$. Such fields are termed Galois and denoted $GF(p^k)$.

The proof of this theorem is beyond the scope of this book. It is given in most advanced books on linear algebra. What is important to note about the theorem is that it is simply stating a necessary condition which applies to the elements of the field. That is to say, a structure without p^k elements cannot possibly be a field. A structure with p^k elements, though, can only be a field if the operators acting upon those p^k elements satisfy the field axioms.

We have already examined one finite field, namely $GF(p)$. Here, the elements are numbers and the operators are addition and multiplication modulo p. Can we construct a field $GF(p^k)$ on the same basis? Some of the numbers of this field are of the form p^i where $i \leq k$ and we note that:

$$\langle p^i \cdot x \rangle_{p^k} = 1$$

has no solution. Thus the field conditions are not satisfied. The problem, though, is not with the operations as it can be shown the only finite field with numbers as elements is $GF(p)$. Consequently, we must consider more complex elements for $GF(p^k)$.

Before proceeding, it is important to note the following:

THEOREM   Any field $GF(p^k)$ is isomorphic to any other field $GF(p^k)$

While several mathematical structures can be considered as field elements, here we will consider just two — polynomials and matrices. For convenience, polynomials are usually described by tuples. Tuples can represent more than polynomials but it is generally true that fields based on these elements use the same operations as polynomial fields.

Consider the problem of constructing a polynomial/tuple field. First, p^k elements need to be selected. The set of polynomials:

$$a_{k-1} \cdot z^{k-1} + a_{k-2} \cdot z^{k-2} + \cdots + a_1 \cdot z + a_0$$

where the coefficients are drawn from $GF(p)$ has p^k members and satisfies this. Next, two operations are needed. Let $A(z)$ and $B(z)$ be two $(k-1)$ degree polynomials with coefficients given by the k-tuples $(a_{k-1}, a_{k-2}, \cdots, a_0)$ and $(b_{k-1}, b_{k-2}, \cdots, b_0)$ respectively. Then define operations as follows:

Addition
A new element of the field $C(z)$ is formed via:

$$C(z) = < A(z) + B(z) >_{P(z)}$$

where the coefficients c_i of $C(z)$ are determined from:

$$c_i = < a_i + b_i >_p$$

Multiplication
This is a little more complex. For the field axioms to be satisfied, the product of two elements must be another element, in this case a polynomial of order $(k-1)$. That can only be achieved via some polynomial modulo operation and for that we need a modulus. For the moment, let the modulus be described as $P(z)$. Then derive:

$$T(z) = < A(z) \cdot B(z) >_{P(z)}$$

where the coefficients of the product are derived from the coefficients t_i of $T(z)$ according to:

$$c_i = < t_i >_p$$

There is no difficulty proving these operations do indeed satisfy all the field conditions.

These operations are quite logical as the condition on the elements of the polynomial/tuples is that they are drawn from $GF(p)$. In turn, this suggests the structure can only be a field if $P(z)$ is the equivalent of a prime, that is, if it is an irreducible polynomial. This is indeed the case but its proof is beyond the scope of this book. Then consider a set of p^k elements in the form of polynomials/tuples of degree $(k-1)$. We shall take a **polynomial field** to be a structure with this set of elements and with operations of addition as given above and multiplication modulo an irreducible polynomial of degree k defined on the field GF(p). Similarly, a **polynomial ring** will be taken to mean this set of elements and the operations of addition as above and multiplication modulo a reducible polynomial of degree k defined on the field GF(p).

There is, of course, no difference in the basic mathematical concepts governing polynomial structures and any other. So, a polynomial field has properties such that, if:

$$< A(z) \cdot B(z) >_{P(z)} = 0$$

then it means either of $A(z)$ or $B(z)$ must be zero modulo $P(z)$. In addition, every element has an inverse. Equally, a ring is distinguished by not necessarily having these properties.

Example

Of all the polynomial rings and fields that can be considered, one stands out as an example of some of these concepts. It is the field GF(p^2) with operations modulo a polynomial $(z^2 + \alpha)$. Consider two elements expressed in polynomial form:

$$(a_1 z + a_0) \qquad \text{and} \qquad (b_1 z + b_0)$$

Addition is simply:

$$C(z) = (< a_1 + b_1 >_p \cdot z + < a_0 + b_0 >_p)$$

and multiplication is:

$$
\begin{aligned}
C(z) &= < A(z) \cdot B(z) >_{P(z)} \\
&= < (a_1 z + a_0) \cdot (b_1 z + b_0) >_{P(z)} \\
&= < a_1 \cdot b_1 \cdot z^2 + (a_0 \cdot b_1 + a_1 \cdot b_0) \cdot z + a_0 \cdot b_0 >_{P(z)} \\
&= (a_0 \cdot b_1 + a_1 \cdot b_0) \cdot z + (a_0 \cdot b_0 - a_1 \cdot b_1 \cdot \alpha)
\end{aligned}
$$

If $(z^2 + 1)$ is irreducible for this structure, then this multiplication gives an expression exactly like complex multiplication.

This field is a candidate for any task requiring the processing of complex data. Constructing algorithms, though, is a little difficult as p must be a prime. Suitable choices are Mersenne or Fermat primes and these allow development of algorithms along the lines of the number theoretic algorithms of the last Part. It is also possible, of course, to use some of the algorithms to be developed in this Part as the problem is a polynomial problem.

Finding irreducible polynomials to form a field can be a difficult task. The requirement is to find which of the p^k monic polynomials of degree k are not equal to the product of elements within GF(p^k). Not all the products of field elements can produce polynomials of this degree, but an exhaustive test still requires of the order of p^{k+1} products to be formed. Unfortunately, there are no general rules allowing an easier approach although there are specific rules for specific cases.

One of those applies to the field GF(p^2). Here, an irreducible polynomial of degree 2 has the form ($z^2 + a_1 \cdot z + a_0$) and elements of this field are of the form ($b_1 \cdot z + b_0$). If this monic polynomial does not equal some product ($b_1 \cdot z + b_0$)·($c_1 \cdot z + c_0$) then it is irreducible. Clearly, we can express this product as $K \cdot (z + B_0) \cdot (z + C_0)$ where the terms are taken modulo p. If we let z be ($-B_0$) then this product is zero (modulo p). This, though, is merely a reduction modulo ($z + B_0$) of ($z^2 + a_1 \cdot z + a_0$) and we can do the same for C_0. Hence for this field, we can state a polynomial F(z) is irreducible if:

$$< F(a) >_p \neq 0 \quad \text{for } a = 0, 1, \cdots, p\text{--}1$$

In the field GF(p), $\Phi(p+1)$ elements exist whose powers generate the multiplicative group. That is to say, the non-zero elements. Here:

THEOREM   Every finite field GF(p^k) has an element ω such that ω^k generates all elements of the field except the additive identity for $0 \leq k \leq p^k - 1$.

Such elements are termed the primitive roots of GF(p^k). Their properties follow from the earlier work on primitive roots in general number sequences. In particular:

1. there are $\Phi(p^k\text{--}1)$ such roots;
2. if g is a root of GF(p^k) then the complete set of roots is given by { g^r } where the indices r are the set of numbers, $2 \leq r \leq p^k - 1$, mutually prime to ($p^k - 1$).

Programming polynomial modulo reductions and some other polynomial field operations is quite difficult and tedious. Taking advantage of the fact any field GF(p^k) is isomorphic to any other, it is often better to perform these tasks in a computationally more appealing structure and then map the results back into the polynomial field. One possibility for this is a field based on matrices.

There are several ways of creating such a field but most offer no computational advantages. However, consider a polynomial field GF(p^k) with operations defined modulo P(z) where P(z) is some irreducible monic polynomial over that field given by:

$$P(z) = r_0 + r_1 \cdot z + r_2 \cdot z^2 + \cdots + r_{k-1} \cdot z^{k-1} + z^k$$

An obvious matrix to link to this polynomial is its k×k companion matrix C where C is:

$$
\begin{bmatrix}
0 & 0 & 0 & \cdot & \cdot & \cdot & 0 & -r_0 \\
1 & 0 & 0 & & & & 0 & -r_1 \\
0 & 1 & 0 & & & & 0 & -r_2 \\
\cdot & & & \cdot & & & & \cdot \\
\cdot & & & & \cdot & & & \cdot \\
\cdot & & & & & \cdot & & \cdot \\
0 & 0 & 0 & & & & 0 & -r_{k-2} \\
0 & 0 & 0 & \cdot & \cdot & \cdot & 1 & -r_{k-1}
\end{bmatrix}
$$

The vectors $\{ I, C, C^2, \cdots, C^{k-1} \}$, with operations performed modulo p between the elements, form a set of k linearly independent vectors. There are p^k vectors of the form:

$$v = a_0 \cdot I + a_1 \cdot C + \cdots + a_{k-1} \cdot C^{k-1}$$

where the scalars $\{ a_i \}$ are drawn from GF(p). Hence this vector space is isomorphic to the polynomial field. Indeed, the tuple $(a_0, a_1, \cdots, a_{k-1})$ describes either a polynomial or a vector.

A vector is of the form:

$$a_0 \cdot I + a_1 \cdot C + a_2 \cdot C^2 + \cdots + a_{k-1} \cdot C^{k-1}$$

If we calculate this expression and collect all the terms, it is found the first column of the resulting matrix is $[a_0\, a_1\, a_2\, \cdots\, a_{k-1}]^T$. Therefore, if we wish to multiply two elements U and V of the field, all that is really necessary to do is form the vector:

$$[u_0 \cdot I + u_1 \cdot C + u_2 \cdot C^2 + \cdots + u_{k-1} \cdot C^{k-1}][v_0\ v_1\ \cdots\ v_{k-1}]^T$$

and this will uniquely define the product. This is very easily computed.

Example

We will consider the field GF(7^2). Elements of this field are given by $(a_0 + a_1 \cdot z)$ or (a_0, a_1) where $0 \le a_0 \le 6$ and $0 \le a_1 \le 6$. A (monic) irreducible polynomial is required to construct the field of the form $q_0 + q_1 \cdot z + z^2$ where $0 \le q_0 \le 6$ and $0 \le q_1 \le 6$. In this case, the polynomial can be determined by selecting values for the coefficients that obey:

$$< q_0 + q_1 \cdot z + z^2 >_7 \ 0$$

for all z, $0 \le z \le 6$. Testing, the complete set of polynomials which satisfy this are:

$z^2 + 1$	$z^2 + 2z + 3$	$z^2 + 4z + 6$
$z^2 + 2$	$z^2 + 2z + 5$	$z^2 + 5z + 2$
$z^2 + 4$	$z^2 + 3z + 1$	$z^2 + 5z + 3$
$z^2 + z + 3$	$z^2 + 3z + 5$	$z^2 + 5z + 5$
$z^2 + z + 4$	$z^2 + 3z + 6$	$z^2 + 6z + 3$
$z^2 + z + 6$	$z^2 + 4z + 1$	$z^2 + 6z + 4$
$z^2 + 2z + 2$	$z^2 + 4z + 5$	$z^2 + 6z + 6$

Next, consider primitive roots. These will depend on the particular irreducible polynomial chosen. Searching for polynomial primitive roots follows exactly the same procedure as in a field GF(p). That is, we only need to find which elements satisfy:

$$< (a_0 + a_1 \cdot z)^r >_{P(z)} \ne 1$$

for powers r equal to the divisors of the period. In this case, that is $p^k - 1 = 49 - 1 = 48$, and the divisors are 1, 2, 3, 4, 6, 8, 12, 16, and 24. We search for $\Phi(48)$ or 16 roots.

Consider all the primitive roots of $(z^2 + 1)$. A candidate root is $(z + 1)$. Testing:

$$
\begin{aligned}
(z + 1)^2 &= 2z & (z + 1)^8 &= 2 \\
(z + 1)^3 &= 2z + 5 & (z + 1)^{12} &= 6 \\
(z + 1)^4 &= 3 & (z + 1)^{16} &= 4 \\
(z + 1)^6 &= 6z & (z + 1)^{24} &= 1
\end{aligned}
$$

hence this is not a root. On the other hand, examining $(z + 2)$:

$$
\begin{aligned}
(z + 2)^2 &= 4z + 3 & (z + 2)^8 &= 5 \\
(z + 2)^3 &= 4z + 2 & (z + 2)^{12} &= z \\
(z + 2)^4 &= 3z & (z + 2)^{16} &= 4 \\
(z + 2)^6 &= 2z + 2 & (z + 2)^{24} &= 6
\end{aligned}
$$

and so this is a primitive root. The remaining 15 primitive roots under $(z^2 + 1)$ could be found by testing each possibility. However, we choose an easier approach. Having discovered that $(z + 2)$ is a root, then we know the others are $(z + 2)^r$ where r is a number mutually prime to 48. That is, r is one of:

1	13	25	37
5	17	29	41
7	19	31	43
11	23	35	47

Testing these, we find:

$$
\begin{aligned}
(z + 2)^1 &= z + 2 & (z + 2)^{13} &= 2z + 6 & (z + 2)^{25} &= 6z + 5 & (z + 2)^{37} &= 5z + 1 \\
(z + 2)^5 &= 6z + 4 & (z + 2)^{17} &= 4z + 1 & (z + 2)^{29} &= z + 3 & (z + 2)^{41} &= 3z + 6 \\
(z + 2)^7 &= 6z + 2 & (z + 2)^{19} &= 2z + 1 & (z + 2)^{31} &= z + 5 & (z + 2)^{43} &= 5z + 6 \\
(z + 2)^{11} &= 6z + 3 & (z + 2)^{23} &= 3z + 1 & (z + 2)^{35} &= z + 4 & (z + 2)^{47} &= 4z + 6
\end{aligned}
$$

Consider all the irreducible polynomials for which z is a primitive root. This requires finding whether $z^2, z^3, z^4, z^6, z^8, z^{12}, z^{16}$, and z^{24} are different from unity (modulo p) modulo the given polynomial. Take that polynomial as $(z^2 + 4)$. Then:

$$
\begin{aligned}
< z^2 >_{z^2+4} &= 3 & < z^6 >_{z^2+4} &= 6 \\
< z^3 >_{z^2+4} &= 3z & < z^8 >_{z^2+4} &= 4 \\
< z^4 >_{z^2+4} &= 2 & < z^{12} >_{z^2+4} &= 1
\end{aligned}
$$

and so z is not a root for this polynomial. Consider $(z^2 + z + 3)$. Then:

$$
\begin{aligned}
< z^2 >_{z^2+z+3} &= 6z + 4 & < z^8 >_{z^2+z+3} &= 3 \\
< z^3 >_{z^2+z+3} &= 5z + 3 & < z^{12} >_{z^2+z+3} &= z + 4 \\
< z^4 >_{z^2+z+3} &= 5z + 6 & < z^{16} >_{z^2+z+3} &= 2 \\
< z^6 >_{z^2+z+3} &= 5z + 4 & < z^{24} >_{z^2+z+3} &= 6
\end{aligned}
$$

hence z is a root for this polynomial. Similarly, it is a root for:

$$z^2 + 2z + 3 \qquad\qquad z^2 + 5z + 3$$
$$z^2 + 2z + 5 \qquad\qquad z^2 + 5z + 5$$
$$z^2 + 3z + 5 \qquad\qquad z^2 + 6z + 3$$
$$z^2 + 4z + 5$$

Consider the irreducible polynomial $(z^2 + 3z + 1)$ in $GF(7^2)$. The product of two elements (a,b) and (x,y) is given by:

$$< (a + bz)\cdot(x + yz) >z^2 3z + 1 = < ax + (bx + ay)z + byz^2 >z^2 3z + 1$$
$$= (ax - by) + (bx + ay - 3by)z$$

Now consider the matrix representation of the field modulo this polynomial. The companion matrix to the polynomial $(z^2 + 3z + 1)$ is:

$$C = \begin{bmatrix} 0 & -1 \\ 1 & -3 \end{bmatrix} = \begin{bmatrix} 0 & 6 \\ 1 & 4 \end{bmatrix}$$

The basis vectors are I and C, hence a vector is described by the tuple (a,b) or:

$$a\begin{bmatrix} 1 & 0 \\ 0 & 1 \end{bmatrix} + b\begin{bmatrix} 0 & 6 \\ 1 & 4 \end{bmatrix} = \begin{bmatrix} a & 6b \\ b & a+4b \end{bmatrix}$$

The product of two vectors is:

$$\begin{bmatrix} a & 6b \\ b & a+4b \end{bmatrix}\begin{bmatrix} x & 6y \\ y & x+4y \end{bmatrix} = \begin{bmatrix} ax + 6by & 6ay + 6bx + 3by \\ bx + ay + 4by & by + ax + 4bx + 4ay \end{bmatrix}$$

The column of the result is the required vector and this could have been achieved by just multiplying the first matrix by the column matrix $[\ x\ \ y\ ]^T$.

The polynomial Chinese remainder theorem

THEOREM   Let $P_i(z)$ be a set of polynomials mutually prime to one another. Then there exists a unique polynomial $Y(z)$ satisfying the polynomial congruences:

$$< Y(z) >_{P_i(z)} = Y_i(z) \qquad\qquad\qquad i = 1, 2, \cdots, k$$

where $$\qquad Deg[Y(z)] \le \sum_{i=1}^{k} Deg[Y_i(z)]$$

Proof of the theorem follows from a simple adaptation of the proof given in Part 3. Assume polynomials $D_i(z)$ exist where:

$$< D_i(z) >_{P_j(z)} = 1 \text{ if } i = j$$
$$= 0 \text{ otherwise}$$

then the solution is just:

$$Y(z) = \left\langle \sum_{i=1}^{k} D_i(z) \cdot Y_i(z) \right\rangle_{P(z)}$$

where $P(z) = \prod_{i=1}^{k} P_i(z)$, and so the problem is one of proving these polynomials $D_i(z)$ exist. From the nature of the congruence relations which define them, they must be of the form:

$$D_i(z) = G_i(z) \cdot Q_i(z) \qquad\qquad \text{where } Q_i(z) = \frac{P(z)}{P_i(z)}$$

and solving these congruences gives polynomials:

$$G_i(z) \cdot Q_i(z) + M_i(z) \cdot P_i(z) = 1$$

Solving these must be via a polynomial form of Euclid's algorithm which we can create by suitably modifying the numeric version of Part 3. Ignoring the subscripts of the last equation for reasons of clarity and just considering two monic polynomials $P(z)$ and $Q(z)$, we find:

$$r_0(z) = < P(z) >_{Q(z)}$$
$$r_1(z) = < Q(z) >_{r_0(z)}$$
$$r_2(z) = < r_0(z) >_{r_1(z)}$$
$$\cdot$$
$$\cdot$$
$$\cdot$$
$$r_m(z) = < r_{m-2}(z) >_{r_{m-1}(z)}$$

which means we can produce the set of equations:

$$\frac{P(z)}{Q(z)} = C_0(z) + \frac{r_0(z)}{Q(z)} \qquad\qquad \text{Deg}[\, r_0(z)\,] < \text{Deg}[\, Q(z)\,]$$

$$\frac{Q(z)}{r_0(z)} = C_1(z) + \frac{r_1(z)}{r_0(z)} \qquad\qquad \text{Deg}[\, r_0(z)\,] < \text{Deg}[\, r_0(z)\,]$$

$$\cdot$$
$$\cdot$$
$$\cdot$$

$$\frac{r_{m-2}(z)}{r_{m-1}(z)} = C_m(z) + 0$$

and from this, the continuing fraction:

$$\frac{P(z)}{Q(z)} = C_0(z) + \cfrac{1}{C_1(z) + \cfrac{1}{C_2(z) + \cdot}}$$

$$+ \cfrac{1}{C_m(z)}$$

This suggests a series of convergents of the form:

$$\frac{U_0(z)}{V_0(z)} = C_0(z)$$

$$\frac{U_1(z)}{V_1(z)} = C_0(z) + \frac{1}{C_1(z)}$$

$$\cdot$$
$$\cdot$$
$$\cdot$$

$$\frac{U_m(z)}{V_m(z)} = \frac{P(z)}{Q(z)}$$

and it follows that a recurrence relation for these is:

$$U_i(z) = C_i(z) \cdot U_{i-1}(z) + U_{i-2}(z)$$
$$V_i(z) = C_i(z) \cdot V_{i-1}(z) + V_{i-2}(z)$$

where $i \geq 2$. An identity easily shown is:

$$V_i(z) \cdot U_{i-1}(z) - U_i(z) \cdot V_{i-1}(z) = (-1)^i \qquad\qquad i = 1, 2, \cdots, m$$

which for i equal to m is:

$$V_m(z) \cdot U_{m-1}(z) - U_m(z) \cdot V_{m-1}(z) = (-1)^m$$

This implies $V_m(z)$ and $U_m(z)$ have no common factors and as this is also true of P(z) and Q(z), the convergents imply:

$$U_m(z) = K \cdot P(z)$$
$$V_m(z) = K \cdot Q(z)$$

where K is a constant. Since P(z) and Q(z) are monic then K, in fact, must be the first coefficient of $U_m(z)$ and $V_m(z)$. Substituting these two terms into the last equation:

$$Q(z) \cdot \{ (-1)^m \cdot K \cdot U_{m-1}(z) \} - P(z) \cdot \{ (-1)^m \cdot K \cdot V_{m-1}(z) \} = 1$$

which means, returning to the original problem, that:

$$G_i(z) = (-1)^m \cdot K \cdot U_{m-1}(z)$$
$$M_i(z) = (-1)^{m+1} \cdot K \cdot V_{m-1}(z)$$

and this must be repeated for each of the k sets of terms.

Example

Let the mutually prime polynomials $P_i(z)$ be:

$$P_1(z) = z - 1 \qquad P_2(z) = z + 1 \qquad P_3(z) = z^4 + z^3 + z^2 + z + 1$$

In essence, the polynomial Chinese remainder theorem reduces to a means of finding polynomials $D_i(z)$ such that:

$$
\begin{array}{lll}
D_1(z) = 1 \bmod P_1(z) & D_2(z) = 0 \bmod P_1(z) & D_3(z) = 0 \bmod P_1(z) \\
\quad = 0 \bmod P_2(z) & \quad = 1 \bmod P_2(z) & \quad = 0 \bmod P_2(z) \\
\quad = 0 \bmod P_3(z) & \quad = 0 \bmod P_3(z) & \quad = 1 \bmod P_3(z)
\end{array}
$$

The solution is that each $D_i(z)$ equals $G_i(z) \cdot Q_i(z)$ and that:

$$G_i(z) \cdot Q_i(z) + M_i(z) \cdot P_i(z) = 1$$

Thus:

$$
\begin{array}{ll}
D_1(z) \cdot (z+1) \cdot (z^4 + z^3 + z^2 + z + 1) + M_1(z) \cdot (z-1) & = 1 \\
D_2(z) \cdot (z-1) \cdot (z^4 + z^3 + z^2 + z + 1) + M_2(z) \cdot (z+1) & = 1 \\
D_3(z) \cdot (z-1) \cdot (z+1) + M_3(z) \cdot (z^4 + z^3 + z^2 + z + 1) & = 1
\end{array}
$$

Consider the solution for $D_1(z)$. The continuing fraction is:

$$\frac{(z-1)}{(z^5 + 2z^4 + 2z^3 + 2z^2 + 2z + 1)} = 0 + \cfrac{1}{(z^4 + 3z^3 + 5z^2 + 7z + 9) + \cfrac{1}{\cfrac{1}{10}(z-1)}}$$

The convergents are therefore:

$$\frac{U_0(z)}{V_0(z)} = 0$$

$$\frac{U_1(z)}{V_1(z)} = \frac{1}{(z^4 + 3z^3 + 5z^2 + 7z + 9)}$$

$$\frac{U_2(z)}{v_2(z)} = \frac{\dfrac{1}{10} \cdot (z - 1)}{\dfrac{1}{10} \cdot (z^5 + 2z^4 + 2z^3 + 2z + 1)}$$

From the identity:

$$V_m(z) \cdot U_{m-1}(z) - V_{m-1}(z) \cdot U_m(z) = (-1)^m$$

with m equal to 2, we obtain by substitution:

$$\frac{1}{10}(z^5 + 2z^4 + 2z^3 + 2z^2 + 2z + 1) - \frac{1}{10}(z - 1) \cdot (z^4 + 3z^3 + 5z^2 + 7z + 9) = 1$$

Comparing against the original equation, $G_1(z)$ must be $\frac{1}{10}$ and so:

$$D_1(z) = \frac{1}{10}(z + 1) \cdot (z^4 + z^3 + z^2 + z + 1)$$

$$= \frac{1}{10}(z^5 + 2z^4 + 2z^3 + 2z^2 + 2z + 1)$$

By a similar process:

$$D_2(z) = -\frac{1}{2} \cdot (z^5 - 1)$$

$$D_3(z) = \frac{1}{5} \cdot (2z^5 - z^4 - z^3 - 5z^2 - z + 6)$$

Example

A periodic convolution can be described by $Y(z) = < H(z) \cdot X(z) >_{P(z)}$, where Deg[P(z)] must be greater than or equal to Deg[H(z)] + Deg[X(z)]. Assume P(z) is given by:

$$P(z) = \prod_{i=1}^{k} (z - a_i)$$

Then from the polynomial Chinese remainder theorem:

$$Y(z) = < \sum_{i=1}^{k} D_i(z) \cdot Y_i(z) >_{P(z)}$$

where $Y_i(z) = \ <Y(z)>_{(z-a_i)} = \ <H(z)\cdot X(z)>_{(z-a_i)} = H(a_i)\cdot X(a_i)$

Compare this to the Toom-Cook algorithm of Part 1. It is the same algorithm.

Cyclotomic polynomials

Consider the unit circle in the complex plane. That is: $z = e^{j\Omega}$, where $0 \le \Omega \le 2\pi$. The set of points $e^{\left(\frac{j2\pi}{N}\right)i}$, ($i = 0, 1, \cdots, N-1$), on this circle divides it into N equal parts. As well, they form the apexes of a regular N-gon and are also the roots of the Nth degree polynomial ($z^N - 1$). They are not unique in dividing the circle into equal parts or in satisfying this equation. For example, the points $e^{j\left(\frac{2\pi k}{N}\right)i}$ also satisfy these requirements, provided k and N are relatively prime. The ancient Greeks became interested in the problem of dividing a circle into equal parts as they had realized the construction of regular N-gons is the circle division problem. They termed the theory of dividing circles **cyclotomy** and it became one of the longest standing mathematical studies of history.

The equation ($z^N - 1$) is quite important in convolution as polynomial residue reduction modulo this polynomial is cyclic convolution. Now this Part studies convolution in a general manner and, to do that from the point of view of polynomial rings and fields, a problem immediately arises. These structures have elements that are defined with respect to various fields including the complex field. The problem, therefore, is of expressing ($z^N - 1$) as the product of a set of irreducible polynomials on these fields. Perhaps strangely, the problem of dividing a circle into N equal parts is mathematically the same as finding these irreducible polynomials of ($z^N - 1$) for GF(p). Consequently, they are termed **cyclotomic polynomials**.

Any cyclotomic polynomial is the product of terms like $\left(z - e^{\left(\frac{j2\pi}{N}\right)i} \right)$. However, to be irreducible on GF(p) means the coefficients of the polynomial must be integers. Thus the nth cyclotomic polynomial is:

$$C_n(z) = \Pi\left(z - e^{\left(\frac{j2\pi}{N}\right)i} \right)$$

where i must satisfy $i \le n$ and (i,n) $= 1$. Further, it can be shown:

$$z^N - 1 = \prod_{j|N} C_j(z)$$

Appendix 4 lists some cyclotomic polynomials.

Cyclotomic polynomials have a number of interesting properties. Appendix 4 suggests they only have coefficients of 0, –1, and 1. That, however, is only true for polynomials up to the 105th. The degree of the nth polynomial is $\Phi(n)$ and for n prime the polynomial is:

$$C_p(z) = z^{p-1} + z^{p-2} + \cdots + z + 1$$

A useful result for creating tables is that if m and n are integers, then:

$$C_{n \cdot m^k}(z) = C_{n \cdot m}\left(z^{m^k}\right)$$

If n is odd and greater than 3 then:

$$C_{2n}(z) = C_n(-z)$$

Equally useful is the identity:

$$C_{np}(z) = \frac{C_n(z^p)}{C_n(z)}$$

where p is a prime. Finally, it can be shown:

$$
\begin{aligned}
C_n(1) &= p \text{ for } n = p^k \text{ where p is a prime and } k > 0 \\
&= 0 \text{ for } n = 1 \\
&= 1 \text{ otherwise}
\end{aligned}
$$

Tensor analysis

Tensors are mathematical structures important in applications such as crystallography, electromagnetic theory and quantum and relativistic mechanics. For most applications, particular tensors are used, namely Cartesian tensors, and the emphasis is very much on geometric transformations.

A tensor is an abstract mathematical structure fabricated from a number of elements. They are typically denoted T_{ijk} where the number of subscripts defines the **rank** of the tensor. If the extent of the indices i, j, and k in this tensor are N, M, and L, then the **size** of the tensor is NML. If all these elements are non-zero, the tensor is **complete**; otherwise it is incomplete. Many tensors are symmetric, that is, all the indices have an extent of N. In this case, the **dimension** of the tensor is N.

A scalar is a tensor of rank zero, a vector is a tensor of rank one and a matrix is a tensor of rank two. However, the converses are not necessarily true as a tensor of any rank can have any dimension. Interest here will be limited to tensors of rank zero, one, and two. The latter two, and in fact all high order tensors, are conveniently represented by matrices but it must be stressed this is a notational convenience and does not in any way imply that matrix operations apply. Rather, tensor addition and tensor multiplication apply suitably modified for this matrix notation.

Tensor addition, also known as direct addition, is not considered here. Tensor multiplication, also known as the Kroenecker or direct or dyadic or outer product multiplication, is a simple generalization of the usual outer product. Two structures A and B with components (A_1, A_2, A_3) and (B_1, B_2, B_3) respectively, have a tensor product:

$$
A \times B = \begin{bmatrix}
A_1 \cdot B_1 & A_1 \cdot B_2 & A_1 \cdot B_3 \\
A_2 \cdot B_1 & A_2 \cdot B_2 & A_2 \cdot B_3 \\
A_3 \cdot B_1 & A_3 \cdot B_2 & A_3 \cdot B_3
\end{bmatrix}
$$

The obvious generalization is to allow these components to be scalars, vectors, polynomials or any other mathematical structure of interest. The product operation '·' must then be appropriate to those elements.

The tensor product of two tensors creates a new tensor with a rank equal to the sum of the ranks of the tensors in the product and whose dimensions are the products of the corresponding dimensions. The product of a scalar, a tensor of rank zero, with any other tensor then, results in all elements being scaled by that value but there is no change in rank. Equally, the product of two tensors of rank zero remains a scalar. The product of two tensors of rank one – vectors – is given by the above tensor product. The product of two tensors a_{ij} and b_{kl} of rank two results in a tensor $c_{ik'jl}$ which is represented as a matrix with elements $c_{ik'jl}$. It is often more convenient though, to represent this as:

$$\begin{bmatrix} a_{11} \cdot B & a_{12} \cdot B \\ a_{21} \cdot B & a_{22} \cdot B \end{bmatrix}$$

Higher products follow from this. There is clearly a problem in how to represent these products in matrix form. In particular, how to form these matrices to allow a simple way of expressing tensor transformations. Since only tensor products involving tensors up to a rank of two will be considered here, this issue will not be examined.

Provided the tensors are symmetric, tensor products are both commutative and associative. Tensors of rank two then — matrices — must be N×N. A number of useful results derive from combining matrix and tensor operations. It can be shown that tensor and matrix products are distributive over one another if the matrices are square. Hence if $\otimes$ represents the tensor product and A_i and B_i are matrices:

$$(A_0 \cdot A_1 \cdot A_2 \cdot \ \cdots \ \cdot A_n) \otimes (B_0 \cdot B_1 \cdot B_2 \cdot \ \cdots \ \cdot B_n)$$

$$= (A_0 \otimes B_0) \cdot (A_1 \otimes B_1) \cdot (A_2 \otimes B_2) \cdot \ \cdots \ \cdot (A_n \otimes B_n)$$

Similarly, the tensor product is distributive over matrix addition. Two other useful formulae are:

$$(A \otimes B)^T = B^T \otimes A^T$$
$$(A \otimes B)^{-1} = B^{-1} \otimes A^{-1}$$

The value of tensor analysis is quite nicely illustrated by revisiting the prime factors algorithm. The basis of this algorithm is that if the number of points N to be transformed is the product of two mutually prime numbers M and L, then via the Chinese remainder theorem, a set of mappings can be defined which result in the DFT being expressed as:

$$X(k_1 k_2) = \sum_{n_1=0}^{M-1} \sum_{n_2=0}^{L-1} x(n_1, n_2) W_M^{k_1 n_1} W_L^{k_2 n_2}$$

$$= \sum_{n_1=0}^{M-1} W_M^{k_1 n_1} \left(\sum_{n_2=0}^{L-1} x(n_1, n_2) W_L^{k_2 n_2} \right)$$

where $0 \leq k_1 \leq M-1$ and $0 \leq k_2 \leq L-1$. The algorithm describes a mapping from one to two dimensions. Execution of the algorithm on the array $x(n_1, n_2)$ proceeds by taking DFTs along one dimension then along the other. The attraction of the algorithm is that it requires no twiddling.

Let Λ be the forward transformation matrix of an L-point DFT. That is:

$$
\begin{bmatrix}
1 & 1 & 1 & \cdot & \cdot & \cdot & 1 \\
1 & W_L & W_L^2 & & & & W_L^{L-1} \\
1 & W_L^2 & W_L^4 & & & & W_L^{2L-2} \\
\cdot & & & \cdot & & & \cdot \\
\cdot & & & & \cdot & & \cdot \\
\cdot & & & & & \cdot & \cdot \\
1 & W_L^{L-1} & W_L^{2L-2} & \cdot & \cdot & \cdot & W_L^{(L-1)(L-1)}
\end{bmatrix}
$$

Let the rows of this matrix be denoted Λ_i. Now define two sets. Each has M elements and those elements are L-point vectors, $\mathbf{X}_{k_1}$ and $\mathbf{x}_{n_1}$, where these vectors are given by:

$$
\begin{aligned}
\mathbf{X}_{k_1} &= X(k_1, k_2) & \text{for } k_2 = 0, 1, \cdots, L-1 \\
\mathbf{x}_{n_1} &= x(n_1, n_2) & \text{for } n_2 = 0, 1, \cdots, L-1
\end{aligned}
$$

where $0 \leq k_1, n_1 \leq M-1$. Substituting into the prime factors algorithm equation:

$$
\mathbf{X}_{k_1} = \sum_{n_1=0}^{M-1} W_M^{k_1 n_1} (\Lambda_{n_1} \cdot \mathbf{x}_{n_1}) \qquad k_1 = 0, 1, \cdots, M-1
$$

Any DFT in matrix form can be expressed as:

$$
\mathbf{X} = \Omega \cdot \mathbf{x}
$$

In this case then, the matrix Ω has elements derived from $W_M^{k_1 n_1} \cdot \Lambda_{n_1}$ and this describes the tensor product of Λ, the forward transformation matrix for an L-point DFT with Ξ, the forward transformation matrix of an M-point DFT. That is:

$$
\Omega = \Lambda \otimes \Xi
$$

This relationship is used later in this Part to enable new variations of the prime factors algorithm to be developed.

THEORY

Winograd's minimal complexity theorem

Winograd's minimal complexity theorem is deceptively simple but it is a key theorem of digital signal processing. It is:

THEOREM   Let F be some polynomial field. Let $X(z)$ and $H(z)$ be two polynomials defined on that field of degree N. Then:

$$Y(z) \ = \ < \ H(z) \cdot X(z) \ >_{P(z)}$$

requires at least ($2N - k$) multiplications where k is the number of irreducible factors of $P(z)$ over F.

A little point. The theorem states that k is the number of irreducible factors not the number of prime divisors. If $P(z)$ is prime, then k is one.

   A proof of the theorem is beyond the scope of this book. However, a sketch can be given which shows that there is at least one algorithm which achieves the minimum. This sketch also serves a useful purpose in highlighting some important implications of the theorem. Assume $P(z)$ can be factored into irreducible polynomials $P_i(z)$ of the field F, so that:

$$P(z) \ = \ \prod_{i=1}^{k} P_i(z)$$

From the Chinese remainder theorem, we may express the polynomial multiplication as:

$$Y(z) \ = \ \left\langle \sum_{i=1}^{k} G_i(z) \cdot Y_i(z) \right\rangle_{P(z)}$$

where $Y_i(z) = < \ Y(z) \ >_{P_i(z)} \ = < \ H(z) \cdot X(z) \ >_{P_i(z)}$

$< \ G_i(z) \ >_{P_j(z)} \ = 1 \ \ \text{for } i{=}j$

$= 0 \ \ \text{otherwise}$

Note carefully what is achieved by this action. The Chinese remainder theorem describes how a set of simple operations may be combined via the equation given to create a more complex result. The combination procedure involves multiplying each of the $Y_i(z)$ by polynomials $G_i(z)$. From the proof of the Chinese remainder theorem, we know a good deal about these polynomials. They are constructed from the $P_i(z)$ and are purely a function of those polynomials and so the field. Let us stress that again. In no way do these polynomials depend on the actual values of the polynomials $H(z)$ or $X(z)$ but solely on the field and the $P_i(z)$. As such, they never vary and so are similar to constants. Certainly, they are not of the same character as those operations which are dependent in some way on the input terms as these represent any potential polynomial of the field. An engineering analogy to multiplication by these polynomials is scaling of signals which can be a hardwired operation. In contrast, operations dependent on input terms would require multipliers because it is never known what the value of those terms may be.

   The distinction between these terms is quite important. The analogy is deliberately chosen because it illustrates such distinctions are significant in both a theoretical and practical sense. Now Winograd's theorem could, in more vernacular style, be phrased as:

'Let H(z) and X(z) be two polynomials of degree N of some polynomial field F. Let:

$$Y(z) = < H(z) \cdot X(z) >_{P(z)}$$

Let an essential multiplication be a multiplication which involves the coefficients of H(z) or X(z) in some way. Then computing Y(z) requires at least ($2N - k$) essential multiplications where k is the number of irreducible factors of P(z).'

Thus Winograd's theorem recognizes no multiplications in the combination stage because any that exist are by 'constant' terms, that is, by terms purely dependent on the field.

Without the Chinese remainder theorem, operations would be modulo P(z) and since that is reducible, we would therefore have a ring not a field. However, employment of the theorem re-expresses the problem into a set of problems, each of which only involves irreducible polynomials and so each enjoys the advantages of a polynomial field.

With this clarification, we see that Winograd's theorem is counting just those multiplications due to the terms $Y_i(z)$ and they are given by:

$$Y_i(z) = < H(z) \cdot X(z) >_{P_i(z)} = << H(z) >_{P_i(z)} \cdot < X(z) >_{P_i(z)} >_{P_i(z)}$$

Reduction any polynomial leaves a result equal to at most the degree of that polynomial, hence this operation must involve the product of two polynomials of degree Deg[$P_i(z)$] and that requires ($2 \cdot$ Deg[$P_i(z)$] $- 1$) multiplications of the form of interest. From the reconstruction equation:

$$\begin{aligned} \text{number of multiplications} &= \sum_{i=1}^{k} (2 \cdot \text{Deg}\big[P_i(z)\big] - 1) \\ &= 2 \cdot \text{Deg[} P(z) \text{]} - k \\ &= 2N - k \end{aligned}$$

A minimal algorithm, therefore, cannot involve any multiplications in the reductions of H(z) and X(z) modulo $P_i(z)$, nor in the final reduction of their product. Does such an algorithm exist? The answer is yes as the Toom-Cook algorithm satisfies these requirements. Here, the $P_i(z)$ are of the form ($z - a_j$) and so a term like $< H(z) >_{P_i(z)}$ becomes $H(a_i)$. All multiplications in this term are by fixed constants of the field and this is true of all other reductions by such a polynomial. There is, therefore, at least one algorithm that satisfies Winograd's theorem.

The theorem doesn't answer all questions one would like to pose of a convolution problem. It doesn't give the minimal number of additions, nor does it indicate the minimal number of multiplications by fixed constants of the field. What it does show, though, is that the choice of the field deserves the very closest attention. Consider cyclic convolution. Here, the polynomial P(z) is given by:

$$P(z) = z^N - 1$$

The irreducible factors of this vary significantly from field to field. Consider ($z^8 - 1$). In the complex field, there are eight factors ($z - W_8^i$) where W is $e^{-j\frac{2\pi}{8}}$ and $0 \le i \le 7$. The minimal algorithm therefore has (2N – N) or 8 (complex) multiplications. In the real number field, the minimal algorithm requires 11 multiplications as ($z^N - 1$) factors into:

$$(z-1) \cdot (z+1) \cdot (z^2+1) \cdot \left(z^2 - 2\cos\left(\frac{\pi}{4}\right) \cdot z + 1\right) \cdot \left(z^2 - 2\cos\left(\frac{3\pi}{4}\right) \cdot z + 1\right)$$

In the field of Gaussian integers, it factors to:

$$(z-1) \cdot (z+1) \cdot (z-j) \cdot (z+j) \cdot (z^4 + 1)$$

which also gives a minimal algorithm of 11 multiplications. In the field of rationals, the minimal algorithm has 12 multiplications as the factorization is:

$$(z-1) \cdot (z+1) \cdot (z^2+1) \cdot (z^4 + 1)$$

The structure of minimal multiplication algorithms

Expressed in matrix form, the minimal Winograd algorithm becomes the equation:

$$\mathbf{y} = \mathbf{C} \cdot (\mathbf{Bh} \cdot \mathbf{Ax})$$

where

A = a (2N – k)×N matrix of fixed constants of the field
B = a (2N – k)×N matrix of fixed constants of the field
C = a N×(2N – k) matrix of fixed constants of the field
• = pointwise multiplication of elements
y = an N-dimensional vector of output elements
h = an N-dimensional vector of input elements
x = an N-dimensional vector of input elements

Matrices A and B are the results of all the reductions modulo the $P_i(z)$ and matrix C is due to Chinese remainder theorem reconstruction. Multiplications and additions within these are by constant elements of the coefficient fields concerned. While they may be fixed operations, they are nonetheless of considerable practical importance. At this stage, it is not possible to make definite conclusions on what they should be — that is a concern of DEVELOPMENTS — but we can note some broad characteristics.

The A and B matrix elements are drawn from the irreducible polynomials $P_i(z)$. Note these from the example of the previous section. For the complex field, these terms are W^{kn} and involve complex multiplications. The real field is less complicated but still involves a number of significant multiplications by these 'constants'. For the Gaussian and rational field, however, the constants are typically 0, 1, or –1 and require no multiplications at all. Therefore, while they may involve more general multiplications than the other fields, practically they are better choices. This suggests very strongly that our interest should not be so much in the minimal algorithm or even in the field but rather in the nature of the operations required by these fixed constants of the field.

In Winograd's investigation of minimal algorithms, he was able to show a result of some practical importance. Its value will become more apparent in DEVELOPMENTS but the result itself is an alternative structure for the minimal algorithm. The individual terms of the vector **y** in the matrix form can be expressed as:

$$y_n = \sum_{i=0}^{2N-k-1} c_{ni} \cdot V_i \qquad\qquad n = 0, 1, \cdots, N-1$$

where c_{ni} = the elements of matrix C
           V_i = the $(2N-k)$ general multiplication terms

Equally:

$$V_i = \left(\sum_{r=0}^{N-1} b_{ir} \cdot h_r \right) \cdot \left(\sum_{s=0}^{N-1} a_{is} \cdot x_s \right)$$

Hence:

$$y_n = \sum_{i=0}^{2N-k-1} c_{ni} \cdot \left(\sum_{r=0}^{N-1} b_{ir} \cdot h_r \right) \left(\sum_{s=0}^{N-1} a_{is} \cdot x_s \right)$$

$$= \sum_{r=0}^{N-1} \cdot \sum_{s=0}^{N-1} h_r \cdot x_s \cdot \left(\sum_{i=0}^{2N-k-1} c_{ni} \cdot b_{ir} \cdot a_{is} \right)$$

Consider the specific case of cyclic convolution. This has the form:

$$y_n = \sum_{j=0}^{N-1} h_j \cdot x_{n-j} \qquad\qquad \text{where } j = 0, 1, \cdots, N-1$$

Thus it must be true in this equation that:

$$\sum_{i=0}^{2N-k-1} c_{ni} \cdot b_{ir} \cdot a_{is} \quad = 1 \text{ when } <r+s-n>_N = 0$$

$$= 0 \text{ otherwise}$$

Denote this as $\delta(n,r,s)$. The equation becomes:

$$y_n = \sum_{r=0}^{N-1} \cdot \sum_{s=0}^{N-1} h_r \cdot x_s \cdot \delta(n,r,s) \qquad\qquad n = 0, 1, \cdots, N-1$$

Let θ be an N-dimensional vector with (non-zero) elements θ_i. Then multiply y_n in this equation by the corresponding element θ_n of θ and sum over all values of n. This gives:

$$\sum_{n=0}^{N-1} y_n \cdot \theta_n = \sum_{n=0}^{N-1} \sum_{r=0}^{N-1} \sum_{s=0}^{N-1} \theta_n \cdot h_r \cdot x_s \cdot \delta(n,r,s)$$

$$= \sum_{r=0}^{N-1} h_r \cdot \sum_{n=0}^{N-1} \sum_{s=0}^{N-1} \theta_n \cdot x_s \cdot \delta(n,r,s)$$

We can follow a similar procedure for the original set of equations to obtain:

$$\sum_{n=0}^{N-1} y_n \cdot \theta_n = \sum_{i=0}^{2N-k-1} \left(\sum_{n=0}^{N-1} c_{ni} \cdot \theta_n \right) \cdot \left(\sum_{r=0}^{N-1} b_{ir} \cdot h_r \right) \left(\sum_{s=0}^{N-1} a_{is} \cdot x_s \right)$$

$$= \sum_{i=0}^{2N-k-1} \left(\sum_{r=0}^{N-1} b_{ir} \cdot h_r \right) \cdot \left(\sum_{n=0}^{N-1} c_{ni} \cdot \theta_n \right) \cdot \left(\sum_{s=0}^{N-1} a_{is} \cdot x_s \right)$$

$$= \sum_{r=0}^{N-1} h_r \sum_{i=0}^{2N-k-1} b_{ir} \cdot \left(\sum_{n=0}^{N-1} c_{ni} \cdot \theta_n \right) \cdot \left(\sum_{s=0}^{N-1} a_{is} \cdot x_s \right)$$

Equating the coefficients of the elements h_r in these two equations gives:

$$\sum_{n=0}^{N-1} \sum_{s=0}^{N-1} \theta_n \cdot x_s \cdot \delta(n,r,s) = \sum_{i=0}^{2N-k-1} b_{ir} \cdot \left(\sum_{n=0}^{N-1} c_{ni} \cdot \theta_n \right) \cdot \left(\sum_{s=0}^{N-1} a_{is} \cdot x_s \right)$$

The indices do not matter in this equation. Similarly, there is no reason why we cannot choose the elements θ_n as h_n. Therefore, we can state:

$$\sum_{r=0}^{N-1} \sum_{s=0}^{N-1} h_r \cdot x_s \cdot \delta(r,n,s) = \sum_{i=0}^{2N-k-1} b_{in} \cdot \left(\sum_{r=0}^{N-1} c_{ri} \cdot h_r \right) \cdot \left(\sum_{s=0}^{N-1} a_{is} \cdot x_s \right)$$

How are we to interpret this? From the earlier definition of $\delta(n, r, s)$, we note $\delta(r, n ,s)$ is non-zero only when $< n + s - r >_N$ is zero. Thus the left-hand side is equivalent to:

$$y_k = \sum_{n=0}^{N-1} h_{n+k} \cdot x_n$$

and this describes cyclic correlation. The entire equation is equivalent to:

$$\mathbf{y'} = \mathbf{B}^T \cdot (\mathbf{C}^T \mathbf{h} \bullet \mathbf{Ax})$$

which is similar to the original equation and so is termed the **dual minimal multiplication algorithm** or the **transpose algorithm**.

The dual algorithm describes a cyclic correlation. A cyclic correlation, though, is transformed into a cyclic convolution simply by reversing one of the sequences involved. That could be done here but given the form of this vector equation it is more convenient to adjust the matrices. If all the rows of the C matrix in the original vector equation, except the first, and all the columns of the B matrix except the first are reversed, that equation is converted to a correlation. That is, if:

$$y_s = C_s \cdot (B_s h \cdot Ax)$$

describes a cyclic correlation then a cyclic convolution is given by:

$$y_s' = B_s^T \cdot (C_s^T h \cdot Ax)$$

Reversing is a permutation achieved by exchanging a_{ij} with $a_{i(N-j)}$. This is not the only possibility. A method described ahead enables other transformations to be devised.

In general terms, C in these vector equations is a far more complex matrix than either A or B. In the practical situation, one of the input sequences would be known, thus all the terms up to the general multiplications can be pre-calculated. The dual algorithm allows the most involved calculations to be those pre-calculations — a considerable advantage.

Although just one case was considered here, there is no difficulty proving a similar result for swapping matrices C and A. Also, since in almost all cases of practical importance matrices A and B are identical, it is not unreasonable to restate the equations as:

$$y = C \cdot (Ax \cdot Ah)$$

or for the transpose form:

$$y = A_s^T \cdot (Ax \cdot C_s^T h)$$

Note too, that if one of x or h is pre-computed, a very likely event, then the vector equation can be expressed as:

$$y = C \cdot D \cdot Ax$$

where D is a diagonal matrix formed from Bh. In the case where A equals B and where the transpose form is used, this reduces to the form:

$$y = A_s^T \cdot D \cdot Ax$$

In these cases, all (general) multiplications are due to the matrix D.

Winograd's theorem is quite general but the transpose form above was derived solely for cyclic convolution. For a more general case, consider a polynomial $(z^N - \alpha)$. This is cyclic convolution for α equal to 1. If α is different from 1, then we would expect a scale factor to be introduced into the equations. If it is -1, then it is possible the polynomial is prime. If it is some other negative term, the same possibility occurs and there can also be scale factors in the equation.

The product of two polynomials $H(z)$ and $X(z)$ modulo this $P(z)$ gives a matrix equation:

$$
\begin{bmatrix} y(0) \\ y(1) \\ \cdot \\ \cdot \\ \cdot \\ y(N-1) \end{bmatrix}
=
\begin{bmatrix}
h(0) & \alpha h(N-1) & \alpha h(N-2) & . & . & \alpha h(1) \\
h(1) & h(0) & \alpha h(N-1) & & & \alpha h(2) \\
\cdot & & & & & \cdot \\
\cdot & & & & & \cdot \\
\cdot & & & & & \cdot \\
h(N-1) & h(N-2) & & . & . & h(0)
\end{bmatrix}
\begin{bmatrix} x(0) \\ x(1) \\ \cdot \\ \cdot \\ \cdot \\ x(N-1) \end{bmatrix}
$$

This is a Toeplitz matrix but not a circulant matrix. A question here is whether this can be expressed in a matrix form like the previous equations. That is, whether it is possible to factor the matrix into the equivalents of A, B, and C. The answer is yes in the sense that one technique at least exists to do this, namely Toom-Cook. It also satisfies Winograd's minimum complexity algorithm. In Toom-Cook, matrices A and B are the same and are formed by reducing the polynomials H(z) and X(z) by substituting the different interpolation points for z. Matrix C is then a matrix of coefficients which derives from the Lagrange interpolation polynomials. If Y(z) is reduced modulo P(z) where P(z) is of the form given, then that need only affect matrix C. In turn this suggests, and it can be shown to be true, that this matrix factorization exists for any Toeplitz matrix.

Considering the derivation of the transpose equation, the term $\delta(r,n,s)$ must be:

$$
\begin{aligned}
\delta(r,n,s) \quad &= 1 \quad \text{for } r+n-s=0 \\
&= \alpha \quad \text{for } r+n-s=kN \text{ where } k>0 \\
&= 0 \quad \text{otherwise}
\end{aligned}
$$

If we wish a matrix factorization using $(z^N - \alpha)$, then Toom-Cook defines matrices A and B and the multiplications and this relation could be used to derive matrix C.

An important problem is to derive the transpose form for this particular P(z). While this can be quite easily done by reworking the previous equations, a simpler derivation is as follows. For any Toeplitz matrix A, the counter identity matrix J given by:

$$
\begin{bmatrix}
0 & 0 & \cdot & \cdot & 0 & 1 \\
0 & 0 & & & & 0 \\
\cdot & & & \cdot & & \cdot \\
\cdot & & \cdot & & & \cdot \\
0 & 1 & & & 0 & 0 \\
1 & 0 & \cdot & \cdot & 0 & 0
\end{bmatrix}
$$

has the property that:

$$
A^T = J \cdot A \cdot J
$$

Note that JJ is the identity. Then consider some algorithm expressed as:

$$
y = Tx = C \cdot D \cdot Ax
$$

where T is some Toeplitz matrix and D is a diagonal matrix given in this instance by Bh. This second form holds for cyclic convolution, which is reduction modulo $(z^N - 1)$, and for the more general case of $(z^N - \alpha)$. Thus:

$$
T^T = J \cdot C \cdot D \cdot A \cdot J = C^* \cdot D \cdot A^*
$$

where C* and A* are matrices C and A with their rows and columns exchanged respectively. Therefore:

$$T = (C^* \cdot D \cdot A^*)^T = (A^*)^T \cdot D \cdot (C^*)^T$$

and this is the desired result.

Polynomial transforms

In Part 2, we studied transforms on an infinite field, namely the DFT and its fast algorithms. In Part 3, we considered the application of number theory to the transform problem, including the development of transforms within finite fields. This Part has again been far-ranging by looking at the applications of polynomial theory to the convolution problem. Now consider transforms defined on polynomial rings and fields.

Polynomial rings and fields have elements described by tuples. That makes them quite different from the previous rings and fields studied where the elements were always scalar values. They are still rings and fields though and so transforms exist provided they meet the requirements of transforms on rings and fields. These requirements are essentially the same as for NTTs as the usual polynomial structure considered is a finite polynomial structure. Therefore, operations are defined with respect to some irreducible polynomial $P(z)$. A length N **polynomial transform** exists on some finite polynomial structure if:

(a) there exists a (polynomial) element of the structure $\omega(z)$ such that $< \omega(z)^N >_{P(z)} = 1$;
(b) both N and $\omega(z)$ have inverses modulo $P(z)$;
(c) the following holds:

$$\left\langle \sum_{n=0}^{N-1} \omega(z)^{kn} \right\rangle_{P(z)} = N \text{ for } < k >_N = 0$$

$$= 0 \text{ otherwise}$$

If these conditions hold, the transform pair is given by:

$$X(k) = \left\langle \sum_{n=0}^{N-1} x(n) \cdot \omega(z)^{kn} \right\rangle_{P(z)} \qquad k = 0, 1, \cdots, N-1$$

$$x(n) = \left\langle N^{-1} \sum_{k=0}^{N-1} X(k) \cdot \omega(z)^{-nk} \right\rangle_{P(z)} \qquad n = 0, 1, \cdots, N-1$$

where X and x are elements of the structure and so are tuples and $P(z)$ is irreducible.

These conditions are for a transform to exist with the cyclic convolution property. Thus, the existence of the transform means the congruence:

$$y(n) = \left\langle \sum_{m=0}^{N-1} h(m) \cdot x(n-m) \right\rangle_{P(z)}$$

on the structure concerned can in fact be determined as:

$$Y(k) = \ < \ H(k) \cdot X(k) \ >_{P(z)} \qquad\qquad k = 0, 1, \ \cdots, N-1$$

Now $y(n)$, $h(n)$ and $x(n)$ are all polynomials (or tuples). Expressed in the form:

$$y(n) \ = \ y_n(z) \ = \ y_{n,(N-1)} \cdot z^{N-1} + y_{n,(N-2)} \cdot z^{N-2} + \ \cdots \ + y_{n,1} \cdot z + y_{n,0}$$

and similarly for the other terms, they are recognized as a z-transform. Following the notation used here and considering the inverse z-transform, then the polynomial convolution can be seen to be equivalent to the two-dimensional convolution:

$$y(i,j) \ = \ \sum_{u=0}^{N-1} \sum_{v=0}^{N-1} h(u,v) \cdot x(i-u, j-v) \qquad\qquad \text{where } i, j = 0, 1, \ \cdots, N-1$$

The conditions on the existence of a transform are not very restrictive and a number could be expected. The elements of the polynomials are usually drawn from a field and if so then the inverse to N exists. Thus determining transforms reduces to finding a root in the polynomial field and proving the properties given. Note that as with NTTs, this is equivalent to proving the polynomials ($\omega(z)^i - 1$) for $1 \le i \le N-1$ are not congruent to zero modulo P(z).

An important potential application for polynomial transforms is in computing cyclic convolutions. However, these involve polynomial products modulo ($z^N - 1$) and this polynomial is reducible. What can be done is to use the Chinese remainder theorem to express the problem as a set of problems, each modulo one of the cyclotomic factors of the polynomial ($z^N - 1$), and use polynomial transforms in solving each of these. What this suggests though, is that a possibility for the root of those polynomial transforms is z^k. The reason is that any root raised to the Nth power must be congruent to 1 modulo P(z). For z^k to be a root means that:

$$z^{Nk} - 1 \ = \ Q(z) \cdot P(z)$$

where $Q(z)$ is some other polynomial of the structure, must hold true. Provided (N, k) is one, ($z^{Nk} - 1$) can always be factored to:

$$z^{Nk} - 1 \ = \ (z^N - 1) \cdot (z^{N(k-1)} + z^{N(k-2)} + \ \cdots \ + z + 1)$$

and so this condition is satisfied if ($z^N - 1$) has P(z) as one of its (cyclotomic) factors. Proving an inverse exists and also that the ($\omega(z)^i - 1$) are not zero modulo P(z) is trivial. Therefore, as z^k is a root of order N for any of the cyclotomic factors of ($z^N - 1$), it may be used in polynomial transforms for computing all the subproblems of this cyclic convolution problem.

One root of the form z^k that always exists is z^1. With this root, the polynomial transform becomes:

$$X(k) \ = \ \left\langle \sum_{n=0}^{N-1} x(n) \cdot z^{kn} \right\rangle_{P(z)} \qquad\qquad k = 0, 1, \ \cdots, N-1$$

Multiplications by z are just shifts. This transform therefore only involves shifts and additions and that makes it similar to NTTs but without many of their restrictions.

Consider the computation of cyclic convolution via polynomial transforms. As those polynomial transforms only involve shifts and additions, the only multiplications must be when the two polynomial transforms are multiplied. Since they are polynomials and as the product of two polynomials modulo a third is one of the principal concerns of this Part, we know this can be done very efficiently. Hence the polynomial transform approach offers a useful addition to the other techniques studied.

DEVELOPMENTS

Polynomial theory and the DFT

Polynomial theory is a very useful tool to show the close links between the DFT and convolution. It is appropriate to begin the section, therefore, by using polynomial theory in a simple way to compute the DFT. As an added bonus, this work will also point to some very attractive digital filter implementations for the DFT.

The DFT is:

$$X(k) = \sum_{n=0}^{N-1} x(n) \cdot W^{kn} \qquad\qquad k = 0, 1, 2, \cdots, N-1$$

Arranging the input values { $x(n)$ } into a polynomial:

$$\Psi(z) = x(0) + x(1) \cdot z + x(2) \cdot z^2 + \cdots + x(N-1) \cdot z^{N-1}$$

then it can be expressed as:

$$X(k) = <\Psi(z)>_{z-W^k} \qquad\qquad k = 0, 1, 2, \cdots, N-1$$

Provided ($z - W^k$) is a divisor of $Q(z)$, this is unchanged if it is expressed as:

$$X(k) = << \Psi(z)>_{Q(z)}>_{z-W^k} \qquad\qquad k = 0, 1, 2, \cdots, N-1$$

Taking the product of all N of these terms gives the polynomial ($z^N - 1$). Choosing $Q(z)$ as this product, the DFT can be described as:

$$X(k) = << \Psi(z)>_{z^N-1}>_{z-W^k} \qquad\qquad k = 0, 1, 2, \cdots, N-1$$

This result can be extended to develop algorithms which minimize the number of operations. They are based on expressing the DFT as a nested set of polynomial reductions modulo various polynomials each divisible by ($z - W^k$). The number of operations is minimized by ensuring most of these polynomials have real coefficients so giving a real-valued result. At best, only one stage of reduction by a complex value is required, hence these algorithms need around half as many multiplications as conventional DFT algorithms.

We will investigate two algorithms. Both are similar and similar in turn to the Rader-Brenner algorithm. They both require N to be 2^α and just differ in their factorization of the polynomial ($z^N - 1$). Like the Rader-Brenner algorithm, they substantially reduce the number of operations by field constants.

The first is based on an algorithm developed by Bruun. If N is a power of 2, then:

$$z^N - 1 = (z^{\frac{N}{2}} + 1) \cdot (z^{\frac{N}{2}} - 1)$$

The product of all the terms ($z - W^k$) when k is even equals ($z^{\frac{N}{2}} - 1$) and the product when k is odd gives ($z^{\frac{N}{2}} + 1$). Therefore, the DFT can be expressed as:

$$X(k) = \; << \psi(z) >_{z^{\frac{N}{2}} - 1} >_{z - W^k} \qquad\qquad \text{for } k = 0, 2, 4, \cdots, N{-}2$$

$$= \; << \psi(z) >_{z^{\frac{N}{2}} + 1} >_{z - W^k} \qquad\qquad \text{for } k = 1, 3, 5, \cdots, N{-}1$$

Now provided all the input terms { x(n) } are real:

$$\langle \psi(z) \rangle_{z^{\frac{N}{2}} - 1} = \sum_{n=0}^{\frac{N}{2} - 1} \left(x(n) + x\left(\frac{N}{2} + n\right) \right) \cdot z^n$$

involves no complex operations at all and neither does:

$$\langle \psi(z) \rangle_{z^{\frac{N}{2}} + 1} = \sum_{n=0}^{\frac{N}{2} - 1} \left(x(n) - x\left(\frac{N}{2} + n\right) \right) \cdot z^n$$

Consider the even values of k. The calculation here can be described as:

$$X(k) = \; << \psi(z) >_{z^{\frac{N}{2}} - 1} >_{z - W^{2j}} \qquad\qquad j = 0, 1, 2, \cdots, \frac{N}{2} - 1$$

Now:

$$(z^{\frac{N}{2}} - 1) = (z^{\frac{N}{4}} + 1) \cdot (z^{\frac{N}{4}} - 1)$$

and the product of ($z - W^j$) for all the even terms of j gives ($z^{\frac{N}{4}} - 1$) and equally, the product of the odd terms gives ($z^{\frac{N}{4}} + 1$). Hence we can express this stage as:

$$X(k) = \; <<< \psi(z) >_{z^{\frac{N}{2}} - 1} >_{z^{\frac{N}{4}} - 1} >_{z - W^k} \qquad\qquad \text{for } k = 0, 4, 8, \cdots$$

$$= \; <<< \psi(z) >_{z^{\frac{N}{2}} - 1} >_{z^{\frac{N}{4}} + 1} >_{z - W^k} \qquad\qquad \text{for } k = 2, 6, 10, \cdots$$

Consider the odd terms of k which give the product ($z^{\frac{N}{2}} + 1$). It can be shown:

$$z^{4\beta} + \gamma z^{2\beta} + 1 = (z^{2\beta} + \sqrt{2-\gamma} \cdot z^{\beta} + 1) \cdot (z^{2\beta} - \sqrt{2-\gamma} \cdot z^{\beta} + 1)$$

Then:

$$z^{\frac{N}{2}} + 1 = (z^{\frac{N}{4}} + \sqrt{2} \cdot z^{\frac{N}{8}} + 1) \cdot (z^{\frac{N}{4}} - \sqrt{2} \cdot z^{\frac{N}{8}} + 1)$$

and so these odd terms can be defined as:

$$X(k) = \; <<< \psi(z) >_{z^{\frac{N}{2}}+1} \quad >_{z^{\frac{N}{4}} + \sqrt{2} \cdot z^{\frac{N}{8}} +1} \quad >_{z-W^k}$$

for those values of $(z-W^k)$ which are factors of $(z^{\frac{N}{4}} + \sqrt{2} \cdot z^{\frac{N}{8}} + 1)$ and:

$$X(k) = \; <<< \psi(z) >_{z^{\frac{N}{2}}+1} \quad >_{z^{\frac{N}{4}} - \sqrt{2} \cdot z^{\frac{N}{8}} +1} \quad >_{z-W^k}$$

for those values of $(z-W^k)$ which are factors of $(z^{\frac{N}{4}} - \sqrt{2} \cdot z^{\frac{N}{8}} + 1)$. This is a little more complex than the previous step but not unnecessarily so.

This procedure can be repeated. At each stage, the set of values of X(k) being treated is divided into two and a new reduction polynomial formed for each of these two sets. That polynomial has an order half that of the previous reduction polynomial, hence overall, there must be $(\alpha-1)$ stages involving real operations only.

Example

The simplest example of merit that we can consider is N equal to 8. Thus:

$$X(k) = \; << x(z) >_{z^8-1} >_{z-W^k} \qquad \text{for } k = 0, 1, 2, \cdots 7$$

The algorithm will have three stages based on the factorization of (z^8-1), namely:

$$
\begin{aligned}
z^8-1 &= (z^4-1) \cdot (z^4+1) \\
&= (z^2-1) \cdot (z^2+1) \cdot (z^2 + \sqrt{2} \cdot z + 1) \cdot (z^2 - \sqrt{2} \cdot z + 1) \\
&= (z-1) \cdot (z+1) \cdot (z+j) \cdot (z-j) \cdot (z-W^3) \cdot (z-W^5) \cdot (z-W) \cdot (z-W^7)
\end{aligned}
$$

Graphically:

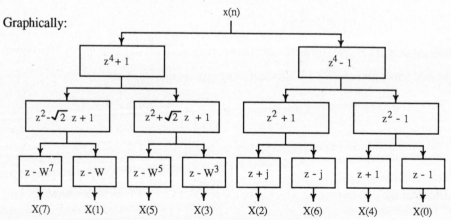

The formulation here is slightly different from Bruun's but captures the essence of his procedure. It is an attractive way of deriving a digital filter realization of a DFT and is similar to the Rader-Brenner algorithm in that all products up to the last stage are by real numbers. It has the disadvantage though, of involving transcendental terms in the multiplications. For real-valued data sequences, further efficiencies can be made based on the properties of those sequences and this gives an especially attractive algorithm for both software and hardware realizations.

The second algorithm is based on the fact that ($z^N - 1$) can always be expressed as the product of cyclotomic polynomials when N is even. In this case, because N is 2^α:

$$z^N - 1 = (z-1) \cdot \prod_{i=0}^{R-1} \left(z^{2^i} + 1 \right)$$

An identical formulation to the first algorithm can be made, giving:

$$X(0) \qquad = \ < X(z) >_{z-1}$$
$$X((\ 2k+1 \ )2^{R-1-i}) = \ << X(z) >_{z^{2^i}+1} \ >_{z-W^{(2k+1)2^{R-1-i}}}$$

We require a factorization for the cyclotomic polynomials ($z^{2i} + 1$). One is:

$$z^{2i} + 1 = (\ z^{\frac{2i}{2}} - j \ ) \cdot (\ z^{\frac{2i}{2}} + j \ )$$

Substituting and arranging terms, the resulting algorithm is found to be the split radix algorithm.

In both these algorithms, a procedure of reducing the initial polynomial formulation of the DFT first by a higher order polynomial then by the final reducing polynomial has gained practical benefits. We could include here the Rader-Brenner algorithm as it can be expressed in similar fashion. These algorithms are quite efficient, largely requiring just real multiplications, and are easily implemented in both hardware and software. In spite of the more sophisticated tacks that will be taken in the rest of this section, these remain among the best (in the sense of minimal multiplications and minimal additions) and most easily implemented DFT algorithms. Note, too, that this procedure can be used to derive any of the DFT formulations given earlier; it is just a question of the polynomial factorization used.

Short-length convolution algorithms

Winograd's theorem does not specify how to construct algorithms for digital convolution with a minimum number of multiplications; it merely states that minimum number. Nevertheless, it does hint at approaches. The sketch proof given of the theorem is expressed in terms of reducing a convolution modulo P(z) to a series of simple polynomial congruences, each modulo an irreducible polynomial divisor of P(z), and combining these results via the Chinese remainder theorem to give the final result. The simple congruences require an efficient algorithm and in the sketch Lagrange interpolation was suggested. Therefore, our initial efforts need to be directed at examining the effectiveness and limitations of Lagrange interpolation and of the remainder theorem as a combination procedure.

For cyclic convolution in a structure with elements drawn from the complex field:

$$P(z) = z^N - 1 = \prod_{i=0}^{N-1} \left(z - W_N^i\right) \qquad \text{where } W = e^{-j\frac{2\pi}{N}}$$

and the irreducible polynomials $P_i(z)$ are $(z - W_N^i)$. As k is N here, the minimal number of multiplications is $(2N - N)$ or N multiplications. Following the outline given in theory, constructing an algorithm means first forming:

$$\langle X(z) \rangle_{P_i(z)} = X\left(W_N^i\right) = \sum_{n=0}^{N-1} x(n) W_N^{ni}$$

$$\langle X(z) \rangle_{P_i(z)} = X\left(W_N^i\right) = \sum_{n=0}^{N-1} x(n) W_N^{ni}$$

to construct an algorithm and then the polynomials:

$$Y_i(z) = \langle H(z) \cdot X(z) \rangle_{P_i(z)} = H(W_N^i) \cdot X(W_N^i)$$

The convolution — the polynomial $Y(z)$ — is formed from the polynomials $Y_i(z)$ according to the Chinese remainder theorem reconstruction. This requires polynomials:

$$\langle G_i(z) \cdot Q_i(z) \rangle_{P_j(z)} \begin{aligned} &= 1 \text{ for } i = j \\ &= 0 \text{ otherwise} \end{aligned}$$

where
$$Q_i(z) = \prod_{j=0, j \neq i}^{N-1} P_j(z) = \prod_{j=0, j \neq i}^{N-1} \left(z - W_N^j\right) = \frac{\left(z^N - 1\right)}{\left(z - W_N^i\right)}$$

As $(z - W_N^i)$ is a factor of $Q_i(z)$, then $\langle Q_i(z) \rangle_{P_i(z)}$ is zero. Also, since:

$$Q(z) = \left(z^{N-1} + W_N^i \cdot z^{N-2} + W_N^{2i} \cdot z^{N-3} + \cdots + W_N^{(N-1)i} \right)$$

then:

$$Q(W_N^i) = \left(W_N^{(N-1)i} + W_N^{(N-1)i} + W_N^{(N-1)i} + \cdots + W_N^{(N-1)i} \right)$$

$$= N \cdot W_N^{(N-1)i}$$

$$= N \cdot W_N^{(-i)}$$

This implies $G_i(z)$ must be some constant α_i. In fact:

$$\begin{aligned} \alpha_0 &= N^{-1} \\ \alpha_i &= N^{-1} \cdot W_N^i \end{aligned} \qquad i = 1, 2, \cdots, N-1$$

Therefore:

$$Y(z) = \left\langle \sum_{i=0}^{N-1} \left(N^{-1} \cdot W_N^i \cdot Q_i(z) \right) \cdot H\left(W_N^i\right) \cdot X\left(W_N^i\right) \right\rangle_{P(z)}$$

$$= \left\langle \sum_{i=0}^{N-1} \left(N^{-1} \cdot W_N^i \cdot \sum_{j=0}^{N-1} W^{i(N-1-j)} \cdot z^j \right) \cdot H\left(W_N^i\right) \cdot X\left(W_N^i\right) \right\rangle_{P(z)}$$

$$= \left\langle \sum_{j=0}^{N-1} z^j \left\{ N^{-1} \cdot \sum_{i=0}^{N-1} W^{i(N-j)} \cdot H\left(W_N^i\right) \cdot X\left(W_N^i\right) \right\} \right\rangle_{P(z)}$$

$$= \left\langle \sum_{j=0}^{N-1} z^j \cdot y(j) \right\rangle_{P(z)}$$

where $\qquad P(z) = z^N - 1$

$$y(j) = N^{-1} \cdot \sum_{i=0}^{N-1} H\left(W_N^i\right) \cdot X\left(W_N^i\right) \cdot W_N^{-ij}$$

This is an interesting result. The algorithm with minimal multiplications for computing convolutions in the complex field is the DFT. In the matrix form:

$$\mathbf{y} = \mathbf{C} \cdot (\mathbf{Bh} \bullet \mathbf{Ax})$$

we recognize A and B as being the forward DFT transformation and C as the inverse.

While it is pleasing to derive a result like this, practical considerations temper any enthusiasm. The DFT does have just N general multiplications but the multiplications involving constants from the field are hardly insignificant. Part 2, after all, was devoted to means of circumventing problems caused by these. Nevertheless, what this theorem is also stating is that the best FFT, in whatever sense best is defined, is indeed the optimum approach for computing convolutions in the complex field.

This result clearly demonstrates the need to carefully consider the coefficient field in order to minimize these operations with constants. Of the cases briefly examined in THEORY, the field of real numbers offers few practical advantages but both the rational field and the field of Gaussian integers do. In both cases, irreducible polynomials mainly involve terms of the form 0, 1, −1, j, and −j. Since the former only involves real integers, we shall concentrate on that.

To illustrate the construction of a minimal algorithm in the rational field, we will take a four point convolution as an example. Let:

$$X(z) = x(0) + x(1) \cdot z + x(2) \cdot z^2 + x(3) \cdot z^3$$
$$H(z) = h(0) + h(1) \cdot z + h(2) \cdot z^2 + h(3) \cdot z^3$$

with $\qquad P(z) = z^4 - 1 = (z-1) \cdot (z+1) \cdot (z^2+1) = P_1(z) \cdot P_2(z) \cdot P_4(z)$

As P(z) factors into three irreducible polynomials in this field, the minimal number of multiplications must be five. The approach taken in constructing the minimal algorithm is exactly as for the complex field. Form the polynomials $X_i(z)$ and $H_i(z)$ where these are given by $< X(z) >_{P_i(z)}$ and $< H(z) >_{P_i(z)}$. From these, construct polynomials $Y_i(z)$ given by $X_i(z) \cdot H_i(z)$. The required result $Y(z)$ is formed from these polynomials via Chinese remainder theorem reconstruction.

Consider the polynomials $X_i(z)$. Then:

$$
\begin{aligned}
X_1(z) &= < X(z) >_{z-1} \\
&= < x(0) + x(1) \cdot z + x(2) \cdot z^2 + x(3) \cdot z^3 >_{z-1} \\
&= x(0) + x(1) + x(2) + x(3) \\
&= x_{10}
\end{aligned}
$$

$$
\begin{aligned}
X_2(z) &= < X(z) >_{z+1} \\
&= < x(0) + x(1) \cdot z + x(2) \cdot z^2 + x(3) \cdot z^3 >_{z+1} \\
&= x(0) - x(1) + x(2) - x(3) \\
&= x_{20}
\end{aligned}
$$

$$
\begin{aligned}
X_4(z) &= < X(z) >_{z^2+1} \\
&= < x(0) + x(1) \cdot z + x(2) \cdot z^2 + x(3) \cdot z^3 >_{z^2+1} \\
&= (x(0) - x(2)) + (x(1) - x(3)) \cdot z \\
&= x_{40} + x_{41} \cdot z
\end{aligned}
$$

Similarly:

$$
\begin{aligned}
H_1(z) &= < H(z) >_{z-1} &&= h_{10} \\
H_2(z) &= < H(z) >_{z+1} &&= h_{20} \\
H_4(z) &= < H(z) >_{z^2+1} &&= h_{40} + h_{41} \cdot z
\end{aligned}
$$

The polynomials $Y_i(z)$ are given by:

$$
Y_i(z) = < H_i(z) \cdot X_i(z) >_{P_i(z)}
$$

hence:

$$
\begin{aligned}
Y_1(z) &= y_{10} = h_{10} \cdot x_{10} \\
Y_2(z) &= y_{20} = h_{20} \cdot x_{20} \\
Y_4(z) &= < (h_{40} + h_{41} \cdot z) \cdot (x_{40} + x_{41} \cdot z) >_{z^2+1} \\
&= < h_{40} \cdot x_{40} + (h_{40} \cdot x_{41} + x_{40} \cdot h_{41}) \cdot z + h_{41} \cdot x_{41} \cdot z^2 >_{z^2+1} \\
&= (h_{40} \cdot x_{40} - h_{41} \cdot x_{41}) + (h_{40} \cdot x_{41} + x_{40} \cdot h_{41}) \cdot z \\
&= y_{40} + y_{41} \cdot z
\end{aligned}
$$

These three equations require six multiplications whereas the minimum number is five. This is not unexpected. A minimal algorithm requires the use of an efficient algorithm such as Toom-Cook and that has not been used here. However, because this is a simple problem, such an algorithm is easily derived by heuristic means. For example, let:

$$m_1 = h_{40} \cdot x_{40}$$
$$m_2 = h_{41} \cdot x_{41}$$
$$m_3 = (h_{40} + h_{41}) \cdot (x_{40} + x_{41})$$

Then:

$$y_{40} = m_1 - m_2$$
$$y_{41} = m_3 - m_2 - m_1$$

This requires only five multiplications and additions. Another algorithm is to let:

$$T_1 = h_{40} + h_{41}$$
$$T_2 = x_{40} + x_{41}$$
$$T_3 = -x_{40} + x_{41}$$

and:

$$m_1 = T_1 \cdot x_{40}$$
$$m_2 = T_2 \cdot h_{41}$$
$$m_3 = T_3 \cdot h_{40}$$

Then:

$$y_{40} = m_1 - m_2$$
$$y_{41} = m_1 + m_3$$

This algorithm also requires five multiplications and additions. It will be the one used.

The Chinese remainder theorem reconstruction gives:

$$Y(z) = \langle D_1(z) \cdot Y_1(z) + D_2(z) \cdot Y_2(z) + D_4(z) \cdot Y_4(z) \rangle_{P(z)}$$

where

$$P(z) = z^4 - 1$$
$$\langle D_i(z) \rangle_{P_j(z)} = 1 \text{ for } i = j$$
$$= 0 \text{ otherwise}$$

The form of the polynomials $D_i(z)$ is $G_i(z) \cdot Q_i(z)$ where $Q_i(z)$ is $\prod_{j \neq i} P_j(z)$. Since $Q_i(z)$ has the $P_i(z)$ as factors, then the $G_i(z)$ must be constants α_i. Now:

$$Q_1(z) = z^3 + z^2 + z + 1$$
$$Q_2(z) = z^3 - z^2 + z - 1$$
$$Q_4(z) = z^2 - 1$$

Thus computing $\langle \alpha_i \cdot Q_i(z) \rangle_{P_j(z)}$ to find the constants α_i, we discover:

$$D_1(z) = \frac{1}{4} \cdot (z^3 + z^2 + z + 1)$$

$$D_2(z) = -\frac{1}{4} \cdot (z^3 + z^2 + z - 1)$$

$$D_4(z) = -\frac{1}{2} \cdot (z^2 - 1)$$

Then:

$$Y(z) = < \frac{1}{4}(z^3 + z^2 + z + 1) \cdot y_{10} - \frac{1}{4}(z^3 - z^2 + z - 1) \cdot y_{20} - \frac{1}{2}(z^2 - 1) \cdot (y_{41} \cdot z + y_{40}) > P(z)$$

Defining:

$$Y(z) = y(0) + y(1) \cdot z + y(2) \cdot z^2 + y(3) \cdot z^3$$

we obtain:

$$y(0) = \frac{1}{4} \cdot (y_{10} + y_{20} + 2 \cdot y_{40})$$

$$y(1) = \frac{1}{4} \cdot (y_{10} - y_{20} + 2 \cdot y_{41})$$

$$y(2) = \frac{1}{4} \cdot (y_{10} + y_{20} - 2 \cdot y_{40})$$

$$y(3) = \frac{1}{4} \cdot (y_{10} - y_{20} - 2 \cdot y_{41})$$

This completes the algorithm.

To summarize, we list the equations. The factor $\frac{1}{4}$ will be included with the terms h_{10}, h_{20}, h_{40}, and h_{41} as in practice these terms are likely to be pre-calculated and this minimizes the number of products by constants:

$$x_{10} = x(0) + x(1) + x(2) + x(3)$$
$$x_{20} = x(0) - x(1) + x(2) - x(3)$$
$$x_{40} = x(0) - x(2)$$
$$x_{41} = x(1) - x(3)$$
$$h_{10} = \frac{1}{4} \cdot (h(0) + h(1) + h(2) + h(3))$$
$$h_{20} = \frac{1}{4} \cdot (h(0) - h(1) + h(2) - h(3))$$
$$h_{40} = \frac{1}{4} \cdot (h(0) - h(2))$$
$$h_{41} = \frac{1}{4} \cdot (h(1) - h(3))$$
$$T_1 = h_{40} + h_{41}$$
$$T_2 = x_{40} + x_{41}$$
$$T_3 = -x_{40} + x_{41}$$
$$y_{10} = h_{10} \cdot x_{10}$$
$$y_{20} = h_{20} \cdot x_{20}$$
$$m_1 = T_1 \cdot x_{40}$$
$$m_2 = T_2 \cdot h_{41}$$

$$m_3 = T_3 \cdot h_{40}$$
$$y_{40} = m_1 - m_2$$
$$y_{41} = m_1 + m_3$$
$$y(0) = y_{10} + y_{20} + 2 \cdot y_{40}$$
$$y(1) = y_{10} - y_{20} + 2 \cdot y_{41}$$
$$y(2) = y_{10} + y_{20} - 2 \cdot y_{40}$$
$$y(3) = y_{10} - y_{20} - 2 \cdot y_{41}$$

These equations require five general multiplications, eight multiplications by constants and 29 real additions. The number of additions can be reduced to 23 by combining terms. A systematic procedure for accomplishing this will be discussed later in this section.

These equations can be alternatively expressed in the matrix form:

$$\mathbf{y} = \mathbf{C} \, (\, \mathbf{Bh} \bullet \mathbf{Ax} \,)$$

The five minimal multiplications are those forming y_{10}, y_{20}, m_1, m_2, and m_3. These intermediate terms are the products of x_{10}, x_{20}, x_{40}, $(x_{40} + x_{41})$, and $(-x_{40} + x_{41})$ by h_{10}, h_{20}, $(h_{40} + h_{41})$, h_{41}, and h_{40}. As the input terms are $\{ x(0), x(1), x(2), x(3) \}$ and $\{ h(0), h(1), h(2), h(3) \}$, then:

$$A = \begin{bmatrix} 1 & 0 & 0 & 0 \\ 0 & 1 & 0 & 0 \\ 0 & 0 & 1 & 0 \\ 0 & 0 & 1 & 1 \\ 0 & 0 & -1 & 1 \end{bmatrix} \begin{bmatrix} 1 & 1 & 1 & 1 \\ 1 & -1 & 1 & -1 \\ 1 & 0 & -1 & 0 \\ 0 & 1 & 0 & -1 \end{bmatrix} = \begin{bmatrix} 1 & 1 & 1 & 1 \\ 1 & -1 & 1 & -1 \\ 1 & 0 & -1 & 0 \\ 1 & 1 & -1 & -1 \\ -1 & 1 & 1 & -1 \end{bmatrix}$$

In this case, matrices A and B are not the same due to the asymmetric heuristic algorithm chosen. Thus:

$$B = \frac{1}{4} \begin{bmatrix} 1 & 0 & 0 & 0 \\ 0 & 1 & 0 & 0 \\ 0 & 0 & 1 & 1 \\ 0 & 0 & 0 & 1 \\ 0 & 0 & 1 & 0 \end{bmatrix} \begin{bmatrix} 1 & 1 & 1 & 1 \\ 1 & -1 & 1 & -1 \\ 1 & 0 & -1 & 0 \\ 0 & 1 & 0 & -1 \end{bmatrix} = \frac{1}{4} \begin{bmatrix} 1 & 1 & 1 & 1 \\ 1 & -1 & 1 & -1 \\ 1 & 1 & -1 & -1 \\ 0 & 1 & 0 & -1 \\ 1 & 0 & -1 & 0 \end{bmatrix}$$

Determining matrix C follows in much the same manner. It is a case of relating the output terms to the product terms, hence:

$$C = \begin{bmatrix} 1 & 1 & 2 & 0 \\ 1 & -1 & 0 & 2 \\ 1 & 1 & -2 & 0 \\ 1 & -1 & 0 & -2 \end{bmatrix} \begin{bmatrix} 1 & 0 & 0 & 0 & 0 \\ 0 & 1 & 0 & 0 & 0 \\ 0 & 0 & 1 & -1 & 0 \\ 0 & 0 & 1 & 0 & 1 \end{bmatrix} = \begin{bmatrix} 1 & 1 & 2 & -2 & 0 \\ 1 & -1 & 2 & 0 & 2 \\ 1 & 1 & -2 & 2 & 0 \\ 1 & -1 & -2 & 0 & -2 \end{bmatrix}$$

Although there are only five general multiplications in this equation, there are 36 additions among the constants, eight multiplications by 2 and scaling of some terms by $\frac{1}{4}$. Matrix A only has elements drawn from $\{\,0,-1,1\,\}$. Matrix B is similar but it has the factor $\frac{1}{4}$. If Bh is pre–calculated, that presents no practical problems. Matrix C is more complicated than either of A or B and, if the factor $\frac{1}{4}$ in B is ignored, accounts for all of the multiplications by constants. Its elements though, are only drawn from $\{\,0,1,-1,2,-2\,\}$.

The number of multiplications by constants can be reduced by reverting to the dual algorithm. There are several possibilities here but only the case described earlier will be considered. That is, to simply reverse the order of all rows or columns except the first of the respective matrices and transpose. Take B and C as the matrices. Then form:

$$B_s = \begin{bmatrix} 1 & 1 & 1 & 1 \\ 1 & -1 & 1 & -1 \\ 1 & -1 & -1 & 1 \\ 0 & -1 & 0 & 1 \\ 1 & 0 & -1 & 0 \end{bmatrix} \qquad C_s = \frac{1}{4}\begin{bmatrix} 1 & 1 & 2 & -2 & 0 \\ 1 & -1 & -2 & 0 & -2 \\ 1 & 1 & -2 & 2 & 0 \\ 1 & -1 & 2 & 0 & 2 \end{bmatrix}$$

The dual vector equation is:

$$y = B_s^{\,T}\!\cdot(\,C_s^{\,T}h\!\cdot\!Ax\,)$$

This can be alternatively expressed as the set of equations:

$$
\begin{aligned}
x_1 &= x(0) + x(1) + x(2) + x(3) \\
x_2 &= x(0) - x(1) + x(2) - x(3) \\
x_3 &= x(0) \qquad\;\; - x(2) \\
x_4 &= x(0) + x(1) - x(2) - x(3) \\
x_5 &= -x(0) + x(1) + x(2) - x(3)
\end{aligned}
$$

$$
\begin{aligned}
h_1 &= \tfrac{1}{4}\,(\; h(0) + h(1) + h(2) + h(3)\;) \\
h_2 &= \tfrac{1}{4}\,(\; h(0) - h(1) + h(2) - h(3)\;) \\
h_3 &= \tfrac{1}{2}\,(\; h(0) - h(1) - h(2) + h(3)\;) \\
h_4 &= \tfrac{1}{2}\,(- h(0) \qquad\quad + h(2)\qquad\;\;) \\
h_5 &= \tfrac{1}{2}\,(\qquad\;\; - h(1) \qquad\quad + h(3)\;)
\end{aligned}
$$

$$
\begin{aligned}
y_1 &= x_1\!\cdot\! h_1 \\
y_2 &= x_2\!\cdot\! h_2 \\
y_3 &= x_3\!\cdot\! h_3 \\
y_4 &= x_4\!\cdot\! h_4 \\
y_5 &= x_5\!\cdot\! h_5
\end{aligned}
$$

$$
\begin{aligned}
y(0) &= y_1 + y_2 + y_3 + \qquad y_5 \\
y(1) &= y_1 - y_2 - y_3 - \qquad y_4 \\
y(2) &= y_1 + y_2 - y_3 - \qquad y_5 \\
y(3) &= y_1 - y_2 + y_3 + \qquad y_4
\end{aligned}
$$

This has five general multiplications, five multiplications by constants, and 36 additions. The latter can be reduced to 25 additions by combining terms. If the h_i terms are pre-calculated, then by combining terms the algorithm can be reduced to five general multiplications and 18 additions.

How does this approach to computing convolutions compare to an FFT approach? Assume real input data and that one set of input terms is pre-calculated. A four-point optimal FFT requires two multiplications by j, two complex additions and six real additions to form its complex result. The product of the two Fourier transforms gives the Fourier transform of the convolution. As two of the terms are real, this requires just two complex multiplications and two real. A complex inverse transform then generates the required result with eight complex additions, two multiplications by j plus four real multiplications for the scaling operation. If the conventional complex multiplication algorithm is used, the total number of operations is 14 real multiplications and 30 real additions. If the minimal multiplication complex multiplication algorithm is used, the number is 12 real multiplications and 32 real additions. The polynomial approach therefore compares quite favorably.

Although this appears to be a very attractive means of executing convolutions, this section is labeled 'short-length convolutions' for good reason. As N grows two problems arise. First, the number of additions begins to increase very substantially. It is quite difficult to give an explicit formula for this as the number fluctuates quite strongly with N. Second, problems arise with the algorithms for forming the products $H(z_i) \cdot X(z_i)$. In Lagrange polynomials, the coefficients are only 0, –1 or 1 for up to third order polynomials. For nth degree polynomials, the coefficients can be numbers up to $\pm\left(\frac{n-1}{2}\right)$. These numbers will appear in the matrices A and B, so they will no longer be simple.

As a result of these two problems, the FFT begins to gain an advantage for relatively small N. Of course, there is no need to restrict the algorithm to the minimum number of multiplications as any inefficiency in the number of general multiplications can be more than compensated for by the simplicity of the A, B, and C matrices. Earlier work, too, has shown there is scope for minimizing the number of additions by defining new terms. Once N exceeds about 20 though, the FFT approach is better in all respects. Thus Chinese remainder theorem decomposition is a short-length convolution procedure only.

Can anything be done about the number of additions? Or, to be more precise, can anything systematic be done? To lead into this question, consider the first of these two 4–point convolution algorithms. Define terms:

$$
\begin{aligned}
s_1 &= x(0) + x(2) & s_2 &= x(1) + x(3) \\
s_3 &= x(0) - x(2) & s_4 &= x(1) - x(3)
\end{aligned}
$$

so that:

$$
\begin{aligned}
X_{10} &= s_1 + s_2 \\
X_{20} &= s_1 - s_2 \\
X_{40} &= s_3 \\
X_{41} &= s_4
\end{aligned}
$$

These reduce the number of additions from eight to six. A similar result applies to the h terms. Using these results, matrix A can be expressed as:

$$A = \begin{bmatrix} 1 & 1 & 0 & 0 \\ 1 & -1 & 0 & 0 \\ 0 & 0 & 1 & 0 \\ 0 & 0 & 1 & 1 \\ 0 & 0 & -1 & 1 \end{bmatrix} \begin{bmatrix} 1 & 1 & 0 & 0 \\ 1 & -1 & 0 & 0 \\ 0 & 0 & 1 & 0 \\ 0 & 0 & 0 & 1 \end{bmatrix} \begin{bmatrix} 1 & 0 & 1 & 0 \\ 0 & 1 & 0 & 1 \\ 1 & 0 & -1 & 0 \\ 0 & 1 & 0 & -1 \end{bmatrix}$$

In a similar fashion, matrix C can be expressed as:

$$C = \begin{bmatrix} 1 & 0 & 1 & 0 \\ 0 & 1 & 0 & 1 \\ 1 & 0 & -1 & 0 \\ 0 & 1 & 0 & -1 \end{bmatrix} \begin{bmatrix} 1 & 1 & 0 & 0 \\ 1 & -1 & 0 & 0 \\ 0 & 0 & 2 & 0 \\ 0 & 0 & 0 & 2 \end{bmatrix} \begin{bmatrix} 1 & 0 & 0 & 0 & 0 \\ 0 & 1 & 0 & 0 & 0 \\ 0 & 0 & 1 & -1 & 0 \\ 0 & 0 & 1 & 0 & 1 \end{bmatrix}$$

Thus minimizing the number of additions can be viewed as a problem of matrix factorization.

It would be remiss not to mention a quite effective and flexible approach to devising short-length convolution algorithms. This is to create them simply by inspection. Some obvious difficulties arise as N increases but there are also some advantages. For one, it is often the best way of finding a balance between the number of additions and multiplications in an algorithm. It has the disadvantage though of not being optimal. To illustrate the approach, consider a four-point aperiodic convolution. Listing the equations:

$$\begin{aligned} Y(z) \ = \ & h(0) \cdot x(0) + (\, h(1) \cdot x(0) + h(0) \cdot x(1) \,) \cdot z \\ & + (\, h(2) \cdot x(0) + h(1) \cdot x(1) + h(0) \cdot x(2) \,) \cdot z^2 \\ & + (\, h(3) \cdot x(0) + h(2) \cdot x(1) + h(1) \cdot x(2) + h(0) \cdot x(3) \,) \cdot z^3 \\ & + (\, h(3) \cdot x(1) + h(2) \cdot x(2) + h(1) \cdot x(3) \,) \cdot z^4 \\ & + (\, h(3) \cdot x(2) + h(2) \cdot x(3) \,) \cdot z^5 + h(3) \cdot x(3) \cdot z^6 \\[4pt] = \ & y(0) + y(1) \cdot z + y(2) \cdot z^2 + y(3) \cdot z^3 + y(4) \cdot z^4 + y(5) \cdot z^5 + y(6) \cdot z^6 \end{aligned}$$

A totally naive approach to constructing an algorithm is inadvisable, thus a procedure will be adopted which includes a measure of systematic development together with some intuition. The systematic part uses the fact that:

$$\begin{aligned} x_i \cdot h_j + x_j \cdot h_i \ & = \ (\, x_i + x_j \,) \cdot (\, h_i + h_j \,) - x_i \cdot h_i - x_j \cdot h_j \\ & = \ x_i \cdot h_i + x_j \cdot h_j - (\, x_i - x_j \,) \cdot (\, h_i - h_j \,) \end{aligned}$$

Substituting these alternative forms and taking advantage of the resulting symmetry can result in algorithms with fewer than N^2 multiplications and $(\, N - 1 \,)^2$ additions. Intuition comes into play in deciding the exact form and number of these substitutions. To illustrate, expand all terms in the convolution and re-group into common, symmetrical, factors. This gives:

$$y(0) \ = \ x(0) \cdot h(0)$$

$$y(1) \ = \ (\, x(0) + x(1) \,) \cdot (\, h(0) + h(1) \,) - x(0) \cdot h(0) - x(1) \cdot h(1)$$

$$y(2) = (x(0) + x(1) + x(2))\cdot(h(0) + h(1) + h(2)) - (x(0) + x(1))\cdot(h(0) + h(1))$$
$$- (x(1) + x(2))\cdot(h(1) + h(2)) + 2\cdot x(1)\cdot h(1)$$

$$y(3) = (x(0) + x(1) + x(2) + x(3))\cdot(h(0) + h(1) + h(2) + h(3))$$
$$- (x(0) + x(1) + x(2))\cdot(h(0) + h(1) + h(2))$$
$$- (x(1) + x(2) + x(3))\cdot(h(1) + h(2) + h(3))$$
$$+ 2\cdot(x(1) + x(2))\cdot(h(1) + h(2)) - x(1)\cdot h(1) - x(2)\cdot h(2)$$

$$y(4) = (x(1) + x(2) + x(3))\cdot(h(1) + h(2) + h(3)) - (x(2) + x(3))\cdot(h(2) + h(3))$$
$$- (x(1) + x(2))\cdot(h(1) + h(2)) + 2\cdot x(2)\cdot h(2)$$

$$y(5) = (x(2) + x(3))\cdot(h(2) + h(3)) - x(3)\cdot h(3) - x(2)\cdot h(2)$$

$$y(6) = x(3)\cdot h(3)$$

i.e.

$$x_0 = x(0) \qquad\qquad h_0 = h(0)$$
$$x_1 = x(1) \qquad\qquad h_1 = h(1)$$
$$x_2 = x(2) \qquad\qquad h_2 = h(2)$$
$$x_3 = x(3) \qquad\qquad h_3 = h(3)$$
$$x_4 = x(0) + x(1) \qquad\qquad h_4 = h(0) + h(1)$$
$$x_5 = x(1) + x(2) \qquad\qquad h_5 = h(1) + h(2)$$
$$x_6 = x(2) + x(3) \qquad\qquad h_6 = h(2) + h(3)$$
$$x_7 = x(0) + x(1) + x(2) \qquad\qquad h_7 = h(0) + h(1) + h(2)$$
$$x_8 = x(1) + x(2) + x(3) \qquad\qquad h_8 = h(1) + h(2) + h(3)$$
$$x_9 = x(0) + x(1) + x(2) + x(3) \qquad\qquad h_9 = h(0) + h(1) + h(2) + h(3)$$

$$m_i = x_i\cdot h_i \qquad\qquad i = 0, 1, \cdots, 9$$

$$y(0) = m_0$$
$$y(1) = m_4 - m_0 - m_1$$
$$y(2) = m_7 - m_4 - m_5 + 2\cdot m_1$$
$$y(3) = m_9 - m_7 - m_8 + 2\cdot m_5 - m_1 - m_2$$
$$y(4) = m_8 - m_6 - m_5 + 2\cdot m_2$$
$$y(5) = m_6 - m_3 - m_2$$
$$y(6) = m_3$$

Only ten essential multiplications, three multiplications by constants and 25 essential additions are required here. Essential in this case meaning additions in the h_i terms are ignored as it is assumed they can be pre-computed. The number of additions can be reduced to 18 by expressing the input terms as :

$$x_7 = x_4 + x(2)$$
$$x_8 = x_6 + x(1)$$
$$x_9 = x_4 + x_6$$

and defining new terms for the output:

$$b_0 = m_4 - m_1 \qquad b_3 = m_6 - m_2$$
$$b_1 = m_7 - m_5 \qquad b_4 = m_1 + b_1$$
$$b_2 = m_2 - m_5 \qquad b_5 = m_8 + b_2$$

These also eliminate the multiplications by constants.

Different expansions of the terms lead to different algorithms. For example:

$$x_0 = x(0) \qquad\qquad\qquad h_0 = h(0)$$
$$x_1 = x(1) \qquad\qquad\qquad h_1 = h(1)$$
$$x_2 = x(2) \qquad\qquad\qquad h_2 = h(2)$$
$$x_3 = x(3) \qquad\qquad\qquad h_3 = h(3)$$
$$x_4 = x(0) + x(1) \qquad\quad h_4 = h(0) + h(1)$$
$$x_5 = x(0) + x(2) \qquad\quad h_5 = h(0) + h(2)$$
$$x_6 = x(1) + x(3) \qquad\quad h_6 = h(1) + h(3)$$
$$x_7 = x(2) + x(3) \qquad\quad h_7 = h(2) + h(3)$$
$$x_8 = x_4 + x_7 \qquad\qquad\quad h_8 = h_4 + h_7$$

$$m_i = x_i \cdot h_i \qquad\qquad i = 0, 1, \cdots, 8$$

$$Y_0 = m_1 - m_2$$
$$y(0) = m_0$$
$$y(1) = m_4 - m_0 - m_1$$
$$y(2) = m_5 - m_0 + Y_0$$
$$y(3) = m_8 - y(0) - y(1) - y(2) - y(4) - y(5) - y(6)$$
$$y(4) = m_6 - m_3 - Y_0$$
$$y(5) = m_7 - m_3 - m_2$$
$$y(6) = m_3$$

This algorithm requires nine essential multiplications, no multiplications by constants and 20 essential additions. This compares quite well against the optimal Winograd algorithm. This and similar algorithms are developed by noting that the cross diagonal terms in:

$$\begin{array}{cccc} x_0 \cdot h_0 & x_1 \cdot h_0 & x_2 \cdot h_0 & x_3 \cdot h_0 \\ x_0 \cdot h_1 & x_1 \cdot h_1 & x_2 \cdot h_1 & x_3 \cdot h_1 \\ x_0 \cdot h_2 & x_1 \cdot h_2 & x_2 \cdot h_2 & x_3 \cdot h_2 \\ x_0 \cdot h_3 & x_1 \cdot h_3 & x_2 \cdot h_3 & x_3 \cdot h_3 \end{array}$$

are the desired product terms and the alternative forms merely generate subarrays of this structure. Experimenting with these alternatives will generate quite a range of algorithms.

Although this approach does not lead to very efficient algorithms, they are not totally unattractive. The algorithms have the same matrix form as the optimal algorithm, namely:

$$y = C \cdot (Ax \bullet Bh)$$

Matrices A and B, and so (usually) matrix C, consist of 1's. The algorithms are easily manipulated and so form a useful approach when a convolution is required modulo some general P(z). Two additional points. First, if this technique is used for cyclic convolution then the transpose algorithm can be applied leading to a further reduction in the number of additions. Second, it is clear that factorization of the matrices is possible and that, too, can lead to a significant reduction in the number of additions.

Convolution algorithms and the DFT

There is one very important application of these short-length algorithms. Rader's result showed prime length DFTs can be converted to cyclic convolutions and here is a very attractive method for implementing short-length convolutions. The DFT is defined for the complex field not the rational and, consequently, we are not employing a Winograd-minimal algorithm for the structure by using these algorithms. However, we gain substantially by having all the field constant terms as very simple terms.

Example

Consider a 5-point DFT. The DFT equations are:

$$X(0) = x(0) + x(1) + x(2) + x(3) + x(4)$$
$$X(k) = x(0) + V(k) \qquad\qquad k = 1, 2, 3, 4$$

where
$$V(k) = \sum_{n=1}^{4} x(n) \cdot W_5^{kn}$$

Define a new set of variables:

$$Y(k) = X(k) - x(0) \qquad\qquad k = 0, 1, 2, 3, 4$$

i.e.
$$Y(0) = x(1) + x(2) + x(3) + x(4)$$
$$Y(k) = V(k) \qquad\qquad k = 1, 2, 3, 4$$

Now define a further set of variables $\zeta(k)$ where:

$$\zeta(k) = Y(k) - Y(0) \qquad\qquad k = 1, 2, 3, 4$$

Substituting:

$$\zeta(k) \quad = \sum_{n=1}^{4} \left(W_5^{kn} - 1 \right) \cdot x(n) \qquad\qquad k = 1, 2, 3, 4$$

$$= \sum_{n=1}^{4} w(kn) \cdot x(n) \qquad\qquad k = 1, 2, 3, 4$$

A primitive root of 5 is 2 as:

$$< 2^0 >_5 = 1 \quad < 2^1 >_5 = 2 \quad < 2^2 >_5 = 4 \quad < 2^3 >_5 = 3 \quad < 2^4 >_5 = 1$$

Defining mappings:

$$n = < 2^{-u} >_5 \qquad\qquad\qquad\qquad k = < 2^v >_5$$

and substituting:

$$\zeta(2^v) = \sum_{u=1}^{4} w\left(2^{v-u}\right) \cdot x\left(2^{-u}\right) \qquad\qquad \text{where } v = 0, 1, 2, 3$$

gives a result describing the convolution of the two sequences { x(1), x(3), x(4), x(2) } and { w(1), w(2), w(4), w(3) } to give the sequence { ζ(1), ζ(2), ζ(4), ζ(3) }. Any algorithm can be used to compute this but the transpose algorithm given earlier will be used. The notation will follow that of the transpose algorithm example given earlier.

Let h(i) represent the set { w(1), w(2), w(4), w(3) }. Thus:

$$
\begin{aligned}
h(0) &= \cos\ 72^\circ - j \cdot \sin\ 72^\circ - 1 \\
h(1) &= \cos 144^\circ - j \cdot \sin 144^\circ - 1 \\
h(2) &= \cos 288^\circ - j \cdot \sin 288^\circ - 1 \\
h(3) &= \cos 216^\circ - j \cdot \sin 216^\circ - 1
\end{aligned}
$$

Substituting into the transpose algorithm equations:

$$
\begin{aligned}
h_1 &= -1.250000 \\
h_2 &= 0.559017 \\
h_3 &= -j\,0.363271 \\
h_4 &= j\,0.951057 \\
h_5 &= j\,0.587785
\end{aligned}
$$

The x_i terms here are obtained by substituting from the set { x(1), x(3), x(4), x(2) }:

$$
\begin{aligned}
x_1 &= x(1) \;+\; x(2) \;+\; x(3) \;+\; x(4) \\
x_2 &= x(1) \;-\; x(2) \;-\; x(3) \;+\; x(4) \\
x_3 &= x(1) \;-\; x(4) \\
x_4 &= x(1) \;-\; x(2) \;+\; x(3) \;-\; x(4) \\
x_5 &= -x(1) \;-\; x(2) \;+\; x(3) \;+\; x(4)
\end{aligned}
$$

The five general multiplications are therefore:

$$
\begin{aligned}
y_1 &= h_1 \cdot x_1 = -1.250000 \cdot x_1 \\
y_2 &= h_2 \cdot x_2 = 0.559017 \cdot x_2 \\
y_3 &= h_3 \cdot x_3 = -j\,0.363271 \cdot x_3 \\
y_4 &= h_4 \cdot x_4 = j\,0.951057 \cdot x_4 \\
y_5 &= h_5 \cdot x_5 = j\,0.587785 \cdot x_5
\end{aligned}
$$

and so the DFT is given by:

$$X(0) = x(0) + x(1) + x(2) + x(3) + x(4)$$
$$X(1) = X(0) + y_1 + y_2 + y_3 + 0 + y_5$$
$$X(2) = X(0) + y_1 - y_2 - y_3 - y_4 + 0$$
$$X(3) = X(0) + y_1 - y_2 + y_3 + y_4 - 0$$
$$X(4) = X(0) + y_1 + y_2 - y_3 + 0 - y_5$$

It is useful to express this in matrix form. To begin, note that the form of the initial DFT matrix is:

$$
\begin{bmatrix} X(0) \\ X(1) \\ X(2) \\ X(3) \\ X(4) \end{bmatrix} =
\begin{bmatrix}
1 & 1 & 1 & 1 & 1 \\
1 & W_5^1 & W_5^2 & W_5^3 & W_5^4 \\
1 & W_5^2 & W_5^4 & W_5^1 & W_5^3 \\
1 & W_5^3 & W_5^1 & W_5^4 & W_5^2 \\
1 & W_5^4 & W_5^3 & W_5^2 & W_5^1
\end{bmatrix}
\begin{bmatrix} X(0) \\ X(1) \\ X(2) \\ X(3) \\ X(4) \end{bmatrix}
$$

Consider the matrix equation:

$$
\begin{bmatrix}
1 & 0 & 0 & 0 & 0 \\
1 & 1 & 0 & 0 & 0 \\
1 & 0 & 1 & 0 & 0 \\
1 & 0 & 0 & 1 & 0 \\
1 & 0 & 0 & 0 & 1
\end{bmatrix}
\begin{bmatrix}
1 & 0 & 0 & 0 & 0 \\
-1 & 1 & 0 & 0 & 0 \\
-1 & 0 & 1 & 0 & 0 \\
-1 & 0 & 0 & 1 & 0 \\
-1 & 0 & 0 & 0 & 1
\end{bmatrix}
=
\begin{bmatrix}
1 & 0 & 0 & 0 & 0 \\
0 & 1 & 0 & 0 & 0 \\
0 & 0 & 1 & 0 & 0 \\
0 & 0 & 0 & 1 & 0 \\
0 & 0 & 0 & 0 & 1
\end{bmatrix}
$$

Denote this as:

$$[\,P_0\,] \cdot [\,P_1\,] = I$$

The DFT equation can be pre-multiplied by the identity without change, hence substituting:

$$
\begin{bmatrix} X(0) \\ X(1) \\ X(2) \\ X(3) \\ X(4) \end{bmatrix} =
\begin{bmatrix}
1 & 0 & 0 & 0 & 0 \\
1 & 1 & 0 & 0 & 0 \\
1 & 0 & 1 & 0 & 0 \\
1 & 0 & 0 & 1 & 0 \\
1 & 0 & 0 & 0 & 1
\end{bmatrix}
\begin{bmatrix}
1 & 1 & 1 & 1 & 1 \\
0 & w(1) & w(2) & w(3) & w(4) \\
0 & w(2) & w(4) & w(1) & w(3) \\
0 & w(3) & w(1) & w(4) & w(2) \\
0 & w(4) & w(3) & w(2) & w(1)
\end{bmatrix}
\begin{bmatrix} x(0) \\ x(1) \\ x(2) \\ x(3) \\ x(4) \end{bmatrix}
$$

where $w(i)$ is $(W_5^i - 1)$ as above. A second matrix equation is:

$$
\begin{bmatrix}
1 & -1 & -1 & -1 & -1 \\
0 & 1 & 0 & 0 & 0 \\
0 & 0 & 1 & 0 & 0 \\
0 & 0 & 0 & 1 & 0 \\
0 & 0 & 0 & 0 & 1
\end{bmatrix}
\begin{bmatrix}
1 & 1 & 1 & 1 & 1 \\
0 & 1 & 0 & 0 & 0 \\
0 & 0 & 1 & 0 & 0 \\
0 & 0 & 0 & 1 & 0 \\
0 & 0 & 0 & 0 & 1
\end{bmatrix}
=
\begin{bmatrix}
1 & 0 & 0 & 0 & 0 \\
0 & 1 & 0 & 0 & 0 \\
0 & 0 & 1 & 0 & 0 \\
0 & 0 & 0 & 1 & 0 \\
0 & 0 & 0 & 0 & 1
\end{bmatrix}
$$

Denote this as:

$$[P_2]{\cdot}[P_3] = I$$

The DFT equation can be post-multiplied by the identity without change, thus:

$$\begin{bmatrix} X(0) \\ X(1) \\ X(2) \\ X(3) \\ X(4) \end{bmatrix} = \begin{bmatrix} 1 & 0 & 0 & 0 & 0 \\ 1 & 1 & 0 & 0 & 0 \\ 1 & 0 & 1 & 0 & 0 \\ 1 & 0 & 0 & 1 & 0 \\ 1 & 0 & 0 & 0 & 1 \end{bmatrix} \begin{bmatrix} 1 & 0 & 0 & 0 & 0 \\ 0 & w(1) & w(2) & w(3) & w(4) \\ 0 & w(2) & w(4) & w(1) & w(3) \\ 0 & w(3) & w(1) & w(4) & w(2) \\ 0 & w(4) & w(3) & w(2) & w(1) \end{bmatrix} \begin{bmatrix} 1 & 1 & 1 & 1 & 1 \\ 0 & 1 & 0 & 0 & 0 \\ 0 & 0 & 1 & 0 & 0 \\ 0 & 0 & 0 & 1 & 0 \\ 0 & 0 & 0 & 0 & 1 \end{bmatrix} \begin{bmatrix} x(0) \\ x(1) \\ x(2) \\ x(3) \\ x(4) \end{bmatrix}$$

or:

$$X = [P_0]{\cdot}[Q]{\cdot}[P_3]\, x$$

The terms w(i) in matrix Q form a minor about the main diagonal, hence we can convert this matrix to a convolution matrix form without any real difficulty. The first step is to permute the terms. The identity matrix with rows or columns permuted always has an inverse and when another matrix is pre- or post-multiplied by such a matrix, its rows or columns will also be permuted. Therefore, we can factor Q to:

$$Q = \begin{bmatrix} 1 & 0 & 0 & 0 & 0 \\ 0 & 1 & 0 & 0 & 0 \\ 0 & 0 & 1 & 0 & 0 \\ 0 & 0 & 0 & 0 & 1 \\ 0 & 0 & 0 & 1 & 0 \end{bmatrix} \begin{bmatrix} 1 & 0 & 0 & 0 & 0 \\ 0 & w(1) & w(3) & w(4) & w(2) \\ 0 & w(2) & w(1) & w(3) & w(4) \\ 0 & w(4) & w(2) & w(1) & w(3) \\ 0 & w(3) & w(4) & w(2) & w(1) \end{bmatrix} \begin{bmatrix} 1 & 0 & 0 & 0 & 0 \\ 0 & 1 & 0 & 0 & 0 \\ 0 & 0 & 0 & 1 & 0 \\ 0 & 0 & 0 & 0 & 1 \\ 0 & 0 & 1 & 0 & 0 \end{bmatrix}$$

Denote this as:

$$[Q] = [T_0]{\cdot}[\Psi]{\cdot}[T_1]$$

The minor of Ψ is now in the form of a convolution and the matrix expression for a convolution can be directly substituted. Using the transpose algorithm given earlier, Ψ becomes:

$$\begin{bmatrix} 1 & 0 & 0 & 0 & 0 & 0 \\ 0 & 1 & 1 & 1 & 0 & 1 \\ 0 & 1 & -1 & -1 & -1 & 0 \\ 0 & 1 & 1 & -1 & 0 & -1 \\ 0 & 1 & -1 & 1 & 1 & 0 \end{bmatrix} = \begin{bmatrix} 1 & 0 & 0 & 0 & 0 & 0 \\ 0 & m_1 & 0 & 0 & 0 & 0 \\ 0 & 0 & m_2 & 0 & 0 & 0 \\ 0 & 0 & 0 & m_3 & 0 & 0 \\ 0 & 0 & 0 & 0 & m_4 & 0 \\ 0 & 0 & 0 & 0 & 0 & m_5 \end{bmatrix} \begin{bmatrix} 1 & 0 & 0 & 0 & 0 \\ 0 & 1 & 1 & 1 & 1 \\ 0 & 1 & -1 & 1 & -1 \\ 0 & 1 & 0 & -1 & 0 \\ 0 & 1 & 1 & -1 & -1 \\ 0 & -1 & 1 & 1 & -1 \end{bmatrix}$$

Collecting all the terms, this DFT algorithm can therefore be expressed in matrix form as:

$$
\begin{bmatrix} X(0) \\ X(1) \\ X(2) \\ X(3) \\ X(4) \end{bmatrix}
=
\begin{bmatrix}
1 & 0 & 0 & 0 & 0 & 0 \\
1 & 1 & 1 & 1 & 0 & 1 \\
1 & 1 & -1 & -1 & -1 & 0 \\
1 & 1 & -1 & 1 & 1 & 0 \\
1 & 1 & 1 & 1 & 0 & 0
\end{bmatrix}
\begin{bmatrix}
1 & 0 & 0 & 0 & 0 & 0 \\
0 & m_1 & 0 & 0 & 0 & 0 \\
0 & 0 & m_2 & 0 & 0 & 0 \\
0 & 0 & 0 & m_3 & 0 & 0 \\
0 & 0 & 0 & 0 & m_4 & 0 \\
0 & 0 & 0 & 0 & 0 & m_5
\end{bmatrix}
\begin{bmatrix}
1 & 1 & 1 & 1 & 1 \\
0 & 1 & 1 & 1 & 1 \\
0 & 1 & -1 & -1 & 1 \\
0 & 1 & 0 & 0 & -1 \\
0 & 1 & -1 & 1 & -1 \\
0 & -1 & -1 & 1 & -1
\end{bmatrix}
\begin{bmatrix} X(0) \\ X(1) \\ X(2) \\ X(3) \\ X(4) \end{bmatrix}
$$

where
$$
\begin{aligned}
m_1 &= -1.250000 \\
m_2 &= 0.559017 \\
m_3 &= -j\,0.363271 \\
m_4 &= j\,0.951057 \\
m_5 &= j\,0.587785
\end{aligned}
$$

This matrix is general. Any prime length DFT can be expressed as a matrix of this form. The terms m_i are always either real or imaginary but never in the general complex form.

The matrix equation can be expressed as:

$$\mathbf{X} = \mathbf{F \cdot M \cdot R}\, \mathbf{x}$$

where
$\mathbf{X}$ = an N-dimensional vector, N prime
$\mathbf{x}$ = an N-dimensional vector, N prime
$\mathbf{F}$ = a N×K matrix
$\mathbf{M}$ = a K×K diagonal matrix
$\mathbf{R}$ = a K×N matrix
$K = 2N - d(N-1) - 1$

If it derives from Winograd's transpose form for the convolution, then:

$$\mathbf{F} = \mathbf{R}^T$$

Matrices F and R also have elements drawn only from { 0, 1, –1 }. This is general as they originate from the convolution matrix form and the variations made did not add, nor could they add, any different terms. Hence these two matrices perform additions (and subtractions) only. They are usually referred to as the **post-weave** and **pre-weave** matrices respectively. The only multiplications in this equation are due to the diagonal matrix M. Of the K non-zero terms of this matrix, one is always unity, hence there are in fact only (K – 1) general multiplications. Some of these elements of M are negative and some are imaginary. For practical reasons, it is often convenient to adjust F so these are included in that matrix leaving M as having only positive real values.

Just as in the corresponding convolution algorithms, there is a need here for a systematic means of factoring matrices F and R in order to minimize the number of additions. This can be done as follows. As we have seen, the $(p-1)$ length convolution algorithm at the heart of this DFT can be described as:

$$y = C \cdot (Ax \cdot Bw)$$

where w is a vector of terms based on powers of $e^{-j\frac{2\pi}{N}}$. Since p is prime, $(p-1)$ is composite. In fact, it must be even. Thus the polynomial modulus in the cyclic convolution can be expressed as:

$$z^{p-1} - 1 = (z^{\frac{(p-1)}{2}} - 1) \cdot (z^{\frac{(p-1)}{2}} + 1)$$

The Chinese remainder theorem enables an efficient algorithm to be evolved for computing the convolution based on these two factors. It is easy to show that for this new matrix equation, the matrices A and B are of the form:

$$A_0 = A = B = \begin{bmatrix} I & I \\ I & -I \end{bmatrix}$$

where I is a $\frac{(p-1)}{2} \times \frac{(p-1)}{2}$ identity matrix. Note that this shows why the matrix F above has only real or imaginary terms. The vector w is complex and so Bw must give a diagonal matrix where the upper half has real terms and the lower imaginary.

Now this matrix equation can be expressed as:

$$\begin{bmatrix} x_1(z) \\ x_2(z) \end{bmatrix} = A_0 x$$

where x is the input vector and $x_1(z)$ and $x_2(z)$ are polynomials of order $\frac{(p-1)}{2}$ resulting from the reductions modulo ($z^{\frac{(p-1)}{2}} - 1$). If either of these polynomials has factors, then we may repeat the process and derive another matrix equation of the form:

$$\begin{bmatrix} x_3(z) \\ x_4(z) \end{bmatrix} = A_1 x_1(z)$$

If ($z^{\frac{(p-1)}{2}} - 1$) has factors, they can only be of order $\frac{(p-3)}{4}$. Therefore, $x_3(z)$ and $x_4(z)$ in this equation, at most, must be of this order. Matrix A_1 accordingly, must be $\frac{(p-1)}{2} \times \frac{(p-1)}{2}$. The process can be repeated on these factors and this may be continued until either constants or irreducible polynomials result. Now construct a series of matrices of order $(p-1) \times (p-1)$ from these matrices A_i according to:

$$\begin{bmatrix} A_i & 0 \\ 0 & I_i \end{bmatrix}$$

Matrix A_i is a minor in this matrix of order $\frac{(p+1-2^i)}{2^i}$ and I_i is an identity matrix of order $\frac{(p(1+2^i)-1)}{2^i}$. Let the product of all of these terms form a matrix A' and consider the equation:

$$s = A' x$$

What are the terms of the vector **s**? If this procedure is reviewed, it will be seen they must be the terms resulting from the reduction of x modulo the irreducible (cyclotomic) factors of $(z^{p-1}-1)$. Now the multiplication stage of the convolution will use the Toom-Cook algorithm or some similar technique but it will use these terms. Thus a complete expression for matrix A in the original matrix convolution expression is:

$$A = A_M A'$$

where A' is a square matrix due entirely to the polynomial reduction and A_M is a rectangular matrix due entirely to whatever multiplication algorithm is employed. When A is expressed in this way, the number of additions is minimized. Note the similarity with Bruun's algorithm.

Example

Take the 5-point DFT considered before. This DFT was based on a 4-point cyclic convolution. Note that:

$$z^4 - 1 = (z^2 - 1)\cdot(z^2 + 1) = (z - 1)\cdot(z + 1)\cdot(z^2 + 1)$$

Let:

$$X(z) = x(0) + x(1)\cdot z + x(2)\cdot z^2 + x(3)\cdot z^3$$

Then:

$$< X(z) >_{z^2-1} = (x(0) + x(2)) + (x(1) + x(3))\cdot z = X_1(z)$$
$$< X(z) >_{z^2+1} = (x(0) - x(2)) + (x(1) - x(3))\cdot z = X_2(z)$$

That is:

$$\begin{bmatrix} X_1(z) \\ X_2(z) \end{bmatrix} = \begin{bmatrix} x(0) + x(2) \\ x(1) + x(3) \\ x(0) - x(2) \\ x(1) - x(3) \end{bmatrix} = \begin{bmatrix} 1 & 0 & \cdot & 1 & 0 \\ 0 & 1 & \cdot & 0 & 1 \\ \cdot & \cdot & \cdot & \cdot & \cdot \\ 1 & 0 & \cdot & -1 & 0 \\ 0 & 1 & \cdot & 0 & -1 \end{bmatrix} \begin{bmatrix} x(0) \\ x(1) \\ x(2) \\ x(3) \end{bmatrix}$$

Since $(z^2 - 1)$ has factors $(z - 1)$ and $(z + 1)$, we can repeat the procedure. Then:

$$< X_1(z) >_{z-1} = (x(0) + x(2)) + (x(1) + x(3))$$
$$< X_2(z) >_{z+1} = (x(0) + x(2)) - (x(1) + x(3))$$

That is:

$$\begin{bmatrix} X_3(z) \\ X_4(z) \end{bmatrix} = \begin{bmatrix} 1 & 1 \\ 1 & -1 \end{bmatrix} \begin{bmatrix} X_1(z) \\ X_2(z) \end{bmatrix}$$

We can factor no more. The matrix A' is therefore:

$$A' = \begin{bmatrix} 1 & 1 & 0 & 0 \\ 1 & -1 & 0 & 0 \\ 0 & 0 & 1 & 0 \\ 0 & 0 & 0 & 1 \end{bmatrix} \begin{bmatrix} 1 & 0 & 1 & 0 \\ 0 & 1 & 0 & 1 \\ 1 & 0 & -1 & 0 \\ 0 & 1 & 0 & -1 \end{bmatrix}$$

From the previous example on the 4-point convolution, the matrix A_M is identified as:

$$A_M = \begin{bmatrix} 1 & 0 & 0 & 0 \\ 0 & 1 & 0 & 0 \\ 0 & 0 & 1 & 0 \\ 0 & 0 & 1 & 1 \\ 0 & 0 & -1 & 1 \end{bmatrix}$$

Long-length convolution algorithms

There are two principal approaches for executing long-length convolution algorithms. Both aim at algorithms for combining short-length convolutions into long which are more efficient in some way than Chinese remainder theorem recombination. One has already been introduced in Part 3. It maps a long convolution into a set of aperiodic convolutions. The other is nesting. Nesting procedures are similar to the first approach but, at the price of some restrictions on sequence length, map to cyclic convolutions.

While the algorithms discussed here are certainly better for long sequence lengths than the remainder theorem approach, it needs to be stressed 'long' only means sequence lengths of up to about 256. For even longer sequence lengths, the FFT approaches discussed in Part 2 will perform the convolutions with fewer additions and subtractions than these techniques. There is an exception to this and it is discussed separately in the next section.

The details of the first technique are given in Part 3 and will not be repeated here. The essence of the approach is that if the data is mapped into an array, the long-length convolution can be expressed as the combination of a set of short-length aperiodic convolutions. These short-length aperiodic convolutions are easily mapped into short-length cyclic convolutions and then the techniques of this Part can be applied to compute these. The fact that only convolutions of up to a length of about 200 can be treated by this approach is an obvious limit but, even so, there is the practical advantage of only real operations being required. There is also no restriction on how the initial array is formed.

A slight increase in the upper limit for this method can be achieved. Consider two sequences { h(0), h(1), h(2), h(3) } and { x(0), x(1), x(2), x(3) }. Their aperiodic convolution is given by:

$$
\begin{aligned}
y(z) \quad = \quad & (\, h(0) \cdot x(0) \,) \\
& + \, (\, h(1) \cdot x(0) + h(0) \cdot x(1) \,) \cdot z \\
& + \, (\, h(2) \cdot x(0) + h(1) \cdot x(1) + h(0) \cdot x(2) \,) \cdot z^2 \\
& + \, (\, h(3) \cdot x(0) + h(2) \cdot x(1) + h(1) \cdot x(2) + h(0) \cdot x(3) \,) \cdot z^3 \\
& + \, (\, h(3) \cdot x(1) + h(2) \cdot x(2) + h(1) \cdot x(3) \,) \cdot z^4 \\
& + \, (\, h(3) \cdot x(2) + h(2) \cdot x(3) \,) \cdot z^5 \\
& + \, (\, h(3) \cdot x(3) \,) \cdot z^6
\end{aligned}
$$

As with any convolution, the bulk of the operations are for the middle terms rather than the initial. Now consider the cyclic convolution $Y(z) = \, < H(z) \cdot X(z) >_{P(z)}$ where $P(z)$ is, say, $(z^5 - 1)$. This gives:

$$
\begin{aligned}
y(z) \quad = \quad & (\, h(0) \cdot x(0) \,) \; + (\, h(3) \cdot x(2) + h(2) \cdot x(3) \,) \\
& + \, (\, h(1) \cdot x(0) + h(0) \cdot x(1) \,) \cdot z + \, (\, h(3) \cdot x(3) \,) \cdot z \\
& + \, (\, h(2) \cdot x(0) + h(1) \cdot x(1) + h(0) \cdot x(2) \,) \cdot z^2 \\
& + \, (\, h(3) \cdot x(0) + h(2) \cdot x(1) + h(1) \cdot x(2) + h(0) \cdot x(3) \,) \cdot z^3 \\
& + \, (\, h(3) \cdot x(1) + h(2) \cdot x(2) + h(1) \cdot x(3) \,) \cdot z^4
\end{aligned}
$$

This cyclic convolution can be computed very efficiently. Examining the two sets of equations, we see that the aperiodic output terms can be found from the cyclic provided $h(0) \cdot x(0)$ and $h(3) \cdot x(3)$ are separately determined and this only requires two extra multiplications. Thus there is a reduction in operations compared to computing a cyclic convolution equal to the full length of the aperiodic convolution or in computing the aperiodic convolution directly.

Consider two N-point sequences. Aperiodically convolving gives a $(2N - 1)$-point sequence as a result. If the two sequences are padded out with zeroes to $(2N - 1 - \alpha)$ points, but are cyclically convolved, from Winograd's theorem at least $(2 \cdot (2N - 1 - \alpha) - k)$ multiplications will be required where k is the number of irreducible factors of the new sequence length $(2N - 1 - \alpha)$. To compute the extra products needed to determine all the terms of the aperiodic convolution requires M_α multiplications where:

$$
M_\alpha \quad = \frac{(\alpha+1)^2}{4} \qquad\qquad \text{for } \alpha \text{ odd}
$$

$$
= \frac{(\alpha+1)^2}{4} - \frac{1}{4} \qquad\qquad \text{for } \alpha \text{ even}
$$

multiplications. Examining this, it is found an advantage is only gained for N up to 11. For example, consider N equal to 9. The aperiodic sequence length is 17. A 17-point cyclic convolution requires 32 multiplications. Examining the different values of α, it is found a value of 5 leads to only 27 multiplications overall; a quite significant saving.

For N greater than 11, a different approach is needed. For the two cyclic convolutions:

$$
\begin{aligned}
y(z) \; &= \; < H(z) \cdot X(z) >_{P_1(z)} \qquad\qquad \text{where } P_1(z) = z^r - 1 \\
y(z) \; &= \; < H(z) \cdot X(z) >_{P_1(z)} \qquad\qquad \text{where } P_2(z) = z^s - 1
\end{aligned}
$$

where r and s are positive, $(r+s)$ is less than or equal to $(2N-1)$ and GCD$(r,s) <$ min (r,s). Then the number of overlap terms is just:

$$\alpha = 2N - 1 - (r + s) + |r - s|$$

and the number of multiplications is:

$$M = (2r - k(r)) + (2s - k(s)) + M_\alpha$$

where M_α is the appropriate one of the formulae given earlier and $k(r)$ and $k(s)$ are the number of prime factors of r and s respectively. This approach gives efficient algorithms up to N equal to 36. Since for both this and the previous approach there is a significant number of multiplications by zero, pruning can be applied.

Now consider nesting. A common approach to solving any complex problem is to transform it in some way into a connected sequence of simpler problems. Nesting is such a procedure. By regarding a long-length convolution as being N-dimensional, it transforms the problem into an embedded set of simpler convolutions within each of the dimensions. Thus nesting has the property of maintaining the original mathematical structure for each of the subproblems.

One of the most important nesting algorithms is the Agarwal-Cooley algorithm. Assume in the cyclic convolution:

$$y(n) = \sum_{i=0}^{N-1} h(i) \cdot x(n-i) \qquad\qquad n = 0, 1, \cdots, N-1$$

that N has two relatively prime factors M and L. From the Chinese remainder theorem, mappings can be defined to convert this one-dimensional cyclic convolution to two. Following from earlier work on the Chinese remainder theorem and the prime factors algorithm, we know a unique mapping is:

$$n = \; < s_2 \cdot n_2 + s_1 \cdot n_1 >_N$$

where
$$< s_1 >_M \; = 1 \qquad\qquad\qquad < s_2 >_M \; = 0$$
$$< s_1 >_L \; = 0 \qquad\qquad\qquad < s_2 >_L \; = 1$$
$$n_2 = 0, 1, \cdots, L-1$$
$$n_1 = 0, 1, \cdots, M-1$$

A similar expression can be derived for index i. A simplification is possible here. We know that s_1 and s_2 will have the structure:

$$S_1 = g_1 \frac{N}{M} = g_1 \cdot L \qquad\qquad\qquad S_2 = g_2 \frac{N}{L} = g_2 \cdot M$$

We have chosen M and L to be mutually prime and s_1 and s_2 are required to be by these equations. Therefore, the two constants g_1 and g_2 must be mutually prime. A term like $g_1 \cdot L \cdot n_1$ is consequently just a permutation of any one of these terms or their combination. Without any loss of generality then, we may alter this mapping to:

$$n = < M \cdot n_2 + L \cdot n_1 >_N$$

where
$$n_2 = 0, 1, \cdots, L-1$$
$$n_1 = 0, 1, \cdots, M-1$$

A similar mapping for i can be constructed in the same way. Then substituting gives:

$$y(n_1, n_2) = \sum_{i_1=0}^{M-1} \cdot \sum_{i_2=0}^{L-1} h(M \cdot i_2 + L \cdot i_1) \cdot x(M \cdot (n_2 - i_2) + L \cdot (n_1 - i_1))$$

Define polynomials:

$$Y_{n_1}(z) = \sum_{n_2=0}^{L-1} y(M \cdot n_2 + L \cdot n_1) \cdot z^{n_1}$$

$$H_{i_1}(z) = \sum_{i_2=0}^{L-1} h(M \cdot i_2 + L \cdot i_1) \cdot z^{i_1}$$

$$X_{m_1}(z) = \sum_{m_2=0}^{L-1} x(M \cdot m_2 + L \cdot m_1) \cdot z^{m_1}$$

Substituting gives the one-dimensional polynomial equation:

$$Y_{n_1}(z) = \sum_{i_1=0}^{M-1} \Big\langle H_{i_1}(z) \cdot X_{(n_1-i_1)}(z) \Big\rangle_{z^L-1} \qquad n_1 = 0, 1, \cdots, M-1$$

This describes the convolution of the two sequences { $H_0(z)$, $H_1(z)$, $\cdots$, $H_{M-1}(z)$ } with the sequence { $X_0(z)$, $X_1(z)$, $\cdots$, $X_{M-1}(z)$ } where the elements of these sequences are polynomials not scalars. Thus the substitution has retained the original mathematical structure and so is a nesting operation. In computational terms though, a long-length convolution has been expressed as a combination procedure applied to short-length convolutions. If M or L in turn have relatively prime factors, the process can be repeated in which case the elements of the convolution will have the form of matrices.

Assume that an M-point convolution requires P_M multiplications and S_M additions and that an L-point convolution requires P_L multiplications and S_L additions. Since this computation is an M-point convolution (of polynomial elements) then it requires P_M multiplicative operations and S_M additive. Those multiplicative operations, though, are with polynomial elements and are convolutions of length L. The additive operations act on the L-point results of these. Thus the total number of operations must be:

number of multiplications $= P_M \cdot P_L$
number of additions $= P_M \cdot S_L + S_M \cdot L$

What is interesting about this is the number of additions. It would have made no difference to the problem if the order of M and L had been reversed but that would have altered the number of additions to $(P_L \cdot S_M + S_L \cdot M)$ which is quite different to the above. Since M and L are distinct, then order counts in the number of additions. Which of M or L to use for the initial convolutions should therefore be the value which minimizes the term $\frac{(P_K - K)}{S_K}$ where K is either M or L.

A matrix version of this nested algorithm can be determined with moderate difficulty. It is a tensor equation and of the form:

$$y = (C_M{}^T \otimes C_L{}^T)(((A_M \otimes A_L) x) \otimes ((B_M \otimes B_L) h))$$

where $\otimes$ is the tensor product operation. Extending this to higher dimensions is simple. Since the tensor product is distributive over matrix multiplication, the terms of this equation can be rearranged to create a range of different algorithms but none have any great practical advantage.

Although at first glance attractive, this nesting algorithm has a number of problems. It retains the advantage of requiring only real arithmetic but it requires the factors to be relatively prime. If some examples are tested, it is found that the algorithm is less efficient than FFT approaches once N exceeds about 200. This plus the relatively prime condition on the factors, means that the number of practical algorithms of this type is relatively small and most algorithms will have only two factors. The algorithm is also quite difficult to program as separate routines must be written for each of the possible factors.

An algorithm similar to the nesting algorithm but without many of its practical difficulties is the **split-nesting algorithm**. It employs the polynomial Chinese remainder theorem to transform the problem into four subproblems, three of which are quite simple and easily handled and one of which may be treated as a nesting problem. Split nesting achieves no advantage over nesting with respect to multiplications but it is both easier to implement and more economical in the number of additions.

The key feature of the algorithm is the way in which it uses the polynomial Chinese remainder theorem to simplify the problem. For p a prime, it is always true that:

$$\begin{aligned}(z^p - 1) &= (z - 1) \cdot (z^{p-1} + z^{p-2} + \cdots + z + 1)\\ &= (z - 1) \cdot Q(z)\end{aligned}$$

This property and the polynomial Chinese remainder theorem indicate that the p-point cyclic convolution $Y(z)$ of two polynomials $H(z)$ and $X(z)$ may be expressed as:

$$Y(z) = S_1(z) \cdot H(1) \cdot X(1) + S_2(z) \cdot H_2(z) \cdot X_2(z)$$

where $\qquad H_2(z) = <H(z)>_{Q(z)} \qquad$ and $\qquad X_2(z) = <X(z)>_{Q(z)}$

Now this structure depends only on p being prime not on its actual value and that can greatly simplify implementation. Given that, it is always the case that:

$$H(1) = \sum_{n=0}^{p-1} h(n)$$

$$H_2(z) = \sum_{n=0}^{p-2}(h(n) - h(p-1)) \cdot z^n$$

and similarly for X(z). Polynomial multiplications only occur within the second set of terms and so these are the only terms that need to be stored. The reduction in the order of the polynomial may seem a slight change but two points need to be kept in mind. First, this technique is to be used in a multi-dimensional convolution and so a reduction in the number of operations in any one of the factors leads to a significant reduction overall. Second, p prime means (p − 1) is composite and often highly composite. Thus the efficient techniques discussed earlier can be employed in the polynomial multiplication phase leading to additional significant savings.

Assume that the cyclic convolution Y(z) of two N-point polynomials H(z) and X(z) is desired where N is M·L and M and L are prime. (This is more restrictive than the Agarwal-Cooley nesting procedure where the requirement was only that M and L be relatively prime.) Following the same mapping procedure as the Agarwal-Cooley algorithm, we may convert this convolution to a two-dimensional convolution. Thus:

$$H(z_1,z_2) = \sum_{i_1=0}^{M-1}\cdot\sum_{i_2=0}^{L-1}h(M\cdot i_2 + L\cdot i_1)z_1^{i_1}z_2^{i_2}$$

$$X(z_1,z_2) = \sum_{j_1=0}^{M-1}\cdot\sum_{j_2=0}^{L-1}x(M\cdot j_2 + L\cdot j_1)z_1^{j_1}z_2^{j_2}$$

which leads to:

$$Y(z_1,z_2) = \sum_{m_1=0}^{M-1}\cdot\sum_{m_2=0}^{L-1}x(M\cdot m_2 + L\cdot m_1)z_1^{m_1}z_2^{m_2}$$

$$= << H(z_1,z_2)\cdot X(z_1,z_2) >_{z^M-1} >_{z^L-1}$$

Since M and L are prime:

$$z^M - 1 = (z-1)\cdot Q_M(z)$$
$$z^L - 1 = (z-1)\cdot Q_L(z)$$

and from the Chinese remainder theorem:

$$\begin{aligned}
Y(z_1,z_2) = \ & S_{11}(z_1,z_2)\cdot Y_{11}(z_1,z_2) \\
+\ & S_{12}(z_1,z_2)\cdot Y_{1Q}(z_1,z_2) \\
+\ & S_{21}(z_1,z_2)\cdot Y_{Q1}(z_1,z_2) \\
+\ & S_{22}(z_1,z_2)\cdot Y_{QQ}(z_1,z_2)
\end{aligned}$$

where
$$Y_{11}(z_1,z_2) = << H(z_1,z_2)\cdot X(z_1,z_2) >_{z_1-1} >_{z_2-1}$$
$$Y_{1Q}(z_1,z_2) = << H(z_1,z_2)\cdot X(z_1,z_2) >_{z_1-1} >_{Q(z_2)}$$

$$Y_{Q1}(z_1, z_2) = << H(z_1, z_2) \cdot X(z_1, z_2) >_{Q(z_1)} >_{z_2-1}$$
$$Y_{QQ}(z_1, z_2) = << H(z_1, z_2) \cdot X(z_1, z_2) >_{Q(z_1)} >_{Q(z_2)}$$

The polynomials $S_{11}(z_1, z_2)$, $S_{12}(z_1, z_2)$, $S_{21}(z_1, z_2)$, and $S_{22}(z_1, z_2)$ are defined as expected. For $S_{11}(z_1, z_2)$ for example:

$$S_{11}(z_1, z_2) = 1 \text{ modulo } (z_1 - 1) \cdot (z_2 - 1)$$
$$= 0 \text{ otherwise}$$

To execute this algorithm involves determining the four polynomials $Y_{11}(z_1, z_2)$, $Y_{1Q}(z_1, z_2)$, $Y_{Q1}(z_1, z_2)$, and $Y_{QQ}(z_1, z_2)$, and then combining them via the Chinese remainder theorem. The last of these polynomials can be determined via the Agarwal-Cooley algorithm. Therefore, insofar as additions are concerned, this algorithm is order-dependent . However, we can also show it involves fewer additions than that algorithm.

Consider the number of operations involved in split nesting. Assume an M-point convolution can be done with P_M products and, similarly, that an L-point requires P_L. The convolution in both cases is assumed to involve the Chinese remainder theorem as outlined. Because of this, we shall differentiate between the additions. Let S_M^q be the number of additions in that convolution associated with the polynomial term Q(z). Then:

$$S_M = S_M^1 + S_M^q$$

is the total number of additions where S_M^1 covers all the additions in Chinese remainder theorem combination and so forth not covered by S_M^q. Similarly, the number of additions S_L in the L-point convolutions can be described as $(S_L^1 + S_L^q)$.

The reductions leave two scalars; hence forming the polynomial $Y_{11}(z_1, z_2)$ requires just one general multiplication and no additions. The polynomial $Y_{QQ}(z_1, z_2)$ is computed via the Agarwal-Cooley algorithm discussed earlier. This polynomial is formed from $(M-1) \times (L-1)$ arrays. Thus there must be $(P_M - 1) \cdot (P_L - 1)$ multiplications together with $((L-1) \cdot S_M^q + (P_M - 1) \cdot S_L^1)$ additions.

The remaining two terms $Y_{1Q}(z_1, z_2)$ and $Y_{Q1}(z_1, z_2)$ require reductions modulo $(z-1)$ in one dimension and modulo Q(z) in the other. The first convolution involves $(L-1)$-point polynomials, where the coefficients are the result of a reduction by $(z_1 - 1)$, and the second $(M-1)$-point polynomials where the coefficients result from reductions by $(z_2 - 1)$. Thus the first convolution requires $L \cdot S_M^1$ additions to form the coefficients plus S_L^q additions and $(P_L - 1)$ products for the convolution. Similarly, computing $Y_{Q1}(z_1, z_2)$ requires $(P_M - 1)$ products and, in total, $(S_M^q + M \cdot S_L^1)$ additions.

The total number of products in these four convolution subproblems is therefore $P_M \cdot P_L$ and that means split-nesting requires the same number as the Agarwal-Cooley algorithm. The total number of additions though, is:

$$L \cdot (S_M^1 + S_M^P) + M \cdot S_L^1 + P_M \cdot S_L^P$$

and this is $(P_M - M) \cdot S_M^1$ fewer additions. As M increases, so this term increases. Hence there is an advantage using split nesting for long convolutions.

Recursive nesting

Recursive nesting is similar to the other forms of nesting but with the advantage that under certain circumstances the number of operations is comparable to the FFT approaches. Hence it may be considered for large as well as small N.

We will begin by developing the basic nesting procedure of this technique. This does describe recursive nesting but that will not become obvious until later. Consider:

$$H(z) = \sum_{n=0}^{N-1} h(n) \cdot z^n$$

and assume N has factors M and L. Define a mapping:

$$n = M \cdot n_2 + n_1 \qquad\qquad \begin{aligned} n_1 &= 0, 1, \cdots, M-1 \\ n_2 &= 0, 1, \cdots, L-1 \end{aligned}$$

which, when substituted, gives:

$$H(z) = \sum_{n_1=0}^{M-1} \left(\sum_{n_2=0}^{L-1} h(M \cdot n_2 + n_1) \cdot z^{Mn_2} \right) \cdot z^{n_1}$$

Now define $z_1 = z$ and $z_2 = z^M$ which leads to:

$$H(z_1, z_2) = \sum_{n_1=0}^{M-1} \sum_{n_2=0}^{L-1} h(n_1, n_2) \cdot z_2^{n_2} z_1^{n_1}$$

Assume a second polynomial $X(z)$ also of degree N. We may express this in similar fashion as $X(z_1, z_2)$. Aperiodically convolving the two polynomials means forming the order $(2N - 1)$ polynomial $X(z) \cdot H(z)$. With these transformations, this is equivalent to:

$$Y(z_1, z_2) = \sum_{n_1=0}^{M-1} \sum_{m_1=0}^{M-1} \sum_{n_2=0}^{L-1} \sum_{m_2=0}^{L-1} h(n_1, n_2) \cdot x(m_1, m_2) \cdot z_2^{n_2+m_2} z_1^{n_1+m_1}$$

Define polynomial terms:

$$H_{n_1}(z_2) = \sum_{n_2=0}^{L-1} h(n_1, n_2) \cdot z^{n_2}$$

$$X_{m_1}(z_2) = \sum_{m_2=0}^{L-1} x(m_1, m_2) \cdot z^{m_2}$$

and substitute to give:

$$Y(z_1, z_2) = \sum_{n_1=0}^{M-1} \sum_{m_1=0}^{M-1} H_{n_1}(z_2) \cdot X_{m_1}(z_2) z_1^{n_1+m_1}$$

This has the form of an aperiodic convolution between two M-point sequences. However, those sequences are { $H_i(z_2)$ } and { $X_i(z_2)$ } and the elements are polynomials (in z_2) not scalars. Therefore, any multiplication between these elements is the product of two L-point polynomials (in z_2) and that also describes an aperiodic convolution.

As with the other nesting techniques, a long-length convolution has been broken down into the sum of a set of short-length convolutions. Any efficient routines for computing the latter will result in a saving in operations being made. A small difficulty exists in that some of those convolutions involve polynomials elements but that is easily overcome.

Assume that the convolution desired is $< H(z) \cdot X(z) >_{P(z)}$ where P(z) is some polynomial. To compute this, we can transform the aperiodic convolution via this particular nesting procedure and that results in a two-dimensional expression for the polynomial product. However, P(z) is a one-dimensional polynomial. The questions arise of whether P(z) can be transformed so it, too, is two-dimensional and what is the result of doing that.

Let P(z) in this operation be some polynomial:

$$P(z) = z^N + \sum_{n=0}^{N-1} p(n) \cdot z^n$$

and consider the same transformation used for H(z) and X(z) is applied. If it should be the case that $p(M \cdot n_2 + n_1) = 0$ for $n_1 \neq 0$ but is finite otherwise, then:

$$P'(z_2) = \sum_{n=0}^{N-1} p(n) \cdot z^{Mn_2} = \sum_{n=0}^{N-1} p(n) \cdot z_2^{n_2}$$

Consider the reduction of any polynomial Q(z) by P(z). Given the form of P(z), we could do this by first substituting z_2 for z^M in these polynomial expressions and then reduce by $P'(z_2)$. That is equivalent to reduction modulo the two-dimensional polynomial:

$$P(z_1, z_2) = (z_1^M - z_2) \cdot P'(z_2)$$

This choice of P(z) may seem a little restrictive here. However, many members of a very important set of polynomials — the cyclotomic polynomials — do satisfy this condition.

To give an example, consider the cyclotomic polynomial:

$$C_{12}(z) = z^4 - z^2 + 1$$

Making the transformation $z_1 = z^2$, then:

$$C_{12}(z) = z_1^2 - z_1 + 1$$

Hence a reduction using $C_{12}(z)$ would involve first substituting z_1 for z^2 everywhere in the expression and then reducing modulo ($z_1^2 - z_1 + 1$). That is, the overall reduction is:

$$P(z) = (z^2 - z_1) \cdot (z_1^2 - z_1 + 1)$$

Consider this transformed P(z) applied to this multi-dimensional convolution. First the polynomial convolution is purely in terms of z_1 while its elements – the polynomials – are purely functions of z_2. Therefore a computational procedure that can be adopted is to perform all the polynomial (convolution) operations modulo $P'(z_2)$ and the overall convolution modulo ($z_1^M - z_2$). These operations can be executed in any way desired.

Example

Consider the problem of convolving the two polynomials:

$$H(z) = h(0) + h(1) \cdot z + h(2) \cdot z^2 + h(3) \cdot z^3$$
$$X(z) = x(0) + x(1) \cdot z + x(2) \cdot z^2 + x(3) \cdot z^3$$

modulo ($z^4 + 1$). Assume that H(z) is known, so allowing some terms to be pre-computed. Comments on the algorithm will therefore be in terms of the number of operations that cannot be pre-computed.

Using the direct approach of multiplying the polynomials then reducing modulo ($z^4 + 1$) requires N^2 multiplications and $N \cdot (N - 1)$ additions. Here, that is 16 and 12 respectively. Toom-Cook, requires at least ($2N - 1$) multiplications and $2N \cdot (2N - 3)$ additions. Here, that is seven multiplications and 40 additions.

For the nesting procedure, the transformed convolution is:

$$Y(z_1, z_2) = \sum_{i=0}^{M-1} \sum_{j=0}^{M-1} H_i(z_2) \cdot X_j(z_2) \cdot z_1^{i+j}$$

where M and L are 2 and:

$$H_0(z_2) = h(0) + h(2) \cdot z_2 \qquad\qquad X_0(z_2) = x(0) + x(2) \cdot z_2$$
$$H_1(z_2) = h(1) + h(3) \cdot z_2 \qquad\qquad X_1(z_2) = x(1) + x(3) \cdot z_2$$

Also $\qquad P(z_1, z_2) = (z_1^2 - z_2) \cdot (z_2^2 + 1)$

Given the form of these equations, we require an efficient short-length algorithm for convolving first order polynomials modulo ($z^2 - \alpha$). For convenience, describe those polynomials as ($h_0 + h_1 \cdot z$) and ($x_0 + x_1 \cdot z$). Then the algorithm in matrix form is:

$$y = C \cdot (Ax \cdot Bh)$$

where using Toom-Cook we find the matrices are:

$$C = \begin{bmatrix} 1 - \alpha & \dfrac{\alpha}{2} & \dfrac{\alpha}{2} \\ 0 & \dfrac{1}{2} & -\dfrac{1}{2} \end{bmatrix} \qquad\qquad A = B = \begin{bmatrix} 1 & 0 \\ 1 & 1 \\ 1 & -1 \end{bmatrix}$$

Since **h** is known, this is best converted to the transpose form. That results in an algorithm with only three multiplications and five additions. While acceptable, a better algorithm based on the informal methods outlined earlier gives three multiplications and just three additions. Now:

$$
\begin{aligned}
Y(z) &= h_0 \cdot x_0 + (h_1 \cdot x_0 + h_0 \cdot x_1) \cdot z + h_1 \cdot x_1 \cdot z^2 \\
&= h_0 \cdot x_0 + ((h_0 + h_1) \cdot (x_0 + x_1) - h_0 \cdot x_0 - h_1 \cdot x_1) \cdot z + h_1 \cdot x_1 \cdot z^2
\end{aligned}
$$

Reducing this modulo $(z^2 - \alpha)$, we derive an algorithm as above but now:

$$
C = \begin{bmatrix} 1 & \alpha & 0 \\ -1 & -1 & 1 \end{bmatrix}
\qquad\qquad
A = B = \begin{bmatrix} 1 & 0 \\ 0 & 1 \\ 1 & 1 \end{bmatrix}
$$

The form of these matrices suggests a transpose algorithm should be used. By using the matrix $\begin{bmatrix} 0 & 1 \\ 1 & 0 \end{bmatrix}$ to transform the equation, matrices C and B become:

$$
C = \begin{bmatrix} 0 & 1 & 1 \\ 1 & 0 & 1 \end{bmatrix}
\qquad\qquad
B = \begin{bmatrix} -1 & 1 \\ -1 & \alpha \\ 1 & 0 \end{bmatrix}
$$

while A remains the same. In detail then, the algorithm to be used here is:

$$
\begin{aligned}
a_0 &= x_0 & b_0 &= -h_0 + h_1 \\
a_1 &= x_1 & b_1 &= -h_0 + \alpha \cdot h_1 \\
a_2 &= x_0 + x_1 & b_2 &= h_0
\end{aligned}
$$

$$
m_i = a_i \cdot b_i \qquad\qquad i = 0, 1, 2
$$

$$
\begin{aligned}
y_0 &= m_1 + m_2 \\
y_1 &= m_0 + m_2
\end{aligned}
$$

Returning now to the problem, we need to first compute:

$$
Y(z_1, z_2) = < (H_0(z_2) + H_1(z_2) \cdot z_1) \cdot (X_0(z_2) + X_1(z_2) \cdot z_1) >_M
$$

where M is $(z_1^2 - z_2)$. Applying our algorithm using z_2 for α gives:

$$
\begin{aligned}
A_0(z_2) &= X_0(z_2) & B_0(z_2) &= -H_0(z_2) + H_1(z_2) \\
A_1(z_2) &= X_1(z_2) & B_1(z_2) &= -H_0(z_2) + z_2 \cdot H_1(z_2) \\
A_2(z_2) &= X_0(z_2) + X_1(z_2) & B_2(z_2) &= H_0(z_2)
\end{aligned}
$$

$$
M_i(z_2) = A_i(z_2) \cdot B_i(z_2) \qquad i = 0, 1, 2
$$

$$Y_0(z_2) = M_1(z_2) + M_2(z_2)$$
$$Y_1(z_2) = M_0(z_2) + M_2(z_2)$$

All the elements here are polynomials. Since:

$$Y(z) = Y_0(z_2) + Y_1(z_2) \cdot z_1$$

the final two terms equal:

$$Y_0(z_2) = y(0) + y(2) \cdot z_2$$
$$Y_1(z_2) = y(1) + y(3) \cdot z_2$$

where $y(0)$, $y(1)$, $y(2)$, and $y(3)$ are the desired convolution terms. In order to derive these terms we need to perform the various operations implied in these equations modulo $(z_2^2 + 1)$. To begin this second step, we derive the set of polynomials $A_i(z_2)$ and $B_i(z_2)$:

$$A_0(z_2) = x(0) + x(2) \cdot z_2$$
$$A_1(z_2) = x(1) + x(3) \cdot z_2$$
$$A_2(z_2) = (x(0) + x(1)) + (x(2) + x(3)) \cdot z_2$$

$$B_0(z_2) = (h(1) - h(0)) + (h(3) - h(2)) \cdot z_2$$
$$B_1(z_2) = < -h(0) - h(2) \cdot z_2 + h(1) \cdot z_2 + h(3) \cdot z_2^2 >_{z_2^2+1}$$
$$= -(h(0) + h(3)) + (h(1) - h(2)) \cdot z_2$$
$$B_2(z_2) = h(0) + h(2) \cdot z_2$$

Next we need to compute the (polynomial) products $M_i(z)$ modulo $(z_2^2 + 1)$. We will use the earlier convolution algorithm with -1 substituted for α. Computing the product gives the intermediate values:

$$a_0 = x(0) \qquad\qquad b_0 = -(h(1) - h(0)) + (h(3) - h(2))$$
$$a_1 = x(2) \qquad\qquad b_1 = -(h(1) - h(0)) - (h(3) - h(2))$$
$$a_2 = x(0) + x(2) \qquad\qquad b_2 = (h(1) - h(0))$$

$$a_3 = x(1) \qquad\qquad b_3 = (h(0) - h(3)) + (h(1) - h(2))$$
$$a_4 = x(3) \qquad\qquad b_4 = (h(0) - h(3)) - (h(1) - h(2))$$
$$a_5 = x(1) + x(3) \qquad\qquad b_5 = -(h(0) - h(3))$$

$$a_6 = x(0) + x(1) \qquad\qquad b_6 = -h(0) + h(2)$$
$$a_7 = x(2) + x(3) \qquad\qquad b_7 = -h(0) - h(2)$$
$$a_8 = (x(0) + x(1)) + (x(2) + x(3)) \quad b_8 = h(0)$$

and:

$$m_i = a_i \cdot b_i \qquad\qquad i = 0, 1, \cdots, 8$$

from that algorithm with the final result that:

$$M_0(z) = (m_1 + m_2) + (m_0 + m_2)\cdot z_2$$
$$M_1(z) = (m_4 + m_5) + (m_3 + m_5)\cdot z_2$$
$$M_2(z) = (m_7 + m_8) + (m_6 + m_8)\cdot z_2$$

Therefore, overall we obtain the desired convolution terms as:

$$y(0) = (m_4 + m_5) + (m_7 + m_8)$$
$$y(1) = (m_1 + m_2) + (m_7 + m_8)$$
$$y(2) = (m_3 + m_5) + (m_6 + m_8)$$
$$y(3) = (m_0 + m_2) + (m_7 + m_8)$$

This requires nine multiplications and 15 additions. While this is two more multiplications than Toom-Cook, it is only about one-third as many additions.

In spite of the fact no mention was made of recursion and that it was not applied at any point, this describes recursive nesting. That becomes apparent when these results are extended from two to many dimensions. This is easily done by transforming polynomials to the form $X(z_1, z_2, \cdots, z_n)$ and taking products modulo $P(z_1, z_2, \cdots, z_n)$. The required computation proceeds along each dimension and results in the processing of successively higher degree polynomials.

One of the most important applications of recursive nesting is in computing cyclic convolutions. Assume N has the particular value 2^α. As shown earlier, here:

$$z^N - 1 = (z^{\frac{N}{2}} - 1)\cdot(z^{\frac{N}{2}} + 1)$$
$$= (z^{\frac{N}{4}} - 1)\cdot(z^{\frac{N}{4}} + 1)\cdot(z^{\frac{N}{2}} + 1)$$
$$\vdots$$
$$= (z - 1)\cdot(z + 1)\cdot(z^2 + 1)\cdot \cdots \cdot(z^{\frac{N}{2}} + 1)$$

Since this final result is the product of a set of irreducible polynomials, it suggests this particular convolution should be computed via the Chinese remainder theorem. If so, then most of the computations will involve the product of two polynomials modulo a polynomial of the form $(z^M + 1)$ where M is some power of 2.

The reductions modulo $(z + 1)$ and $(z - 1)$ are trivial. Those involving $(z^M + 1)$ when M is a significant power of 2 are not so simple. Here, the recursive nesting approach has much to offer. As described earlier, a reduction modulo $(z^M + 1)$ can be transformed to:

$$z^M - 1 = (z^2 - z_1)\cdot(z^{\frac{M}{2}} + 1)$$
$$= (z^2 - z_1)\cdot(z_1^2 - z_2)\cdot(z_2^{\frac{M}{4}} + 1)$$
$$\vdots$$
$$= (z^2 - z_1)\cdot(z_1^2 - z_2)\cdot \cdots \cdot(z_2^u + 1)$$

Hence the problem involves successively applying each of these terms to the polynomials resulting from the previous stage. Other factorizations can be considered. What is important, though, is to maintain a common structure, such as ($z_{i-1}^L - z_i$) here, for all of the terms employed.

The overall efficiency of the computation will depend on the quality of the algorithm used to compute the product of two polynomials modulo this common term. A variation on an earlier algorithm offers some benefits here. For two M-point sequences, the Toom-Cook algorithm uses interpolation over ($2M - 1$) points and a set of interpolation polynomials. Although this algorithm is for aperiodic convolution, it is easy to take the result modulo ($z_{i-1}^L - z_i$). For this problem, the elements of the two polynomials in the product are themselves polynomials, thus the interpolation points cannot, in general, be scalars but must also be polynomials (in z_i). A set of distinct 'points' are needed for the interpolation. Provided M is greater than L and given M in this case will be a power of 2, such a set is:

$$0, \pm 1, \pm z_i, \pm z_i^2, \pm z_i^3, \cdots, \pm z_i^{L-1}$$

A Toom-Cook algorithm modified to account for this, then, will be in terms of additions of polynomials in z_i, the product of polynomials in z_i, and the shifting of terms within a polynomial in z_i due to multiplication by 'constants' of the form of z_i^v. Thus the computation is quite efficient due to the lack of scalar multiplications.

If N is large in this recursive nesting procedure, then the computational effort required to compute the cyclic transform by this method will be dominated by the effort to compute the product modulo ($z^{\frac{N}{2}} + 1$). Expressing this as the product of a series of terms of the form ($z_{i-1}^2 - z_i$), the total number of multiplications using the example given earlier is 3^α. As N is 2^α, the ratio of the two is 1.5^α. That is, the number of multiplications depends exponentially on the number of terms and so the FFT approach must be superior to this one beyond some given number of points. The fact that no sinusoids are involved and that recursive nesting is relatively easy to program make the procedure attractive but it still remains a technique for only moderately long sequences.

For certain values of N this need not be true. There is no reason why the factorization of P(z) into a multi-dimensional term must be fixed to terms of the form ($z_{i-1}^2 - z_i$). It is convenient but just as with the FFT there can be advantages in having a mixed radix form. If M has the form 2^β where β is 2^γ then it is possible to factor ($z^M + 1$) to:

$$z^M + 1 = (z^{2^\alpha} - z_1) \cdot \cdots \cdot (z_{\gamma-2}^4 - z_{\gamma-1}) \cdot (z_{\gamma-1}^2 - z_\gamma) \cdot (z_\gamma^2 + 1)$$

The number of multiplications for this factorization is easily calculated. The first computation, which is modulo ($z_\gamma^2 + 1$), requires three. The next computation, which is modulo ($z_{\gamma-1}^2 - z_\gamma$), also requires three. Each successive operation modulo a polynomial of the form ($z_{i-1}^u - z_i$) requires ($2u - 1$) if an optimal algorithm is used. Thus:

$$\text{number of multiplications} = 3 \cdot 3 \cdot 7 \cdot 31 \cdot 255 \cdot \cdots \cdot (4\beta - 1)$$

Noting that each of these terms is of the form ($2^v - 1$), it can be shown that:

$$\text{number of multiplications} < 9 \cdot 2^{\gamma-3} \cdot 2^\beta$$

and this is just $(\frac{9}{8} \cdot \text{Log}_2 N) \cdot N$. This is essentially the same as the FFT. However, there is no involvement with sinusoids or with scrambling and that makes it a more interesting algorithm for practical application. This algorithm, too, has a regular structure and so the programming effort required is similar to that of an FFT.

The requirements on N to achieve this advantage are quite stringent and that is a disadvantage. What it shows though, is that there are advantages for any arbitrary N in factoring to a mixed radix form and then one where the terms follow in some near-exponential progression. Some practical advantage is lost due to the need to program each section and the number of operations can exceed an FFT approach but this can still create algorithms very useful in many practical situations.

Multi-dimensional DFT algorithms

The multi-dimensional discrete Fourier transform is:

$$X(k_1, k_2, \cdots, k_\alpha) = \sum_{n_1=0}^{N_1-1} \sum_{n_2=0}^{N_2-1} \cdots \sum_{n_\alpha=0}^{N_\alpha-1} x(n_1, n_2 \cdots, n_\alpha) W_{N_1}^{k_1 n_1} W_{N_2}^{k_2 n_2} \cdots W_{N_\alpha}^{k_\alpha n_\alpha}$$

where
$$k_1 = 0, 1, \cdots, N_1-1$$
$$k_2 = 0, 1, \cdots, N_2-1$$
$$\cdot$$
$$\cdot$$
$$\cdot$$
$$k_\alpha = 0, 1, \cdots, N_\alpha-1$$

Three broad approaches can be adopted to tackle this computation. Since this is a separable transform, the most direct is to repeatedly apply some one-dimensional algorithm. A two-dimensional algorithm, for example, involves the application of a one-dimensional transform along the rows of the array and then the columns or vice versa. In the general case, this results in an algorithm which requires:

$$\frac{M}{N_1} \quad N_1 \text{ - point DFTs}$$

$$\frac{M}{N_2} \quad N_2 \text{ - point DFTs}$$

$$\cdot$$
$$\cdot$$
$$\cdot$$

$$\frac{M}{N_\alpha} \quad N_\alpha \text{ - point DFTs}$$

where $M = N_1 \cdot N_2 \cdot \cdots \cdot N_\alpha$

If each dimension has the same number of points, then the algorithm requires $\alpha N^{\alpha-1}$ N-point DFTs. Clearly, the great advantage of this approach is its generality and the ease of application.

A second approach has a limited application but it incorporates some interesting ideas. Its restrictions arise from the fact it is an extension of the prime factors algorithm. Now we showed earlier that an ML-point DFT, M, and L mutually prime, is given by:

$$X = \Omega \, x$$

where $\quad \Omega = \Lambda \otimes \Xi$
Λ = the forward transformation matrix for an L-point DFT
Ξ = the forward transformation matrix for an M-point DFT
$\otimes$ is the tensor product

If V_1 is the forward transformation matrix of an N_1-point DFT, V_2 the forward transformation matrix of an N_2-point DFT and so on up to V_α being the forward transformation matrix of an N_α-point DFT where each of $N_1, N_2, \cdots, N_\alpha$ are mutually prime to one another, then by a simple extension of earlier techniques we can show:

$$\Omega = V_1 \otimes V_2 \otimes \cdots \otimes V_\alpha$$

Some interesting new algorithms can be developed by using this result. Earlier we found that a DFT transforming matrix when N is an odd prime can be expressed as $\Omega = C \cdot D \cdot A$. Matrix C describes the post-weave additions and matrix A the pre-weave. Matrix D is diagonal and defines the general multiplications. It is the only matrix of the three that is complex. Now the tensor product is distributive over matrix multiplication. Thus for a tensor product $P \otimes Q$ where P is the matrix product $M \cdot N$ and Q is the matrix product $X \cdot Y$:

$$P \otimes Q = (M \cdot N) \otimes (X \cdot Y) = (M \otimes X) \cdot (N \otimes Y)$$

provided the matrices are of the correct dimension. This extends to the tensor products of terms involving several matrices. Applying that here, a multi-dimensional prime factors algorithm based on these more efficient DFT algorithms is found by substituting $C_i \cdot D_i \cdot A_i$ for V_i and that gives

$$\Omega = (C_1 \otimes C_2 \cdots \otimes C_\alpha) \cdot (D_1 \otimes D_2 \cdots \otimes D_\alpha) \cdot (A_1 \otimes A_2 \cdots \otimes A_\alpha)$$

The tensor product of the D_i matrices will be diagonal as these matrices are diagonal. Further, the number of terms in this central matrix is just $N_1 \cdot N_2 \cdot \cdots \cdot N_\alpha$ and it remains the only complex term in this algorithm. The other two terms are more complicated than before but are easily executed. The tensor product of two matrices can be expressed in vector terms and so a form of nesting algorithm can be derived for each of these.

This particular matrix decomposition is only one of many that could be considered. To illustrate, consider the derivation of this decomposition. Given N is some odd prime p, then the DFT forward transformation matrix Ω can be described by $\Omega = B^T \cdot F \cdot B$ where B is a matrix:

$$\begin{bmatrix} 1 & 1 & 1 & . & . & . & 1 \\ 0 & 1 & 0 & & & & 0 \\ 0 & 0 & 1 & & & & 0 \\ . & & & . & & & . \\ . & & & & . & & . \\ . & & & & & . & . \\ 0 & 0 & 0 & . & . & . & 1 \end{bmatrix}$$

and F is a matrix:

$$\begin{bmatrix} 1 & 0 & . & . & . & 0 \\ 0 & & & & & \\ . & & & & & \\ . & & & F_{p-1} & & \\ . & & & & & \\ 0 & & & & & \end{bmatrix}$$

where F_{p-1} is a matrix with elements ($W_p^{nk} - 1$) where n, k > 0. From Rader's theorem, F_{p-1} can be expressed as a cyclic convolution by scrambling rows and columns. Thus:

$$F_{p-1} = P_1 \cdot \Phi \cdot Q_1$$

where P_1 and Q_1 are permutation matrices. In turn, matrix Φ can be expressed as the product of three matrices $C \cdot D \cdot A$ where D is a diagonal matrix. Then any of these combinations can be substituted into the tensor product to form new sets of transforming matrices with different properties.

The third approach more directly employs the results of this Part. Within a multi-dimensional structure there are additional symmetries which the first approach cannot identify. Via polynomial theory, algorithms can be developed which exploit this so achieving an improvement in efficiency at a cost of a little more complexity. We will examine three such algorithms in some depth in the following sections. Two of these have some prominent limitations, but introduce and use some important ideas. The third involves polynomial transforms and is more general. However, before we can examine the application of polynomial transforms to the DFT, we need to develop them a little further and examine them with respect to the general convolution problem.

These three techniques decompose a multi-dimensional transform into a smaller set of one-dimensional transforms than that produced by the separable approach. While more efficient in terms of the number of multiplications, in keeping with many other optimal techniques discussed here, the number of additions is usually greater. To a varying degree though, they do have one other advantage. The separable approach is a very difficult parallel computation. These approaches, however, are relatively simple.

Auslander, Feig, and Winograd's method

Auslander, Feig, and Winograd have developed an interesting multi-dimensional algorithm for the situation where the number of points along α dimensions is an odd prime p. The algorithm uses a generalized form of Rader's theorem to convert the multi-dimensional transform matrix into a circulant matrix. Then it factors that circulant matrix into the product of two circulant matrices, one of which only has elements equal to 0 and 1. The result is effectively equal to a set of $\frac{(p^{\alpha}-1)}{(p-1)}$ p-point DFTs. Heideman has recently shown a minimal multiplication algorithm for this multi-dimensional DFT must have this number of p-point DFTs. If those p-point DFTs require M multiplications and A additions each, overall this algorithm requires:

$$\frac{(p^{\alpha}-1)}{(p-1)}\cdot(M-1)+1 \qquad \text{multiplications}$$

$$\frac{(p^{\alpha}-1)}{(p-1)}\cdot A+(p^{\alpha}-1)\cdot(p^{\alpha-1}-1) \qquad \text{additions}$$

This number of multiplications approaches $\frac{1}{\alpha}$ fewer than the separation approach, but has approximately N^2 more additions.

When each dimension has p points, the multi-dimensional discrete Fourier transform is:

$$X(k_1,k_2,\cdots,k_{\alpha}) = \sum_{n_1=0}^{p-1}\sum_{n_2=0}^{p-1}\cdots\sum_{n_{\alpha}=0}^{p-1}x(n_1,n_2\cdots,n_{\alpha})W_p^{k_1 n_1}W_p^{k_2 n_2}\cdots W_p^{k_{\alpha}k_{\alpha}}$$

where $k_1, k_2, \cdots, k_{\alpha} = 0, 1, \cdots, p-1$

The indices form a set of p^{α} tuples. Since p is an odd prime, we will take these as isomorphic to the Galois field $GF(p^{\alpha})$. This field is guaranteed to have primitive roots. Let one of them be G. The set $\{ G^0, G^1, \cdots, G^{M-1} \}$ where M is $(p^{\alpha}-1)$ is a permutation of the non-zero elements of the field. Using this mapping, the DFT becomes:

$$X(0,0,\cdots,0) = x(0,0,\cdots,0)+\sum_{j=0}^{M-1}x\left(G^{-j}\right)$$

$$X\left(G^i\right) = x(0,0,\cdots,0)+\sum_{j=0}^{M-1}x\left(G^{-j}\right)\cdot W^{\beta(i,-j)} \qquad i=0,1,\cdots,M-1$$

Describe the relationship between the powers of the primitive root and the tuple as:

$$G^i \rightarrow (g_i(0), g_i(1), \cdots, g_i(\alpha-1)) \equiv (g_i(0)+g_i(1)\cdot z+\cdots+g_i(\alpha-1)\cdot z^{\alpha-1})$$

Then the exponent $\beta(i,-j)$ in these equations must be given by:

$$\beta(i,-j) \;=\; \sum_{n=0}^{\alpha-1} \big\langle g_i(n) \cdot g_j(n) \big\rangle_p$$

This is not the (polynomial) product $G^i \cdot G^j$ and it is not at all obvious how we may derive it via the usual field operations.

We turn to the matrix description of a field. Field operators are modulo some (monic) irreducible polynomial $P(z)$ and that polynomial will have a companion matrix C. An element G^i of the field is described by the $\alpha \times \alpha$ matrix:

$$\left[\sum_{n=0}^{\alpha-1} g_i(n) \cdot C^n \right] \qquad\qquad C_o = I$$

and the tuple maps into the first column of this matrix. Given two elements G^u and G^v, the element G^{u+v} is formed in this representation by taking the matrix product:

$$\left[\sum_{n=0}^{\alpha-1} g_u(n) \cdot C^n \right]\left[\sum_{n=0}^{\alpha-1} g_v(n) \cdot C^n \right]$$

with the individual element operations performed modulo p. Consider this computation. The first column of the resulting matrix is $\{\, g_{u+v}(0), g_{u+v}(1), \cdots, g_{u+v}(\alpha-1)\,\}$. Let the first row of the first matrix in this expression be the tuple $(\, r_0, r_1, \cdots, r_{\alpha-1})$. Then $g_{u+v}(0)$ is the product of this tuple by the first column of the second matrix which is the tuple $\{\, g_v(n)\,\}$. This suggests a way out of our dilemma. Define a mapping in terms of this matrix representation. If this mapping is applied to some element G^u of the field, the result is the element $(r_0, r_1, \cdots, r_{\alpha-1})$. Denote this as $\overline{G}^u$. Assume this mapping is now used in the DFT and consider the exponent $\beta(i,-j)$. Clearly, it must be:

$$\beta(i,-j) \;=\; \sum_{n=0}^{\alpha-1} r_n \cdot g_j(n) = g_{i+j}(0)$$

Thus the DFT equations can be expressed as:

$$X(0,0,\cdots,0) \;=\; x(0,0,\cdots,0) + \sum_{j=0}^{M-1} x\big(G^{-j}\big)$$

$$X(G^i) \qquad\qquad = x(0,0,\cdots,0) + \sum_{j=0}^{M-1} x\big(G^{-j}\big) \cdot W_p^{g_{i-j}(0)} \qquad i = 0,1,\cdots,M-1$$

There is an important question to answer here. Is the mapping of G^i into $\overline{G}^i$ unique? It is if the first row of the matrix defining G^i is unique. Assume it is not and that there exists at least one other matrix with an identical first row. Then $(G^i - G^k)$ must generate a unique element of the field where k is the corresponding index of G for that matrix and:

$$G^i - G^k = \left[\sum_{n=0}^{\alpha-1} g_i(n) \cdot C^n \right] - \left[\sum_{n=0}^{\alpha-1} g_k(n) \cdot C^n \right] = \left[\sum_{n=0}^{\alpha-1} (g_i(n) - g_k(n)) \cdot C^n \right]$$

The first row, however, can only be zero if indices i and k are the same. Hence the mapping is unique.

Express the DFT equations as:

$$X(0, 0, \cdots, 0) - x(0, 0, \cdots, 0) = \sum_{j=0}^{M-1} x\left(G^{-j}\right)$$

$$X(G^i) - x(0, 0, \cdots, 0) = \sum_{j=0}^{M-1} x\left(G^{-j}\right) \cdot W_p^{g_{i-j}(0)} \quad i = 0, 1, \cdots, M-1$$

The second term may be simplified to the vector equation:

$$\mathbf{T} = [\, \omega(i-j)\,]\, \mathbf{y}$$

or to the polynomial equation:

$$T(z) = <\, \Omega(z) \cdot Y(z)\, >_{z^M-1}$$

where the polynomial $\Omega(z)$ has coefficients $\{\, W_p^{g_i(0)}\, \}$. Thus we have transformed this multi-dimensional DFT into cyclic convolution form and so found a generalized form of Rader's theorem.

Let θ be $\frac{(p^\alpha - 1)}{(p-1)}$. The set of elements $\{\, 0, 1, G^\theta, G^{2\theta}, \cdots, G^{(p-2)\theta}\, \}$ forms a subfield of $GF(p^\alpha)$. These elements represent tuples of the form $(0, 0, \cdots, a)$ where $0 \le a \le p-1$. This subfield, then, is isomorphic to $GF(p)$. In addition, the tuples represent constants in a polynomial field. Define numbers I_k where:

$$I_k = <\, G^{k\theta}\, >_p$$

This describes a unique mapping from the set of integers $\{\, 0, 1, \cdots, p-2\, \}$ into the set of integers $\{\, 1, 2, \cdots, p-1\, \}$. A useful property of these numbers is that the product of I_k with any element x a member of $GF(p^\alpha)$ is just the sum of x taken I_k times. That is:

$$x \cdot G^{k\theta} = \sum_{n=1}^{I_k} x$$

Return briefly to the elements $\omega(i)$ in the vector equation. These represent $W_p^{g_i(0)}$. The index i in $\omega(i)$ is taken from G^i which represents a tuple whose first element is $g_i(0)$. Therefore, as field addition is element by element:

$$\omega(i+j) = W_p^{g_j(0)+g_j(0)}$$

But:

$$W_p^{g_j(0)+g_j(0)} = W_p^{g_j(0)} \cdot W_p^{g_j(0)} = \omega(i) \cdot \omega(j)$$

Assume that j is $k\theta$. Then:

$$G^{i+k\theta} = G^i \cdot G^{k\theta} = \sum_{n=1}^{I_k} G^i$$

Thus:

$$\omega(i+k\theta) = \omega(i)^{I_k} = (W_p^{g_j(0)})^{I_k}$$

As I_k is a unique mapping, then ranging k over the integers $\{0, 1, \cdots, p-2\}$ ranges $\omega(i+k\theta)$ over all the primitive pth roots of unity.

Define sets T_L for $0 \le L \le p-2$ as follows:

$$T_L = \{ j ; \text{ where } g_j(0) = g_{L\theta}(0) \text{ for } 0 \le j \le M-1 \}$$

In addition, define the set T_{-1} as:

$$T_{-1} = \{ j ; \text{ where } g_j(0) = 0 \text{ for } 0 \le j \le M-1 \}$$

Then the union of these sets covers all integers j from 0 to $(M-1)$.

These sets will be used in several ways. To begin, define a class of polynomials:

$$Q_L(z) = \sum z^n \qquad\qquad\qquad \text{where n belongs to } T_L$$

The definition of the sets and the prior mentioned properties of the subfield mean that:

$$Q_L(z) = < Q_0(z) \cdot z^{L\theta} >_{z^M-1}$$

These sets may also be used as a means of dividing polynomial expressions. In particular, the polynomial $\{ W^{g_j(0)} \cdot z^j \}$ can be expressed as the sum of p separate polynomials:

$$\sum_{j=0}^{M-1} W^{g_j(0)} \cdot z^j = \sum_{L=0}^{p-2} W^{I_{L\theta}} Q_L(z) + Q_{-1}(z)$$

Return now to the DFT equations. The second set of these above may be expressed as:

$$X(G^i) = x(0, 0, \cdots, 0) + \sum_{j=0}^{M-1} x(G^{-j}) + \sum_{j=0}^{M-1} \left(W_P^{g_{i-j}(0)} - 1 \right) \cdot x(G^{-j})$$

$$= X(0, 0, \cdots, 0) + \sum_{j=0}^{M-1} \left(W_P^{g_{i-j}(0)} - 1 \right) \cdot x(G^{-j})$$

Thus the multi-dimensional DFT becomes:

$$X(G^i) - X(0, 0, \cdots, 0) = \sum_{j=0}^{M-1} \left(W_P^{g_{i-j}(0)} - 1 \right) \cdot x(G^{-j})$$

where $X(0, 0, \cdots, 0) = x(0, 0, \cdots, 0) + \sum_{j=0}^{M-1} x(G^{-j})$

Let the vector S define the left-hand side of the second equation. That is:

$$\mathbf{S} = [X(G^i) - X(0, 0, \cdots, 0)]$$

and define ($W_P^{g_{i-j}(0)} - 1$) as $h(i{-}j)$. In addition, describe the terms $x(G^{-j})$ by the vector **y** as before. Then, again, we can more compactly describe this by the vector equation:

$$\mathbf{S} = [h(i{-}j)]\, \mathbf{y}$$

or in polynomial form as:

$$S(z) = <H(z) \cdot Y(z)>_{z^{M}-1}$$

where H(z) has coefficients equal to the first column of matrix [$h(i{-}j)$]. Now:

$$H(z) = \sum_{j=0}^{M-1} h(j) \cdot z^j = \sum_{j=0}^{M-1} \left(W_P^{g_{i-j}(0)} - 1 \right) \cdot z^j$$

$$= \sum_{j=0}^{M-1} W_P^{g_{i-j}(0)} \cdot z^j - \sum_{j=0}^{M-1} z^j$$

$$= \left\langle \sum_{L=0}^{P-2} W^{l_{L\theta}} \cdot Q_L(z) - \sum_{L=0}^{P-2} Q_L(z) \right\rangle_{z^{M}-1}$$

$$= \left\langle \sum_{L=0}^{P-2} W^{l_{L\theta}} \cdot Q_0(z) \cdot z^{L\theta} - \sum_{L=0}^{P-2} Q_0(z) \cdot z^{L\theta} \right\rangle_{z^{M}-1}$$

$$= \left\langle \sum_{L=0}^{p-2} \left(W^{lL\theta} - 1 \right) \cdot z^{L\theta} \cdot Q_0(z) \right\rangle_{z^M-1}$$

$$= < A(z) \cdot Q_0(z) >_{z^M-1}$$

This describes a cyclic convolution.

We need to interpret this carefully. We have factored a polynomial into cyclic convolution form when that polynomial itself via another cyclic convolution originates from a circulant matrix. This equation implies then, that this matrix factors into two circulant matrices. The first columns of both are the set of coefficients of $A(z)$ and $Q_0(z)$ respectively and their first rows are those coefficients with all but the first element reversed.

Consider the number of computations in this algorithm. The matrix derived from $Q_0(z)$ has elements consisting only of ones and zeroes. All multiplications are therefore due to the matrix derived from $A(z)$. Let the product of the vector **y** and the matrix derived from $Q_0(z)$ be a vector **d**. Then $S(z)$ is:

$$S(z) = < A(z) \cdot D(z) >_{z^M-1}$$

$$= \left\langle \left(\sum_{L=0}^{p-2} h(L\theta) \cdot z^{L\theta} \right) \cdot \left(\sum_{k=0}^{M-1} d_k \cdot z^k \right) \right\rangle_{z^M-1}$$

where $h(L\theta)$ is $(W^{lL\theta} - 1)$. Use the sets T_L to express this as:

$$S(z) = \left\langle \left(\sum_{L=0}^{p-2} h(L\theta) \cdot z^{L\theta} \right) \cdot \left(\sum_{k=0}^{\theta-1} \sum_{L=0}^{p-2} d_{k+L\theta} \cdot z^{k+L\theta} \right) \right\rangle_{z^M-1}$$

$$= \left\langle \sum_{k=0}^{\theta-1} \left\{ \left(\sum_{L=0}^{p-2} h(L\theta) \cdot z^{L\theta} \right) \cdot \left(\sum_{L=0}^{p-2} d_{k+L\theta} \cdot z^{L\theta} \right) \right\} \cdot z^k \right\rangle_{z^M-1}$$

$$= \left\langle \sum_{k=0}^{\theta-1} \mathbf{hd}(k) \cdot z^k \right\rangle_{z^M-1}$$

The elements of this polynomial are:

$$\mathbf{hd}(k) = \left\langle \left(\sum_{L=0}^{p-2} h(L\theta) \cdot z^{L\theta} \right) \cdot \left(\sum_{L=0}^{p-2} d_{k+L\theta} \cdot z^{L\theta} \right) \right\rangle_{z^M-1}$$

$$= \left\langle \left(\sum_{L=0}^{p-2} h(L\theta) \cdot z_1^{L} \right) \cdot \left(\sum_{L=0}^{p-2} d_{k+L\theta} \cdot z_1^{L} \right) \right\rangle_{z_1^{p-1}-1}$$

where z_1 is z^θ. As these are p-point cyclic convolutions, hence this computation is equivalent to θ or $\frac{(p^\alpha - 1)}{(p - 1)}$ p-point cyclic convolutions.

Each row of the $(M{-}1)\times(M{-}1)$ matrix $Q_0(z)$ has exactly $p^{\alpha-1}$ ones and so the number of additions involved in forming vector **d** is:

$$(M - 1)\cdot(p^{\alpha-1} - 1) = (p^\alpha - 1)\cdot(p^{\alpha-1} - 1)$$

The p-point cyclic convolutions can be computed using p-point DFTs via Winograd's methods. Let the number of additions in each p-point DFT be A_p and the number of multiplications M_p. Then following earlier work, we require $(\theta\cdot(M_p - 1) + 1)$ multiplications to compute the cyclic convolutions, where the extra one is due to the term $X(0, 0, \cdots, 0)$. Hence overall, we require:

$$\frac{(p^\alpha - 1)}{(p-1)}\cdot(M_p - 1) + 1 \qquad \text{multiplications}$$

$$\frac{(p^\alpha - 1)}{(p-1)}\cdot A_p + (p^\alpha - 1)\cdot(p^{\alpha-1} - 1) \qquad \text{additions}$$

To conclude, it is useful to comment on this last manipulation of $S(z)$ showing it is equivalent to a set of p-point convolutions. We can express this result in a vector equation:

$$\mathbf{S'} = [\, B]\, \mathbf{d'}$$

where the terms **S'** and **d'** are used as the equation shows permutation of terms. Now consider matrix B. From the equation, this must consist of a set of $p{\times}p$ blocks on the main diagonal such that when multiplied by the terms of **d'** the result is the set of convolutions. Thus the procedure followed is also a block diagonalizing technique.

Example

We choose the simplest possible, namely 3^2. The first step is to find an irreducible polynomial for the field GF(3^2). There are three possibilities:

$$z^2 + 1$$
$$z^2 + z + 2$$
$$z^2 + 2z + 2$$

For this example, we choose the last. We also require a primitive root for this irreducible polynomial. The four possibilities are:

$$\text{z} \qquad 2z$$
$$z + 2 \quad 2z + 1$$

Although z is an obvious choice, for this example we select $(2z + 1)$.

In matrix form, the irreducible polynomial is:

$$\begin{bmatrix} 0 & -2 \\ 1 & -2 \end{bmatrix} = \begin{bmatrix} 0 & 1 \\ 1 & 1 \end{bmatrix}$$

Therefore, an element of the field is described (in matrix form) by:

$$a\begin{bmatrix} 1 & 0 \\ 0 & 1 \end{bmatrix} + b\begin{bmatrix} 0 & 1 \\ 1 & 1 \end{bmatrix} = \begin{bmatrix} a & b \\ b & a+b \end{bmatrix}$$

From this, if G^i describes the tuple (a,b), then G^4 also describes the tuple (a,b). Here, both this irreducible polynomial and $(z^2 + z + 2)$ map tuples into themselves. If $(z^2 + 1)$ had been chosen however, then a field element in matrix form would be:

$$a\begin{bmatrix} 1 & 0 \\ 0 & 1 \end{bmatrix} + b\begin{bmatrix} 0 & 2 \\ 1 & 0 \end{bmatrix} = \begin{bmatrix} a & 2b \\ b & a \end{bmatrix}$$

Thus (a,b) maps into (a,2b) and the transformed element is in matrix terms:

$$a\begin{bmatrix} 1 & 0 \\ 0 & 1 \end{bmatrix} + 2b\begin{bmatrix} 0 & 2 \\ 1 & 0 \end{bmatrix} = \begin{bmatrix} a & 4b \\ 2b & a \end{bmatrix}$$

The direct expression of the DFT equations can be written as:

$$\begin{bmatrix} X(0,0) \\ X(0,1) \\ X(0,2) \\ X(1,0) \\ X(1,1) \\ X(1,2) \\ X(2,0) \\ X(2,1) \\ X(2,2) \end{bmatrix} = \begin{bmatrix} 1 & 1 & 1 & 1 & 1 & 1 & 1 & 1 & 1 \\ 1 & W & W^2 & 1 & W & W^2 & 1 & W & W^2 \\ 1 & W^2 & W & 1 & W^2 & W & 1 & W^2 & W \\ 1 & 1 & 1 & W & W & W & W^2 & W^2 & W^2 \\ 1 & W & W^2 & W & W^2 & 1 & W^2 & 1 & W \\ 1 & W^2 & W & W & 1 & W^2 & W^2 & W & 1 \\ 1 & 1 & 1 & W^2 & W^2 & W^2 & W & W & W \\ 1 & W & W^2 & W^2 & 1 & W & W & W^2 & 1 \\ 1 & W^2 & W & W^2 & W & 1 & W & 1 & W^2 \end{bmatrix} \begin{bmatrix} x(0,0) \\ x(0,1) \\ x(0,2) \\ x(1,0) \\ x(1,1) \\ x(1,2) \\ x(2,0) \\ x(2,1) \\ x(2,2) \end{bmatrix}$$

We now have the information needed to perform the initial transformation of this. That is:

$$X(0, 0, \ldots, 0) = x(0, 0, \cdots, 0) + \sum_{j=0}^{M-1} x(G^{-j})$$

$$X(G^4) = x(0, 0, \cdots, 0) + \sum_{j=0}^{M-1} x(G^{-j}) \cdot W_p^{g_{i-j}(0)}$$

This creates:

$$
\begin{bmatrix} X(0,0) \\ X(1,0) \\ X(1,2) \\ X(2,2) \\ X(0,1) \\ X(2,0) \\ X(2,1) \\ X(1,1) \\ X(0,2) \end{bmatrix} = \begin{bmatrix} 1 & 1 & 1 & 1 & 1 & 1 & 1 & 1 & 1 \\ 1 & W & 1 & W & W^2 & W^2 & 1 & W^2 & W \\ 1 & W & W & 1 & W & W^2 & W^2 & 1 & W^2 \\ 1 & W^2 & W & W & 1 & W & W^2 & W^2 & 1 \\ 1 & 1 & W^2 & W & W & 1 & W & W^2 & W^2 \\ 1 & W^2 & 1 & W^2 & W & W & 1 & W & W^2 \\ 1 & W^2 & W^2 & 1 & W^2 & W & W & 1 & W \\ 1 & W & W^2 & W^2 & 1 & W^2 & W & W & 1 \\ 1 & 1 & W & W^2 & W^2 & 1 & W^2 & W & W \end{bmatrix} \begin{bmatrix} x(0,0) \\ x(1,0) \\ x(0,2) \\ x(1,1) \\ x(2,1) \\ x(2,0) \\ x(0,1) \\ x(2,2) \\ x(1,2) \end{bmatrix}
$$

For the next stage, form the vectors **S** and **y** and the matrix H, where H is a minor of the above. To do this, subtract the first column from every other column, then the first row from every other row. Letting h_1 be ($W - 1$) and h_2 be ($W^2 - 1$), we get:

$$
\begin{bmatrix} S(0) \\ S(1) \\ S(2) \\ S(3) \\ S(4) \\ S(5) \\ S(6) \\ S(7) \end{bmatrix} = \begin{bmatrix} h_1 & 0 & h_1 & h_2 & h_2 & 0 & h_2 & h_1 \\ h_1 & h_1 & 0 & h_1 & h_2 & h_2 & 0 & h_2 \\ h_2 & h_1 & h_1 & 0 & h_1 & h_2 & h_2 & 0 \\ 0 & h_2 & h_1 & h_1 & 0 & h_1 & h_2 & h_2 \\ h_2 & 0 & h_2 & h_1 & h_1 & 0 & h_1 & h_2 \\ h_2 & h_2 & 0 & h_2 & h_1 & h_1 & 0 & h_1 \\ h_1 & h_2 & h_2 & 0 & h_2 & h_1 & h_1 & 0 \\ 0 & h_1 & h_2 & h_2 & 0 & h_2 & h_1 & h_1 \end{bmatrix} \begin{bmatrix} y_0 \\ y_1 \\ y_2 \\ y_3 \\ y_4 \\ y_5 \\ y_6 \\ y_7 \end{bmatrix}
$$

In polynomial form, this is described by:

$$ S(z) \ = \ < H(z) \cdot Y(z) >_{z^8 - 1} $$

where $H(z) = h_1 + h_1 \cdot z + h_2 \cdot z^2 + h_2 \cdot z^4 + h_2 \cdot z^5 + h_1 \cdot z^6$

We can set about factoring this. As θ, which is $\frac{(p^\alpha - 1)}{(p - 1)}$, is 4, then the polynomial A(z) is given by:

$$ A(z) \ = \ \sum_{L=0}^{p-2} h(\theta L) \cdot z^{\theta L} \ = \ h(0) + h(4) \cdot z^4 \ = \ h_1 + h_2 \cdot z^4 $$

This represents the circulant matrix:

$$\begin{bmatrix} h_1 & 0 & 0 & 0 & h_2 & 0 & 0 & 0 \\ 0 & h_1 & 0 & 0 & 0 & h_2 & 0 & 0 \\ 0 & 0 & h_1 & 0 & 0 & 0 & h_2 & 0 \\ 0 & 0 & 0 & h_1 & 0 & 0 & 0 & h_2 \\ h_2 & 0 & 0 & 0 & h_1 & 0 & 0 & 0 \\ 0 & h_2 & 0 & 0 & 0 & h_1 & 0 & 0 \\ 0 & 0 & h_2 & 0 & 0 & 0 & h_1 & 0 \\ 0 & 0 & 0 & h_2 & 0 & 0 & 0 & h_1 \end{bmatrix}$$

The other part of the factorization is the matrix derived from $Q_0(z)$. This polynomial has powers equal to the members of the set T_0, namely:

$$\{\, j\,;\, \text{where } g_j(0) = g_0(0) = G^0 = 1 \ \text{ for } 0 \le j \le M-1 \,\}$$

The different powers of G are the indices of the vectors X and x in the previous matrix equations, hence:

$$Q_0(z) = 1 + z + z^6$$

Thus the corresponding circulant matrix is:

$$\begin{bmatrix} 1 & 0 & 1 & 0 & 0 & 0 & 0 & 1 \\ 1 & 1 & 0 & 1 & 0 & 0 & 0 & 0 \\ 0 & 1 & 1 & 0 & 1 & 0 & 0 & 0 \\ 0 & 0 & 1 & 1 & 0 & 1 & 0 & 0 \\ 0 & 0 & 0 & 1 & 1 & 0 & 1 & 0 \\ 0 & 0 & 0 & 0 & 1 & 1 & 0 & 1 \\ 1 & 0 & 0 & 0 & 0 & 1 & 1 & 0 \\ 0 & 1 & 0 & 0 & 0 & 0 & 1 & 1 \end{bmatrix}$$

and this completes the algorithm.

This algorithm is equivalent to four 3-point convolutions. To show this, we note:

$$S = [\,A\,][\,Q_0\,]\,y$$

We can take the product of the last two matrices to give:

$$S = [\,A\,]\,d$$

By permuting the elements of S and d, we can transform A to:

$$\begin{bmatrix} h_1 & h_2 & 0 & 0 & 0 & 0 & 0 & 0 \\ h_2 & h_1 & 0 & 0 & 0 & 0 & 0 & 0 \\ 0 & 0 & h_1 & h_2 & 0 & 0 & 0 & 0 \\ 0 & 0 & h_2 & h_1 & 0 & 0 & 0 & 0 \\ 0 & 0 & 0 & 0 & h_1 & h_2 & 0 & 0 \\ 0 & 0 & 0 & 0 & h_2 & h_1 & 0 & 0 \\ 0 & 0 & 0 & 0 & 0 & 0 & h_1 & h_2 \\ 0 & 0 & 0 & 0 & 0 & 0 & h_2 & h_1 \end{bmatrix}$$

which is in block diagonal form where each of the four blocks is the Rader form of a 3-point DFT. The permutation is easily achieved. Let:

$$\alpha = k + L\theta$$
$$\beta = L + k\theta$$

where
$$k = 0, 1, \cdots, \theta - 1$$
$$L = 0, 1, \cdots, p - 2$$

Then the permutation is to swap row α with row β and column α with column β.

Gertner's approach

Gertner has recently proposed an elegant yet simple approach for the computation of two-dimensional discrete Fourier transforms. The N×N DFT given by:

$$X(k_1, k_2) = \sum_{n_1=0}^{N-1} \sum_{n_2=0}^{N-1} x(n_1, n_2) \cdot W_N^{k_1 n_1 + k_2 n_2}$$

where $k_1, k_2 = 0, 1, \cdots, N-1$

describes the transformation of an array of points $x(n_1, n_2)$ whose coordinates define a grid of integers. To use the terminology of mathematics, the coordinates are defined on a **lattice**. Gertner's approach is to identify relationships between points on the lattice and use that to decompose the DFT into a set of one-dimensional DFTs.

We will develop the approach by considering an algorithm for the simplest case: N equal to a prime p. From earlier work, the linear congruence $< m\alpha - n\beta >_p = 0$, where m and n are taken from $0, 1, \cdots, p-1$, has solutions only if m and n are relatively prime. If so, then from Euler's theorem:

$$< \alpha - m^{\Phi(p)-1} \cdot n\beta >_p = 0$$

i.e.
$$\alpha = < m^{\Phi(p)-1} \cdot n\beta >_p = < \theta \cdot \beta >_p$$

Ranging β over $0, 1, \cdots, p-1$ generates p congruent solutions, one of which is $(0,0)$. If either m or n should be zero then the congruence reduces to $< r\delta >_p = 0$. As shown in Part 3, solutions to such a congruence exist only if r and p are mutually prime. If this is the case, then again there will be p incongruent solutions.

Given a particular pair (m, n), then the solution pairs of the congruence will match p points of the coordinate lattice of the DFT. We can clearly find a set of congruences that match all the points of the lattice but can we find a set of p-point congruences that do this and what is the minimum number of such congruences? Following the methods outlined in Part 3, we find that an alternative solution to the congruence is:

$$\alpha = < nt >_p$$
$$\beta = < mt >_p$$
$$t = 0, 1, \cdots, p-1$$

Observe that $\{ nt \}$ and $\{ mt \}$ are permutations of the sequence $\{ t \}$. Therefore, spanning all coordinate pairs (i, j), $i, j = 0, 1, \cdots, p-1$, in a lattice is just a question of choosing appropriate values for the parameter pair (m, n) subject to the condition they are relatively prime to each other or, if either is zero, that they relatively prime to p. A convenient set is:

(a) n a finite value (mutually prime to p) and m ranging over $0, 1, \cdots, p-1$;
(b) n set to 0 and m a finite value (mutually prime to p).

This offers several choices for $(p+1)$ pairs of (m,n) which clearly is the minimum number needed. The most convenient is the set $\{ (0,1), (1,0), (1,1), \cdots, (1,p-1) \}$.

Using this result, we can now decompose the two-dimensional DFT into a set of one-dimensional DFTs; one for each of the congruences. Begin by substituting the congruence solutions into the DFT equations to give:

$$X(\langle nt \rangle_p, \langle mt \rangle_p) = \sum_{n_1=0}^{p-1} \sum_{n_2=0}^{p-1} x(n_1, n_2) \cdot W_p^{n_1 nt + n_2 mt}$$

$$= \sum_{n_1=0}^{p-1} \sum_{n_2=0}^{p-1} x(n_1, n_2) \cdot W_p^{dt}$$

where $\qquad t = 0, 1, 2, \cdots, p-1 \qquad\qquad d = < m \cdot n_2 + n \cdot n_1 >_p$

Since d must range over $0, 1, \cdots, p-1$ then we can express this as:

$$X(\langle nt \rangle_p, \langle mt \rangle_p) = \sum_{d=0}^{p-1} R_d(m, n) \cdot W_p^{dt}$$

where $R_d(m,n) = \sum_{n_1=0}^{p-1} x\left(n_1, \langle m \cdot n_2 + n \cdot n_1 \rangle_p = d\right)$

Hence the result. These terms $R_d(m,n)$ define a transform, the **discrete Radon transform**. It is widely used in areas such as tomography and seismology.

This completes the algorithm and all it requires is:

(a) $p(p^2 - 1)$ additions to form the terms $R_d(m,n)$;
(b) $(p + 1)$ p-point DFTs of the terms $R_d(m,n)$.

This is comparable to Auslander, Feig, and Winograd's method.

Example

We will find the DFT of the array of points

0.000000	0.000000	0.000000	0.000000	0.000000
0.951057	0.293893	−0.769421	−0.769421	0.293893
0.587785	0.181636	−0.475528	−0.475528	0.181636
−0.587785	−0.181636	0.475528	0.475528	−0.181636
−0.951057	−0.293893	0.769421	0.769421	−0.293893

We need a set of six values for (m, n). Here, they are chosen to be (0,1), (1,0), (1,1), (1,2), (1,3), and (1,4). To determine the DFT, we need to find for each of the (p + 1) pairs of (m,n) the terms $R_d(m, n)$ for d equal to 0, 1, $\cdots$, p − 1 and then find the p-point Fourier transform of these values. Thus:

1. $R_d(0,1) = \sum\limits_{j=0}^{4} x(d,j)$

$$R_0(0,1) = x(0,0) + x(0,1) + x(0,2) + x(0,3) + x(0,4) = 0.0$$
$$R_1(0,1) = x(1,0) + x(1,1) + x(1,2) + x(1,3) + x(1,4) = 0.0$$
$$R_2(0,1) = x(2,0) + x(2,1) + x(2,2) + x(2,3) + x(2,4) = 0.0$$
$$R_3(0,1) = x(3,0) + x(3,1) + x(3,2) + x(3,3) + x(3,4) = 0.0$$
$$R_4(0,1) = x(4,0) + x(4,1) + x(4,2) + x(4,3) + x(4,4) = 0.0$$

Taking the Fourier transform:

$$X(0,0) = 0.0$$
$$X(1,0) = 0.0$$
$$X(2,0) = 0.0$$
$$X(3,0) = 0.0$$
$$X(4,0) = 0.0$$

2. $R_d(1,0) = \sum\limits_{j=0}^{4} x(j,d)$

$$R_0(1,0) = x(0,0) + x(1,0) + x(2,0) + x(3,0) + x(4,0) = 0.0$$
$$R_1(1,0) = x(0,1) + x(1,1) + x(2,1) + x(3,1) + x(4,1) = 0.0$$
$$R_2(1,0) = x(0,2) + x(1,2) + x(2,2) + x(3,2) + x(4,2) = 0.0$$
$$R_3(1,0) = x(0,3) + x(1,3) + x(2,3) + x(3,3) + x(4,3) = 0.0$$
$$R_4(1,0) = x(0,4) + x(1,4) + x(2,4) + x(3,4) + x(4,4) = 0.0$$

Taking the Fourier transform:

$$X(0,0) = 0.0$$
$$X(0,1) = 0.0$$
$$X(0,2) = 0.0$$
$$X(0,3) = 0.0$$
$$X(0,4) = 0.0$$

3. $R_d(1,1) = \displaystyle\sum_{j=0}^{4} x\left(j, \langle d-j \rangle_p \right)$

$$R_0(1,1) = x(0,0) + x(1,4) + x(2,3) + x(3,2) + x(4,1) = 0.0$$
$$R_1(1,1) = x(0,1) + x(1,0) + x(2,4) + x(3,3) + x(4,2) = 2.377642$$
$$R_2(1,1) = x(0,2) + x(1,1) + x(2,0) + x(3,4) + x(4,3) = 1.469463$$
$$R_3(1,1) = x(0,3) + x(1,2) + x(2,1) + x(3,0) + x(4,4) = -1.469463$$
$$R_4(1,1) = x(0,4) + x(1,3) + x(2,2) + x(3,1) + x(4,0) = -2.377642$$

Taking the Fourier transform:

$$X(0,0) = 0.0$$
$$X(1,1) = -j6.25$$
$$X(2,2) = 0.0$$
$$X(3,3) = 0.0$$
$$X(4,4) = j6.25$$

4. $R_d(1,2) = \displaystyle\sum_{j=0}^{4} x\left(j, \langle d-2j \rangle_p \right)$

$$R_0(1,2) = x(0,0) + x(1,3) + x(2,1) + x(3,4) + x(4,2) = 0.0$$
$$R_1(1,2) = x(0,1) + x(1,4) + x(2,2) + x(3,0) + x(4,3) = 0.0$$
$$R_2(1,2) = x(0,2) + x(1,0) + x(2,3) + x(3,1) + x(4,4) = 0.0$$
$$R_3(1,2) = x(0,3) + x(1,1) + x(2,4) + x(3,2) + x(4,0) = 0.0$$
$$R_4(1,2) = x(0,4) + x(1,2) + x(2,0) + x(3,3) + x(4,1) = 0.0$$

Taking the Fourier transform:

$$X(0,0) = 0.0$$
$$X(2,1) = 0.0$$
$$X(4,2) = 0.0$$

$$X(1,3) = 0.0$$
$$X(3,4) = 0.0$$

5. $R_d(1,3) = \sum_{j=0}^{4} x\left(j, \langle d - 3j \rangle_p\right)$

$R_0(1,3) = x(0,0) + x(1,2) + x(2,4) + x(3,1) + x(4,3) = 0.0$
$R_1(1,3) = x(0,1) + x(1,3) + x(2,0) + x(3,2) + x(4,4) = 0.0$
$R_2(1,3) = x(0,2) + x(1,4) + x(2,1) + x(3,3) + x(4,0) = 0.0$
$R_3(1,3) = x(0,3) + x(1,0) + x(2,2) + x(3,4) + x(4,1) = 0.0$
$R_4(1,3) = x(0,4) + x(1,1) + x(2,3) + x(3,0) + x(4,2) = 0.0$

Taking the Fourier transform:

$$X(0,0) = 0.0$$
$$X(3,1) = 0.0$$
$$X(1,2) = 0.0$$
$$X(4,3) = 0.0$$
$$X(2,4) = 0.0$$

6. $R_d(1,4) = \sum_{j=0}^{4} x\left(j, \langle d - 4j \rangle_p\right)$

$R_0(1,4) = x(0,0) + x(1,1) + x(2,2) + x(3,3) + x(4,4) = 0.0$
$R_1(1,4) = x(0,1) + x(1,2) + x(2,3) + x(3,4) + x(4,0) = -2.377642$
$R_2(1,4) = x(0,2) + x(1,3) + x(2,4) + x(3,0) + x(4,1) = -1.469463$
$R_3(1,4) = x(0,3) + x(1,4) + x(2,0) + x(3,1) + x(4,2) = 1.469463$
$R_4(1,4) = x(0,4) + x(1,0) + x(2,1) + x(3,2) + x(4,3) = -2.377642$

Taking the Fourier transform:

$$X(0,0) = 0.0$$
$$X(4,1) = j6.25$$
$$X(3,2) = 0.0$$
$$X(2,3) = 0.0$$
$$X(1,4) = j6.25$$

Thus the final result is that the two-dimensional discrete Fourier transform is:

0.00	0.00	0.00	0.00	0.00
0.00	−j6.25	0.00	0.00	−j6.25
0.00	0.00	0.00	0.00	0.00
0.00	0.00	0.00	0.00	0.00
0.00	j6.25	0.00	0.00	j6.25

The basic approach is quite easily extended to arbitrary N. In this case, we begin once more with a linear congruence but now of the form $< m\alpha - n\beta >_N = 0$. This congruence derives from a Diophantine equation. From the work of Part 3, we know a solution exists only if the greatest common divisor of m and n divides N. There are two cases to consider here:

1. m and n are mutually prime. In this case, solutions to the congruence are very similar to the case of N prime.
2. m and n have a common divisor d. In this case, the solutions to the congruence are given by

$$\alpha = < \frac{n}{d} t > N$$

$$\beta = < \frac{m}{d} t > N$$

It is very easy to show these equations have $d \cdot N$ congruent solutions whereas what is required is N congruent solutions.

Then the requirement in the general case is exactly as for the case of N prime; m and n must be mutually prime. Further, if one of m or n is zero, then the other must be mutually prime to N.

Working through the previous equations, we find nothing dependent on N being prime. Thus the same result applies to N composite. The critical question then, is whether a set of congruences can be found, each with N congruent solutions, that span all points of the lattice. The success of the approach in turn is dependent on a minimal set of such congruences being found. This is a moderately difficult problem. The reason is that the requirement m and n are mutually prime does not mean that m and N and n and N are mutually prime. In the solutions to the congruence though, $< mt >_N$ or $< nt >_N$ are only guaranteed to be unique permutations of the sequence $\{t\}$ if m and N are mutually prime or n and N are mutually prime respectively. The result is that there is some overlap between the congruences and so a loss of efficiency. In addition, it is not possible to state a general result for the number of operations. Nevertheless, given that M congruences are found to span the lattice where M depends on N, then the number of operations is:

1. $M \cdot p \cdot (p - 1)$ additions to form the terms $R_d(m,n)$;
2. M p-point DFTs of the terms $R_d(m,n)$.

Example

A simple illustration of the difficulties is for an array of $3^2 \times 3^2$ points. Now the solution to the general congruence $< m\alpha - n\beta >_N = 0$ is:

$$\alpha = < n \cdot t >_N$$
$$\beta = < m \cdot t >_N$$

For this problem, we require a set of pairs (m,n), mutually prime, that cover all pairs in a 9×9 lattice. We proceed as follows:

1. Choose the set of values (1, m) and range m over 0, 1, $\cdots$, N – 1. This spans all points in the lattice bar:

    (1,0) (2,0) (3,0) (4,0) (5,0) (6,0) (7,0) (8,0)
    (1,3) (2,3) (4,3) (5,3) (7,3) (8,3)
    (1,6) (2,6) (4,6) (5,6) (7,6) (8,6)

2. Choosing (0,1) will give a congruence that spans the points (1,0), (2,0), (3,0), (4,0), (5,0), (6,0), (7,0), and (8,0).

3. To span the remaining points, we select (3r,1) and range r over 1, 2 and 3.

We may generalize this result. To span a lattice of $p^\alpha \times p^\alpha$ points, we need to choose congruences with parameters:

1. (1,m) for m = 0, 1, $\cdots$, $p^\alpha - 1$
2. (0,1)
3. (pr,1) for r = 1, 2, $\cdots$, $p^{\alpha-1} - 1$

Thus ($p^\alpha + p^{\alpha-1}$) or $\frac{p+1}{p} p^\alpha$ congruences are needed and so the number of operations is:

1. ($p^2 - 1$)$\cdot p^\alpha$ additions to form the values $R_d(m,n)$;
2. $\frac{p+1}{p} \cdot p^\alpha$ p-point one dimensional DFTs.

Gertner has also analyzed the case of N equal to the product of two primes p_1 and p_2. Here, there are ($N + p_1 + p_2 + 1$) congruences spanning the lattice where the parameter pairs are:

1. (1,m) for m = 0, 1, $\cdots$, N – 1
2. (0,1)
3. (p_1r,1) for r = 1, 2, $\cdots$, $p_2 - 1$
4. (p_2s,1) for s = 1,2, $\cdots$, $p_1 - 1$
5. (p_1,p_2)
6. (p_2,p_1)

It was stated that Gertner's approach is for computing two-dimensional discrete Fourier transforms. It can be extended to other dimensions but with some difficulty. Given a congruence such as $< m\alpha + n\beta + q\gamma >_N = 0$, then it is quite easy to show solutions exist provided the greatest common divisor of m, n and q divides N. However, the problem is to find those solutions as there is no general solution to the multi-dimensional congruence. Some limited approaches exist for dealing with three-dimensional congruences but beyond that trial and error is the major option.

A final comment on this approach is that it creates a very attractive algorithm for parallel computing. For composite N there is some sharing of input data, but the computation of the sets of Radon transforms and their DFTs proceeds completely independently.

Polynomial transform techniques

Polynomial transforms are an attractive means of computing multi-dimensional convolutions. Here, however, we will concentrate on just using polynomial transforms for computing two-dimensional cyclic convolutions as they are the most important of the higher dimensional convolutions.

Assume two arrays $x(i, j)$ and $h(i, j)$ are to be convolved to produce an array $y(i, j)$. Assume further that $h(i, j)$ is known, hence any part of the algorithm purely in terms of these elements can be pre-calculated. Cyclic convolution relates these arrays via:

$$Y(i,j) = \sum_{u=0}^{N-1} \sum_{v=0}^{N-1} h(u,v) \cdot x(i-u, j-v) \qquad\qquad i, j = 0, 1, \cdots, N-1$$

where the exponents are interpreted modulo N. In polynomial terms, this two-dimensional cyclic convolution is given by:

$$Y(z_1, z_2) = << H(z_1, z_2) \cdot X(z_1, z_2) >_{z_1^{N}-1} >_{z_2^{N}-1}$$

where

$$H(z_1, z_2) = \sum_{i=0}^{N-1} \sum_{j=0}^{N-1} h(i,j) \cdot z_1^i z_2^j$$

$$X(z_1, z_2) = \sum_{i=0}^{N-1} \sum_{j=0}^{N-1} x(i,j) \cdot z_1^i z_2^j$$

$$Y(z_1, z_2) = \sum_{i=0}^{N-1} \sum_{j=0}^{N-1} y(i,j) \cdot z_1^i z_2^j$$

Now form a set of polynomials from these equations as follows:

$$H_j(z_1) = \sum_{i=0}^{N-1} h(i,j) \cdot z_1^i$$

$$X_j(z_1) = \sum_{i=0}^{N-1} x(i,j) \cdot z_1^i$$

$$Y_j(z_1) = \sum_{i=0}^{N-1} y(i,j) \cdot z_1^i$$

By substituting into $Y(z_1, z_2)$, it is found that the polynomial elements of the two-dimensional convolution $Y_j(z_1)$ are given by:

$$Y_j(z_1) = \left\langle \sum_{v=0}^{N-1} H_v(z_1) \cdot X_{j-v}(z_1) \right\rangle_{z_1^N - 1} \qquad j = 0, 1, \cdots, N-1$$

That is, the two-dimensional convolution can be expressed as a set of N cyclic convolutions where each of these convolutions is between sequences of polynomial elements. Given the separation between z_1 and z_2 here, there is no need to distinguish between them and so we will omit the subscripts from this point onward.

The simplest approach to solving this problem is to divide it into a set of subproblems via the Chinese remainder theorem. In each of these, operations are defined modulo a cyclotomic polynomial. As these are irreducible, then polynomial transforms can be used to compute the solutions. However, given that each of these subproblems is expressed in terms of N polynomials (each of the same order as the cyclotomic polynomial involved), a root of order N is needed in the polynomial structures concerned. The computation proceeds by taking polynomial transforms of the two arrays using this root. Operations are modulo $Q(z)$ where $Q(z)$ is an (irreducible) cyclotomic polynomial. The resulting polynomials are multiplied modulo $Q(z)$ and the inverse polynomial transform determined, again modulo $Q(z)$. This gives the set of polynomials $Y_j(z)$. The only general multiplications in this procedure will be in the product of the two polynomial transforms. However, if the root is an arbitrary polynomial, then the various reductions will introduce a large number of scaling operations, so creating implementation problems. Therefore, these polynomial transforms need to be based on simple roots such as the set $\{ z^k \}$. This is possible when $Q(z)$ is cyclotomic subject to N and k being relatively prime.

We return to the problem to consider it in detail. Divide it into two specific cases: N prime and N composite. We begin with N an odd prime p. First, the Chinese remainder theorem is used to reduce the problem into two simpler problems. Now:

$$z^p - 1 = (z - 1) \cdot (z^{p-1} + z^{p-2} + \cdots + z + 1)$$
$$= (z - 1) \cdot Q(z)$$

Thus for the set of terms:

$$Y_{1,j}(z) = \langle Y_j(z) \rangle_{z-1} \qquad j = 0, 1, \cdots, p-1$$

$$= \sum_{v=0}^{p-1} H_v(1) \cdot X_{j-v}(1)$$

This is simply a p-point cyclic convolution between scalar elements. Next, form the set of polynomials:

$$Y_{Q,j}(z) = \langle Y_j(z) \rangle_{Q(z)} \qquad j = 0, 1, \cdots, p-1$$

The required convolution is therefore the elements of the set of polynomials:

$$Y_j(z) = S_1(z) \cdot Y_{1,j}(z) + S_2(z) \cdot Y_{Q,j}(z) \qquad\qquad j = 0, 1, \cdots, p-1$$

where
$$< S_1(z) >_{z-1} = 1 \qquad \text{and} \qquad < S_1(z) >_{Q(z)} = 0$$
$$< S_2(z) >_{z-1} = 0 \qquad \text{and} \qquad < S_2(z) >_{Q(z)} = 1$$

Suitable choices for $S_1(z)$ and $S_2(z)$ are:

$$S_1(z) = \frac{1}{p} \cdot Q(z)$$

$$S_2(z) = 1 - \frac{1}{p} \cdot Q(z)$$

Consider in broad terms the execution of this procedure. First, the set of p terms $Y_{1,j}(z)$ form a (scalar) cyclic convolution of the type studied earlier in this Part. There are more additions than usual but using one of the minimal algorithms discussed earlier there need only be $(2p-2)$ general multiplications. The term $\frac{1}{p}$ in $S_1(z)$ can be included in the pre-calculated part of this algorithm to eliminate scaling in the Chinese remainder reconstruction with this term.

The more complex set of terms to compute are the polynomials $Y_{Q,j}(z)$. Since they are expressed in terms of polynomial elements, this can be done via polynomial transforms using, as we have already established, a root of z. Finding the $Y_{Q,j}(z)$ becomes a four-stage exercise. First, the two polynomial sequences { $H_v(z)$ } and { $X_v(z)$ } are reduced modulo $Q(z)$. Second, they are each polynomial transformed. Third, as with the DFT, they are multiplied (polynomial) element by element. Finally, the inverse polynomial transform is taken. As { $H_v(z)$ } is known, the polynomial transform associated with it can be pre-computed. The root of the polynomial transforms here is z, thus the only multiplications of any type are those resulting from the product of the polynomial transforms. These each have p polynomials elements of order $(p-1)$. Since $Q(z)$ is irreducible, a minimal algorithm to multiply these elements requires $(2(p-1)-1)$ general multiplications. There are p such multiplications to perform, hence overall, computing $Y_{1,j}(z)$ and $Y_{Q,j}(z)$ requires:

$$(2p-2) + p \cdot (2(p-1)-1) = 2p^2 - p - 2$$

general multiplications. In comparison, if an optimal Fourier transform algorithm is used, it will require $(2p-2)$ general multiplications for each of the 2p p-point one-dimensional transforms. In all, that is $(4p^2-4p)$. The polynomial transform approach therefore achieves an advantage of almost two in terms of multiplications for large p.

Although these figures are only for general multiplications, the number of scaling terms is quite small. The polynomial transforms involve just additions and shifts, hence scalings can be introduced in one of two ways. One is via the reductions modulo $Q(z)$. However, that just involves the substitution:

$$z^{p-1} = -z^{p-2} - z^{p-3} - \cdots - z - 1$$

and this introduces no scale factors. The other source of scale factors is in the multiplication $S_2(z) \cdot Y_{Q,j}(z)$. Some simple steps can reduce problems here. From the earlier work on the polynomial Chinese remainder theorem:

$$S_2(z) = (z - 1) \cdot T(z)$$

where $T(z)$ is some polynomial. This polynomial is a constant of the polynomial structure within which we are working. Therefore it can be linked in with the (pre-computed) polynomial transform of $H_u(z)$.

What this approach achieves is the mapping of a two-dimensional cyclic convolution problem into a scalar cyclic convolution of p elements plus p polynomial products modulo $Q(z)$ where $Q(z)$ is of order $(p-1)$. These products are between polynomials and so they are similar to scalar cyclic convolutions. Having to perform one convolution and these p products, both of order $(p-1)$, is inconvenient. For practical reasons, therefore, it is useful to adopt a slightly different approach. A reduction modulo $Q(z)$ is the same as a reduction $(z^p - 1)$ followed by reduction modulo $Q(z)$ as $Q(z)$ is cyclotomic and a factor of $(z^p - 1)$. Hence the heart of the problem can be re-expressed as a set of $(p+1)$ p-point scalar cyclic convolutions. The later conversion of p of these modulo $Q(z)$ only involves additions.

Example

This example will both illustrate the technique and serve as a vehicle for a detailed analysis of the number of operations required. Take the two arrays to be cyclically convolved as:

$h(0,0)$	$h(0,1)$	$h(0,2)$	$h(0,3)$	$h(0,4)$		$x(0,0)$	$x(0,1)$	$x(0,2)$	$x(0,3)$	$x(0,4)$
$h(1,0)$	$h(1,1)$	$h(1,2)$	$h(1,3)$	$h(1,4)$		$x(1,0)$	$x(1,1)$	$x(1,2)$	$x(1,3)$	$x(1,4)$
$h(2,0)$	$h(2,1)$	$h(2,2)$	$h(2,3)$	$h(2,4)$	and	$x(2,0)$	$x(2,1)$	$x(2,2)$	$x(2,3)$	$x(2,4)$
$h(3,0)$	$h(3,1)$	$h(3,2)$	$h(3,3)$	$h(3,4)$		$x(3,0)$	$x(3,1)$	$x(3,2)$	$x(3,3)$	$x(3,4)$
$h(4,0)$	$h(4,1)$	$h(4,2)$	$h(4,3)$	$h(4,4)$		$x(4,0)$	$x(4,1)$	$x(4,2)$	$x(4,3)$	$x(4,4)$

As p is 5, therefore:

$$Q(z) = z^4 + z^3 + z^2 + z + 1$$

Assume the first array is fixed and so computations purely in terms of its elements can be pre-computed. Although this is the second simplest problem that can be considered, it is still quite complex. In parts of the problem therefore, we shall merely sketch the actions to be taken.

The first step is to form the arrays into polynomials:

$$H_0(z) = h(0,0) + h(0,1) \cdot z + h(0,2) \cdot z^2 + h(0,3) \cdot z^3 + h(0,4) \cdot z^4$$
$$H_1(z) = h(1,0) + h(1,1) \cdot z + h(1,2) \cdot z^2 + h(1,3) \cdot z^3 + h(1,4) \cdot z^4$$
$$H_2(z) = h(2,0) + h(2,1) \cdot z + h(2,2) \cdot z^2 + h(2,3) \cdot z^3 + h(2,4) \cdot z^4$$
$$H_3(z) = h(3,0) + h(3,1) \cdot z + h(3,2) \cdot z^2 + h(3,3) \cdot z^3 + h(3,4) \cdot z^4$$
$$H_4(z) = h(4,0) + h(4,1) \cdot z + h(4,2) \cdot z^2 + h(4,3) \cdot z^3 + h(4,4) \cdot z^4$$

$$X_0(z) = x(0,0) + x(0,1)\cdot z + x(0,2)\cdot z^2 + x(0,3)\cdot z^3 + x(0,4)\cdot z^4$$
$$X_1(z) = x(1,0) + x(1,1)\cdot z + x(1,2)\cdot z^2 + x(1,3)\cdot z^3 + x(1,4)\cdot z^4$$
$$X_2(z) = x(2,0) + x(2,1)\cdot z + x(2,2)\cdot z^2 + x(2,3)\cdot z^3 + x(2,4)\cdot z^4$$
$$X_3(z) = x(3,0) + x(3,1)\cdot z + x(3,2)\cdot z^2 + x(3,3)\cdot z^3 + x(3,4)\cdot z^4$$
$$X_4(z) = x(4,0) + x(4,1)\cdot z + x(4,2)\cdot z^2 + x(4,3)\cdot z^3 + x(4,4)\cdot z^4$$

To begin the algorithm, compute the scalar convolution — the term resulting from the reduction modulo ($z - 1$). A factor $\frac{1}{5}$ included here comes from $\frac{1}{5}\cdot Q(z)$, where this is the final Chinese remainder combination polynomial for the terms $H_i(1)$. It is convenient to incorporate this scale factor at this point as these terms are pre-calculated. Then form the scalars:

$$H_0(1) = \frac{1}{5}\cdot(h(0,0) + h(0,1) + h(0,2) + h(0,3) + h(0,4))$$

$$H_1(1) = \frac{1}{5}\cdot(h(1,0) + h(1,1) + h(1,2) + h(1,3) + h(1,4))$$

$$H_2(1) = \frac{1}{5}\cdot(h(2,0) + h(2,1) + h(2,2) + h(2,3) + h(2,4))$$

$$H_3(1) = \frac{1}{5}\cdot(h(3,0) + h(3,1) + h(3,2) + h(3,3) + h(3,4))$$

$$H_4(1) = \frac{1}{5}\cdot(h(4,0) + h(4,1) + h(4,2) + h(4,3) + h(4,4))$$

$$X_0(1) = x(0,0) + x(0,1) + x(0,2) + x(0,3) + x(0,4)$$
$$X_1(1) = x(1,0) + x(1,1) + x(1,2) + x(1,3) + x(1,4)$$
$$X_2(1) = x(2,0) + x(2,1) + x(2,2) + x(2,3) + x(2,4)$$
$$X_3(1) = x(3,0) + x(3,1) + x(3,2) + x(3,3) + x(3,4)$$
$$X_4(1) = x(4,0) + x(4,1) + x(4,2) + x(4,3) + x(4,4)$$

Convolving these scalars gives terms:

$$Y_{1,0} = H_0(1)\cdot X_0(1) + H_1(1)\cdot X_4(1) + H_2(1)\cdot X_3(1) + H_3(1)\cdot X_2(1) + H_4(1)\cdot X_1(1)$$
$$Y_{1,1} = H_0(1)\cdot X_1(1) + H_1(1)\cdot X_0(1) + H_2(1)\cdot X_4(1) + H_3(1)\cdot X_3(1) + H_4(1)\cdot X_2(1)$$
$$Y_{1,2} = H_0(1)\cdot X_2(1) + H_1(1)\cdot X_1(1) + H_2(1)\cdot X_0(1) + H_3(1)\cdot X_4(1) + H_4(1)\cdot X_3(1)$$
$$Y_{1,3} = H_0(1)\cdot X_3(1) + H_1(1)\cdot X_2(1) + H_2(1)\cdot X_1(1) + H_3(1)\cdot X_0(1) + H_4(1)\cdot X_4(1)$$
$$Y_{1,4} = H_0(1)\cdot X_4(1) + H_1(1)\cdot X_3(1) + H_2(1)\cdot X_2(1) + H_3(1)\cdot X_1(1) + H_4(1)\cdot X_0(1)$$

In this first stage, the reduction modulo ($z - 1$) requires $p\cdot(p - 1)$ additions. The convolution requires at least $2\cdot(p - 1)$ multiplications but the actual number depends on the algorithm chosen. This is also true for the number of additions.

The second term involves the set of polynomials $Y_{Q,j}(z)$. To compute these, first reduce the initial polynomials modulo $Q(z)$. That gives:

$$U_0(z) = (h(0,0) - h(0,4)) + (h(0,1) - h(0,4))\cdot z + (h(0,2) - h(0,4))\cdot z^2 + (h(0,3) - h(0,4))\cdot z^3$$
$$= u_{00} + u_{01}\cdot z + u_{02}\cdot z^2 + u_{03}\cdot z^3$$
$$U_1(z) = (h(1,0) - h(1,4)) + (h(1,1) - h(1,4))\cdot z + (h(1,2) - h(1,4))\cdot z^2 + (h(1,3) - h(1,4))\cdot z^3$$
$$= u_{10} + u_{11}\cdot z + u_{12}\cdot z^2 + u_{13}\cdot z^3$$
$$U_2(z) = (h(2,0) - h(2,4)) + (h(2,1) - h(2,4))\cdot z + (h(2,2) - h(2,4))\cdot z^2 + (h(2,3) - h(2,4))\cdot z^3$$
$$= u_{20} + u_{21}\cdot z + u_{22}\cdot z^2 + u_{23}\cdot z^3$$

$U_3(z) = (h(3,0) - h(3,4)) + (h(3,1) - h(3,4)) \cdot z + (h(3,2) - h(3,4)) z^2 + (h(3,3) - h(3,4)) \cdot z^3$
$\quad = u_{30} + u_{31} \cdot z + u_{32} \cdot z^2 + u_{33} \cdot z^3$
$U_4(z) = (h(4,0) - h(4,4)) + (h(4,1) - h(4,4)) \cdot z + (h(4,2) - h(4,4)) \cdot z^2 + (h(4,3) - h(4,4)) \cdot z^3$
$\quad = u_{40} + u_{41} \cdot z + u_{42} \cdot z^2 + u_{43} \cdot z^3$

$V_0(z) = (x(0,0) - x(0,4)) + (x(0,1) - x(0,4)) \cdot z + (x(0,2) - x(0,4)) \cdot z^2 + (x(0,3) - x(0,4)) \cdot z^3$
$\quad = v_{00} + v_{01} \cdot z + v_{02} \cdot z^2 + v_{03} \cdot z^3$
$V_1(z) = (x(1,0) - x(1,4)) + (x(1,1) - x(1,4)) \cdot z + (x(1,2) - x(1,4)) \cdot z^2 + (x(1,3) - x(1,4)) \cdot z^3$
$\quad = v_{10} + v_{11} \cdot z + v_{12} \cdot z^2 + v_{13} \cdot z^3$
$V_2(z) = (x(2,0) - x(2,4)) + (x(2,1) - x(2,4)) \cdot z + (x(2,2) - x(2,4)) \cdot z^2 + (x(2,3) - x(2,4)) \cdot z^3$
$\quad = v_{20} + v_{21} \cdot z + v_{22} \cdot z^2 + v_{23} \cdot z^3$
$V_3(z) = (x(3,0) - x(3,4)) + (x(3,1) - x(3,4)) \cdot z + (x(3,2) - x(3,4)) \cdot z^2 + (x(3,3) - x(3,4)) \cdot z^3$
$\quad = v_{30} + v_{31} \cdot z + v_{32} \cdot z^2 + v_{33} \cdot z^3$
$V_4(z) = (x(4,0) - x(4,4)) + (x(4,1) - x(4,4)) \cdot z + (x(4,2) - x(4,4)) \cdot z^2 + (x(4,3) - x(4,4)) \cdot z^3$
$\quad = v_{40} + v_{41} \cdot z + v_{42} \cdot z^2 + v_{43} \cdot z^3$

As the polynomials $U_i(z)$ are pre-calculated, this step requires just $p \cdot (p - 1)$ additions.
Polynomial transforms modulo $Q(z)$ with a root z are formed according to:

$$g_k(z) = \left\langle \sum_{r=0}^{p-1} G(z) \cdot z^{rk} \right\rangle_{Q(z)} \qquad\qquad k = 0, 1, \cdots , p-1$$

The polynomial transforms for the set of polynomials $\{ V_i(z) \}$ are therefore:

$$
\begin{aligned}
X_0(z) &= <\; V_0(z) &+\; V_1(z) &+\; V_2(z) &+\; V_3(z) &+\; V_4(z) &>_{Q(z)} \\
X_1(z) &= <\; V_0(z) &+\; V_1(z) \cdot z &+\; V_2(z) \cdot z^2 &+\; V_3(z) \cdot z^3 &+\; V_4(z) \cdot z^4 &>_{Q(z)} \\
X_2(z) &= <\; V_0(z) &+\; V_1(z) \cdot z^2 &+\; V_2(z) \cdot z^4 &+\; V_3(z) \cdot z^6 &+\; V_4(z) \cdot z^8 &>_{Q(z)} \\
X_3(z) &= <\; V_0(z) &+\; V_1(z) \cdot z^3 &+\; V_2(z) \cdot z^6 &+\; V_3(z) \cdot z^9 &+\; V_4(z) \cdot z^{12} &>_{Q(z)} \\
X_4(z) &= <\; V_0(z) &+\; V_1(z) \cdot z^4 &+\; V_2(z) \cdot z^8 &+\; V_3(z) \cdot z^{12} &+\; V_4(z) \cdot z^{16} &>_{Q(z)}
\end{aligned}
$$

The first of these involves the sum of p polynomials, each of order $(p - 1)$. With no reduction needed, the set just requires $(p - 1) \cdot (p - 1)$ additions.

The other terms involve multiplications by some power of z and that will require reduction. However, this needn't involve the first term, so it may be expressed as:

$$X_i(z) = V_0(z) + <\; V_1(z) \cdot z^i + V_2(z) \cdot z^{2i} + V_3(z) \cdot z^{3i} + V_4(z) \cdot z^{4i} \;>_{Q(z)}$$

for $0 < i < p$. Consider the specific case of i equal to 2. We express the equation as:

$$X_2(z) = V_0(z) + T_2(z)$$

where $T_2(z) = <\; V_1(z) \cdot z^2 + V_2(z) \cdot z^4 + V_3(z) \cdot z^6 + V_4(z) \cdot z^8 \;>_{Q(z)}$

In order to compute $T_2(z)$, it is computationally simpler to reduce modulo $(z^5 - 1)$ and then modulo $Q(z)$. Reducing modulo $(z^5 - 1)$:

$$v_{13} + 0 \cdot z + v_{10} \cdot z^2 + v_{11} \cdot z^3 + v_{12} \cdot z^4 +$$
$$v_{21} + v_{22} \cdot z + v_{23} \cdot z^2 + 0 \cdot z^3 + v_{20} \cdot z^4 +$$
$$0 + v_{30} \cdot z + v_{31} \cdot z^2 + v_{32} \cdot z^3 + v_{33} \cdot z^4 +$$
$$v_{42} + v_{43} \cdot z + 0 \cdot z^2 + v_{40} \cdot z^3 + v_{41} \cdot z^4$$

Collecting the terms:

$$T_2(z) = (v_{13} + v_{21} + v_{42}) + (v_{22} + v_{30} + v_{43}) \cdot z + (v_{10} + v_{23} + v_{31}) \cdot z^2$$
$$+ (v_{11} + v_{32} + v_{40}) \cdot z^3 + (v_{12} + v_{20} + v_{33} + v_{41}) \cdot z^4$$

$$= t_{20} + t_{21} \cdot z + t_{22} \cdot z^2 + t_{23} \cdot z^3 + t_{24} \cdot z^4$$

Reducing modulo $Q(z)$ and adding the terms of $V_0(z)$, we get:

$$X_2(z) = (v_{00} + t_{20} - t_{24}) + (v_{01} + t_{21} - t_{24}) \cdot z + (v_{02} + t_{22} - t_{24}) \cdot z^2$$
$$+ (v_{03} + t_{23} - t_{24}) \cdot z^3$$

$$= x_{20} + x_{21} \cdot z + x_{22} \cdot z^2 + x_{22} \cdot z^3$$

To generalize, forming each of the polynomial terms $T_i(z)$ requires $(p \cdot (p - 3) + 1)$ additions. Reducing these polynomials modulo $Q(z)$ requires $(p - 1)$ additions and finally, adding to the terms of $V_0(z)$ requires another $(p - 1)$ additions. In total, each of these transforms requires $(p^2 - p - 1)$ additions. There are $(p - 1)$ of these terms, so together with the operations for the first term, the transform requires $(p^3 - 3p^2 + 2)$ additions. Denote the overall result as:

$$X_0(z) = x_{00} + x_{01} \cdot z + x_{02} \cdot z^2 + x_{03} \cdot z^3$$
$$X_1(z) = x_{10} + x_{11} \cdot z + x_{12} \cdot z^2 + x_{13} \cdot z^3$$
$$X_2(z) = x_{20} + x_{21} \cdot z + x_{22} \cdot z^2 + x_{23} \cdot z^3$$
$$X_3(z) = x_{30} + x_{31} \cdot z + x_{32} \cdot z^2 + x_{33} \cdot z^3$$
$$X_4(z) = x_{40} + x_{41} \cdot z + x_{42} \cdot z^2 + x_{43} \cdot z^3$$

A similar procedure applies to derive the polynomial transform of the set of polynomials $U_i(z)$. However, as this is normally pre-computed there is an opportunity to reduce the number of scaling operations required later in the computation by modifying this slightly.

The Chinese remainder combination polynomial for the set of polynomials $Y_{Q,j}(z)$ is:

$$S(z) = 1 - \frac{1}{5} \cdot Q(z) = \frac{4}{5} - \frac{1}{5} \cdot z - \frac{1}{5} \cdot z^2 - \frac{1}{5} \cdot z^3 - \frac{1}{5} \cdot z^4$$

This equals $(z - 1) \cdot T(z)$ and hence:

$$T(z) = 1 - \frac{1}{5} \cdot (4 + 3 \cdot z + 2 \cdot z^2 + z^3)$$

We may therefore form polynomials:

$$K_i(z) = \frac{1}{5} < T(z) \cdot \mathbb{H}_i(z) >_{Q(z)} \qquad\qquad i = 0, 1, 2, 3, 4$$

where the polynomials $\mathbb{H}_i(z)$ are the polynomial transforms of the polynomials $U_i(z)$. The constant $\frac{1}{5}$ is included as it is needed in the later inverse polynomial transform but is more conveniently placed here. Then these pre-computed terms may be represented by:

$$\begin{aligned}
K_0(z) &= k_{00} + k_{01} \cdot z + k_{02} \cdot z^2 + k_{03} \cdot z^3 \\
K_1(z) &= k_{10} + k_{11} \cdot z + k_{12} \cdot z^2 + k_{13} \cdot z^3 \\
K_2(z) &= k_{20} + k_{21} \cdot z + k_{22} \cdot z^2 + k_{23} \cdot z^3 \\
K_3(z) &= k_{30} + k_{31} \cdot z + k_{32} \cdot z^2 + k_{33} \cdot z^3 \\
K_4(z) &= k_{40} + k_{41} \cdot z + k_{42} \cdot z^2 + k_{43} \cdot z^3
\end{aligned}$$

Next, the transforms are multiplied modulo $Q(z)$. This gives polynomials:

$$M_i(z) = < K_i(z) \cdot \mathbb{X}_i(z) >_{Q(z)} \qquad\qquad i = 0, 1, 2, 3, 4$$

These polynomials are derived via one of the previously discussed methods. The result is:

$$\begin{aligned}
M_0(z) &= m_{00} + m_{01} \cdot z + m_{02} \cdot z^2 + m_{03} \cdot z^3 \\
M_1(z) &= m_{10} + m_{11} \cdot z + m_{12} \cdot z^2 + m_{13} \cdot z^3 \\
M_2(z) &= m_{20} + m_{21} \cdot z + m_{22} \cdot z^2 + m_{23} \cdot z^3 \\
M_3(z) &= m_{30} + m_{31} \cdot z + m_{32} \cdot z^2 + m_{33} \cdot z^3 \\
M_4(z) &= m_{40} + m_{41} \cdot z + m_{42} \cdot z^2 + m_{43} \cdot z^3
\end{aligned}$$

The inverse polynomial transform is now applied. Since the scale factor $\frac{1}{5}$ has been included in the earlier calculations, this just involves computing:

$$Y_{Q,i}(z) = < M_0(z) + M_1(z) \cdot z^{-i} + M_2(z) \cdot z^{-2i} + M_2(z) \cdot z^{-3i} + M_2(z) \cdot z^{-4i} >_{Q(z)}$$

for $0 \le i \le 4$. The result is $Y_{Q,i}(z)$ as this completes the operations for this stage. Here, z^{-j} is interpreted as z^{4j} or, in general, $z^{(p-1)j}$. The number of operations is exactly the same as before. Then:

$$\begin{aligned}
Y_{Q,0}(z) &= q_{00} + q_{01} \cdot z + q_{02} \cdot z^2 + q_{03} \cdot z^3 \\
Y_{Q,1}(z) &= q_{10} + q_{11} \cdot z + q_{12} \cdot z^2 + q_{13} \cdot z^3 \\
Y_{Q,2}(z) &= q_{20} + q_{21} \cdot z + q_{22} \cdot z^2 + q_{23} \cdot z^3 \\
Y_{Q,3}(z) &= q_{30} + q_{31} \cdot z + q_{32} \cdot z^2 + q_{33} \cdot z^3 \\
Y_{Q,4}(z) &= q_{40} + q_{41} \cdot z + q_{42} \cdot z^2 + q_{43} \cdot z^3
\end{aligned}$$

Finally, Chinese remainder theorem reconstruction follows to combine the outputs of these two stages. This involves forming polynomials:

$$Y_i(z) = Y_{1,i}(z) \cdot Q(z) + Y_{Q,i}(z) \cdot (z - 1) \qquad\qquad i = 0, 1, 2, 3, 4$$

That is:

$$Y_i(z) = Y_{1,i}(z) \cdot (z^4 + z^3 + z^2 + z + 1)$$
$$+ (q_{i3} \cdot z^4 + (q_{i2} - q_{i3}) \cdot z^3 + (q_{i1} - q_{i2}) \cdot z^2 + (q_{i0} - q_{i1}) \cdot z - q_{i0})$$

$$= (Y_{1,i}(z) + q_{i3}) \cdot z^4 + (Y_{1,i}(z) + q_{i2} - q_{i3}) \cdot z^3 + (Y_{1,i}(z) + q_{i1} - q_{i2}) \cdot z^2$$
$$+ (Y_{1,i}(z) \, q_{i0} - q_{i1}) \cdot z + (Y_{1,i}(z) - q_{i0})$$

$$= y_{i0} + y_{i1} \cdot z + y_{i2} \cdot z^2 + y_{i2} \cdot z^3 + y_{i3} \cdot z^4 \qquad\qquad i = 0, 1, 2, 3, 4$$

The required convolution is therefore the array of elements:

$$
\begin{array}{ccccc}
y(0,0) & y(0,1) & y(0,2) & y(0,3) & y(0,4) \\
y(1,0) & y(1,1) & y(1,2) & y(1,3) & y(1,4) \\
y(2,0) & y(2,1) & y(2,2) & y(2,3) & y(2,4) \\
y(3,0) & y(3,1) & y(3,2) & y(3,3) & y(3,4) \\
y(4,0) & y(4,1) & y(4,2) & y(4,3) & y(4,4)
\end{array}
$$

and that completes the problem. This final step requires $2p \cdot (p - 1)$ additions.

Summing all operations, this approach requires $2(p^3 - p^2 - 2p + 2)$ additions, plus the operations required for a p-point cyclic convolution and p p-point polynomial products modulo $Q(z)$. If the latter are executed by first reducing modulo $(z^p - 1)$ then $Q(z)$, the number of operations becomes $(2p^3 - p^2 - 5p + 4)$ additions, plus the number of operations required for $(p + 1)$ p-point cyclic convolutions. The row and column approach requires $2p$ p-point cyclic convolutions. Thus as p grows large the polynomial transform technique requires half as many multiplications per output point but approximately $2p$ more additions per output point.

The general case is the cyclic convolution of two arrays $x(i, j)$ and $h(i, j)$ where each is N×N and N is composite. The algorithm developed will be similar to that derived for N prime and, for that reason, it is useful to review in general terms the approach which was used. A two-dimensional cyclic convolution can be expressed as a one-dimensional cyclic convolution of p-point polynomial elements. The Chinese remainder theorem can be applied to this convolution. As $(z^p - 1)$ factors into two cyclotomic polynomials $(z - 1)$ and $Q(z)$, the effect of that step is to split the problem into two simpler problems. One involves a p-point scalar cyclic convolution problem which can be solved by methods discussed earlier in the Part. The second can be solved by polynomial transforms using a root of z. This results in a set of $(p - 1)$ polynomial products modulo $Q(z)$ which may alternatively be expressed as a set of p-point scalar cyclic convolutions.

For a general N, $(z^N - 1)$ factors into a set of cyclotomic polynomials. While as mentioned earlier the Chinese remainder theorem can be used to split the problem into a set of convolution problems, each of which would involve a one-dimensional convolution problem with elements that are polynomials of the same order as the cyclotomic polynomial, we adopt a different procedure. Form:

$$z^N - 1 = C(z) \cdot R(z)$$

where $C(z)$ is the largest cyclotomic factor of ($z^N - 1$) and $R(z)$ is the product of all other terms. Next, follow a Chinese remainder theorem approach as before with these two terms and determine:

$$< Y_j(z) >_{C(z)} = Y_{C,j}(z) \qquad \text{and} \qquad < Y_j(z) >_{R(z)} = Y_{R,j}(z)$$

with these forming the final convolution via Chinese remainder theorem reconstruction.

The first of these two terms is an N-point, one-dimensional convolution problem modulo the irreducible polynomial $C(z)$ and involves terms which are polynomials of the same order as $C(z)$. As z is a root of order N for a polynomial transform defined modulo $C(z)$, these can be used to compute the result. The second is also an N-point convolution problem but it is modulo the composite $R(z)$ and involves elements that are polynomials of the order of $R(z)$, namely ($N - \Phi(N) - 1$). As $R(z)$ is composite, polynomial transforms cannot be used here.

To overcome this difficulty, return to the original equations developed at the beginning of the section and observe this second convolution is in fact:

$$<< H(z_1, z_2) \cdot X(z_1, z_2) >_{z_2^N - 1} >_{R(z)}$$

This problem can be rephrased to:

$$<< H(z_1, z_2) \cdot X(z_1, z_2) >_{R(z)} >_{z_2^N - 1}$$

without any loss of generality. That is, it can be changed into a ($N - \Phi(N)$)-point convolution problem modulo $R(z)$ where the elements are polynomials of order ($N - 1$). In turn these polynomials may be expressed as the sum of two polynomials via the Chinese remainder theorem and again, one of those will be determined by reducing the polynomial $C(z)$ and the other $R(z)$. The result is an ($N - \Phi(N)$)-point convolution modulo $C(z)$ of ($N - \Phi(N)$)-point polynomial elements and that can be computed via polynomial transforms. There is also a convolution modulo $R(z)$ similar to the original problem which is in effect ($N - \Phi(N)$)$\times$ ($N - \Phi(N)$). Therefore, this algorithm may be recursively applied until we reach a stage similar to the prime algorithm. Thus we have an algorithm for N which has $\Phi(N)$ stages.

Although the algorithm outlined applies to any N, its complexity tends to restrict interest to values where N is p^k. Of these, one very important group is where N is 2^k. Apart from 2^k being a convenient value, these algorithms also have a very regular structure. This, as we showed earlier, is because ($z^N - 1$) factorizes to:

$$z^N - 1 = (z^{\frac{N}{2}} + 1) \cdot (z^{\frac{N}{4}} + 1) \cdot (z^{\frac{N}{8}} + 1) \cdots (z + 1) \cdot (z - 1)$$

Consequently, the algorithm proceeds as follows. First, given a set of polynomials:

$$<< Y_i(z_1, z_2) >_{z_1^N - 1} >_{z_2^N - 1} \qquad\qquad i = 0, 1, \cdots, N–1$$

then via Chinese remainder theorem reconstruction, we compute these as:

$$\ll Y_i(z_1,z_2) \gg_{z_1^{N}-1} \gg_{z_2^{N}-1} = S_{10}(z) \cdot \ll Y_i(z_1,z_2) \gg_{z_1^{\frac{N}{2}}+1} \gg_{z_2^{N}-1}$$

$$+ S_{20}(z) \cdot \ll Y_i(z_1,z_2) \gg_{z_1^{\frac{N}{2}}-1} \gg_{z_2^{N}-1}$$

It is easily shown:

$$S_{10}(z_1) = -\frac{1}{2} \cdot (z_1^{\frac{N}{2}} - 1) \qquad\qquad S_{20}(z_1) = -\frac{1}{2} \cdot (z_1^{\frac{N}{2}} + 1)$$

The first term is reduction modulo an irreducible polynomial in z_1 and can be performed by polynomial transforms of order $\frac{N}{2}$ with a root of z_1. The second term is a two-dimensional convolution and can be expressed as $\ll Y_i(z_1,z_2) \gg_{z_2^{N}-1} \gg_{z_1^{\frac{N}{2}}-1}$. The Chinese remainder theorem can again be applied to give:

$$\ll Y_i(z_1,z_2) \gg_{z_2^{N}-1} \gg_{z_1^{\frac{N}{2}}-1} = S_{11}(z) \cdot \ll Y_i(z_1,z_2) \gg_{z_2^{\frac{N}{2}}+1} \gg_{z_1^{\frac{N}{2}}-1}$$

$$+ S_{21}(z) \cdot \ll Y_i(z_1,z_2) \gg_{z_2^{\frac{N}{2}}-1} \gg_{z_1^{\frac{N}{2}}-1}$$

where again it can be shown:

$$S_{11}(z_2) = -\frac{1}{2} \cdot (z_1^{\frac{N}{2}} - 1) \qquad\qquad S_{21}(z_2) = -\frac{1}{2} \cdot (z_1^{\frac{N}{2}} + 1)$$

The first term is also a reduction modulo an irreducible polynomial and can be executed by a polynomial transform of order $\frac{N}{2}$ but a root of z_2^2. The second term, however, is just a low-order replica of the original problem. Hence it is computed by repeating these steps until a 2×2 term results. As can be seen, each of the k stages of the algorithm incorporate the same form of processing as every other.

Example

We choose the simplest value of N to investigate, namely 4. One array will be assumed to be fixed. In this example, the steps which in practice would be taken to reduce scaling operations will not be described. That can be done by following the same procedures as in the last example. Then the two arrays which are to be convolved will be taken to be:

h(0,0)	h(0,1)	h(0,2)	h(0,3)	x(0,0)	x(0,1)	x(0,2)	x(0,3)
h(1,0)	h(1,1)	h(1,2)	h(1,3)	x(1,0)	x(1,1)	x(1,2)	x(1,3)
h(2,0)	h(2,1)	h(2,2)	h(2,3)	x(2,0)	x(2,1)	x(2,2)	x(2,3)
h(3,0)	h(3,1)	h(3,2)	h(3,3)	x(3,0)	x(3,1)	x(3,2)	x(3,3)

The first step is to form these into polynomials. For reasons of clarity, the subscripts will be dropped as it will always be clear whether reductions are modulo a polynomial in z_1 or one in z_2. Then:

$$H_0(z) = h(0,0) + h(0,1)\cdot z + h(0,2)\cdot z^2 + h(0,3)\cdot z^3$$
$$H_1(z) = h(1,0) + h(1,1)\cdot z + h(1,2)\cdot z^2 + h(1,3)\cdot z^3$$
$$H_2(z) = h(2,0) + h(2,1)\cdot z + h(2,2)\cdot z^2 + h(2,3)\cdot z^3$$
$$H_3(z) = h(3,0) + h(3,1)\cdot z + h(3,2)\cdot z^2 + h(3,3)\cdot z^3$$

and:

$$X_0(z) = x(0,0) + x(0,1)\cdot z + x(0,2)\cdot z^2 + x(0,3)\cdot z^3$$
$$X_1(z) = x(1,0) + x(1,1)\cdot z + x(1,2)\cdot z^2 + x(1,3)\cdot z^3$$
$$X_2(z) = x(2,0) + x(2,1)\cdot z + x(2,2)\cdot z^2 + x(2,3)\cdot z^3$$
$$X_3(z) = x(3,0) + x(3,1)\cdot z + x(3,2)\cdot z^2 + x(3,3)\cdot z^3$$

The algorithm ends with Chinese remainder reconstruction of two sets of polynomials, one derived modulo ($z^{\frac{N}{2}} + 1$) and the other ($z^{\frac{N}{2}} - 1$). The first set are determined through polynomial transforms and are derived from reductions of the above polynomials modulo ($z^{\frac{N}{2}} + 1$). That is, by ($z^2 + 1$). Thus:

$$HO_{0,0}(z) = (h(0,0) - h(0,2)) + (h(0,1) - h(0,3))\cdot z = h0_{00} + h0_{01}\cdot z$$
$$HO_{0,1}(z) = (h(1,0) - h(1,2)) + (h(1,1) - h(1,3))\cdot z = h0_{10} + h0_{11}\cdot z$$
$$HO_{0,2}(z) = (h(2,0) - h(2,2)) + (h(2,1) - h(2,3))\cdot z = h0_{20} + h0_{21}\cdot z$$
$$HO_{0,3}(z) = (h(3,0) - h(3,2)) + (h(3,1) - h(3,3))\cdot z = h0_{30} + h0_{31}\cdot z$$

$$XO_{0,0}(z) = (x(0,0) - x(0,2)) + (x(0,1) - x(0,3))\cdot z = x0_{00} + x0_{01}\cdot z$$
$$XO_{0,1}(z) = (x(1,0) - x(1,2)) + (x(1,1) - x(1,3))\cdot z = x0_{10} + x0_{11}\cdot z$$
$$XO_{0,2}(z) = (x(2,0) - x(2,2)) + (x(2,1) - x(2,3))\cdot z = x0_{20} + x0_{21}\cdot z$$
$$XO_{0,3}(z) = (x(3,0) - x(3,2)) + (x(3,1) - x(3,3))\cdot z = x0_{30} + x0_{31}\cdot z$$

Use a root of z to polynomial transform these to:

$$h0_0(z) = ((h0_{00} + h0_{20}) + (h0_{10} + h0_{30})) + ((h0_{01} + h0_{21}) + (h0_{11} + h0_{31}))\cdot z$$
$$= a0_{00} + a0_{01}\cdot z$$
$$h0_1(z) = ((h0_{00} - h0_{20}) - (h0_{11} - h0_{31})) + ((h0_{01} - h0_{21}) + (h0_{10} - h0_{30}))\cdot z$$
$$= a0_{10} + a0_{11}\cdot z$$
$$h0_2(z) = ((h0_{00} + h0_{20}) - (h0_{10} + h0_{30})) + ((h0_{01} + h0_{21}) - (h0_{11} + h0_{31}))\cdot z$$
$$= a0_{20} + a0_{21}\cdot z$$
$$h0_3(z) = ((h0_{00} - h0_{20}) + (h0_{11} - h0_{31})) + ((h0_{01} - h0_{21}) - (h0_{10} - h0_{30}))\cdot z$$
$$= a0_{30} + a0_{31}\cdot z$$

$$x0_0(z) = ((x0_{00} + x0_{20}) + (x0_{10} + x0_{30})) + ((x0_{01} + x0_{21}) + (x0_{11} + x0_{31}))\cdot z$$
$$= b0_{00} + b0_{01}\cdot z$$
$$x0_1(z) = ((x0_{00} - x0_{20}) - (x0_{11} - x0_{31})) + ((x0_{01} - x0_{21}) + (x0_{10} - x0_{30}))\cdot z$$
$$= b0_{10} + b0_{11}\cdot z$$
$$x0_2(z) = ((x0_{00} + x0_{20}) - (x0_{10} + x0_{30})) + ((x0_{01} + x0_{21}) - (x0_{11} + x0_{31}))\cdot z$$
$$= b0_{20} + b0_{21}\cdot z$$
$$x0_3(z) = ((x0_{00} - x0_{20}) + (x0_{11} - x0_{31})) + ((x0_{01} - x0_{21}) - (x0_{10} - x0_{30}))\cdot z$$
$$= b0_{30} + b0_{31}\cdot z$$

For problems such as these, a very efficient procedure can be employed to derive the polynomial transforms. Generally, the polynomial transform is given by:

$$g_j(z) = \left\langle \sum_{i=0}^{N-1} g_i(z) \cdot z^{ij} \right\rangle_{z^{\frac{N}{2}}+1} \qquad\qquad j = 0, 1, \cdots, N-1$$

We may express this as:

$$g_j(z) = \left\langle \sum_{i=0}^{N-1} g_{2i}(z) \cdot z^{2ij} \right\rangle_{z^{\frac{N}{2}}+1} + \left\langle \sum_{i=0}^{N-1} g_{2i+1}(z) \cdot z^{(2i+1)j} \right\rangle_{z^{\frac{N}{2}}+1}$$

$$= \left\langle \sum_{i=0}^{N-1} g_{2i}(z) \cdot z^{2ij} \right\rangle_{z^{\frac{N}{2}}+1} + \left\langle z^j \cdot \sum_{i=0}^{N-1} g_{2i+1}(z) \cdot z^{2ij} \right\rangle_{z^{\frac{N}{2}}+1}$$

Given the modulus, this in turn can be expressed as the two equations:

$$g_j(z) = \left\langle \sum_{i=0}^{N-1} g_{2i}(z) \cdot z^{2ij} \right\rangle_{z^{\frac{N}{2}}+1} + \left\langle z^j \cdot \sum_{i=0}^{N-1} g_{2i+1}(z) \cdot z^{2ij} \right\rangle_{z^{\frac{N}{2}}+1}$$

$$g_{j+\frac{N}{2}}(z) = \left\langle \sum_{i=0}^{N-1} g_{2i}(z) \cdot z^{2ij} \right\rangle_{z^{\frac{N}{2}}+1} - \left\langle z^j \cdot \sum_{i=0}^{N-1} g_{2i+1}(z) \cdot z^{2ij} \right\rangle_{z^{\frac{N}{2}}+1}$$

$$\text{for } j = 0, 1, \cdots, \frac{N}{2} - 1$$

This describes an algorithm similar to a base-2 FFT. Hence the terms of the polynomial transform require $\frac{N}{2} \cdot \log_2 N$ additions per element or $\frac{1}{2} \cdot N^2 \cdot \log_2 N$ additions for the complete transform.

The two sets of polynomials are multiplied modulo ($z^2 + 1$) to give:

$$\begin{aligned}
MO_0(z) &= (a0_{00} \cdot b0_{00} - a0_{01} \cdot b0_{01}) + (a0_{00} \cdot b0_{01} + a0_{01} \cdot b0_{00}) \cdot z \\
&= m0_{00} + m0_{01} \cdot z \\
MO_1(z) &= (a0_{10} \cdot b0_{10} - a0_{11} \cdot b0_{11}) + (a0_{10} \cdot b0_{11} + a0_{11} \cdot b0_{10}) \cdot z \\
&= m0_{10} + m0_{11} \cdot z \\
MO_2(z) &= (a0_{20} \cdot b0_{20} - a0_{21} \cdot b0_{21}) + (a0_{20} \cdot b0_{21} + a0_{21} \cdot b0_{20}) \cdot z \\
&= m0_{20} + m0_{21} \cdot z \\
MO_3(z) &= (a0_{20} \cdot b0_{20} - a0_{21} \cdot b0_{21}) + (a0_{20} \cdot b0_{21} + a0_{21} \cdot b0_{20}) \cdot z \\
&= m0_{30} + m0_{31} \cdot z
\end{aligned}$$

The inverse polynomial transform is applied. Here, z^{-i} is z^{3i}. Thus:

$$\begin{aligned}
YO_{0,0}(z) &= \frac{1}{4} \cdot ((m0_{00} + m0_{20}) + (m0_{10} + m0_{30})) + \frac{1}{4} \cdot ((m0_{01} + m0_{21}) + (m0_{21} + m0_{31})) \cdot z \\
&= y0_{00} + y0_{01} \cdot z
\end{aligned}$$

$$YO_{0,1}(z) = \frac{1}{4} \cdot ((mO_{00} - mO_{20}) + (mO_{11} - mO_{31})) + \frac{1}{4} \cdot ((mO_{01} - mO_{21}) - (mO_{10} - mO_{30})) \cdot z$$

$$= yO_{10} + yO_{11} \cdot z$$

$$YO_{0,2}(z) = \frac{1}{4} \cdot ((mO_{00} + mO_{20}) - (mO_{10} + mO_{30})) + \frac{1}{4} \cdot ((mO_{01} + mO_{21}) - (mO_{11} + mO_{31})) \cdot z$$

$$= yO_{20} + yO_{21} \cdot z$$

$$YO_{0,3}(z) = \frac{1}{4} \cdot ((mO_{00} - mO_{20}) - (mO_{11} - mO_{31})) + \frac{1}{4} \cdot ((mO_{01} - mO_{21}) + (mO_{10} - mO_{30})) \cdot z$$

$$= yO_{30} + yO_{31} \cdot z$$

These are the required output terms and can now be combined with the reconstruction polynomial $S_{10}(z)$. However, that will be left until later.

To compute the second part of this first stage, reduce the polynomials modulo $(z^2 - 1)$:

$$HO_{1,0}(z) = (h(0,0) + h(0,2)) + (h(0,1) + h(0,3)) \cdot z = hO_{02} + hO_{03} \cdot z$$
$$HO_{1,1}(z) = (h(1,0) + h(1,2)) + (h(1,1) + h(1,3)) \cdot z = hO_{12} + hO_{13} \cdot z$$
$$HO_{1,2}(z) = (h(2,0) + h(2,2)) + (h(2,1) + h(2,3)) \cdot z = hO_{22} + hO_{23} \cdot z$$
$$HO_{1,3}(z) = (h(3,0) + h(3,2)) + (h(3,1) + h(3,3)) \cdot z = hO_{32} + hO_{33} \cdot z$$

$$XO_{1,0}(z) = (x(0,0) + x(0,2)) + (x(0,1) + x(0,3)) \cdot z = xO_{02} + xO_{03} \cdot z$$
$$XO_{1,1}(z) = (x(1,0) + x(1,2)) + (x(1,1) + x(1,3)) \cdot z = xO_{12} + xO_{13} \cdot z$$
$$XO_{1,2}(z) = (x(2,0) + x(2,2)) + (x(2,1) + x(2,3)) \cdot z = xO_{22} + xO_{23} \cdot z$$
$$XO_{1,3}(z) = (x(3,0) + x(3,2)) + (x(3,1) + x(3,3)) \cdot z = xO_{32} + xO_{33} \cdot z$$

Now map these into the z_2 domain, which means the polynomials are rearranged to:

$$H1_0(z) = hO_{02} + hO_{12} \cdot z + hO_{22} \cdot z^2 + hO_{32} \cdot z^3$$
$$H1_1(z) = hO_{03} + hO_{13} \cdot z + hO_{23} \cdot z^2 + hO_{33} \cdot z^3$$

$$X1_0(z) = xO_{02} + xO_{12} \cdot z + xO_{22} \cdot z^2 + xO_{32} \cdot z^3$$
$$X1_1(z) = xO_{03} + xO_{13} \cdot z + xO_{23} \cdot z^2 + xO_{33} \cdot z^3$$

The initial steps are largely repeated. Reduce these polynomials modulo $(z^2 + 1)$ and polynomial transform them and also reduce them modulo $(z^2 - 1)$. Then for the first step:

$$H1_{0,0}(z) = (hO_{02} - hO_{22}) + (hO_{12} - hO_{32}) \cdot z = h1_{00} + h1_{01} \cdot z$$
$$H1_{0,1}(z) = (hO_{03} - hO_{23}) + (hO_{13} - hO_{33}) \cdot z = h1_{10} + h1_{11} \cdot z$$

$$X1_{0,0}(z) = (xO_{02} - xO_{22}) + (xO_{12} - xO_{32}) \cdot z = x1_{00} + x1_{01} \cdot z$$
$$X1_{0,1}(z) = (xO_{03} - xO_{23}) + (xO_{13} - xO_{33}) \cdot z = x1_{10} + x1_{11} \cdot z$$

It is necessary to polynomial transform these with a root of z^2:

$$\ln1_0(z) = (h1_{00} + h1_{10}) + (h1_{01} + h1_{11}) \cdot z = a1_{00} + a1_{01} \cdot z$$
$$\ln1_1(z) = (h1_{00} - h1_{10}) + (h1_{01} - h1_{11}) \cdot z = a1_{10} + a1_{11} \cdot z$$

$$x1_0(z) = (x1_{00} + x1_{10}) + (x1_{01} + x1_{11})\cdot z = b1_{00} + b1_{01}\cdot z$$
$$x1_1(z) = (x1_{00} - x1_{10}) + (x1_{01} - x1_{11})\cdot z = b1_{10} + b1_{11}\cdot z$$

The products are naturally formed modulo $(z^2 + 1)$:

$$M1_0(z) = (a1_{00}\cdot b1_{00} - a1_{01}\cdot b1_{01}) + (a1_{00}\cdot b1_{01} + a1_{01}\cdot b1_{00})\cdot z$$
$$= m1_{00} + m1_{01}\cdot z$$
$$M1_1(z) = (a1_{10}\cdot b1_{10} - a1_{11}\cdot b1_{11}) + (a1_{10}\cdot b1_{11} + a1_{11}\cdot b1_{10})\cdot z$$
$$= m1_{10} + m1_{11}\cdot z$$

The inverse transform is taken. Here, z^{-i} is interpreted as z^i. Thus:

$$Y1_{0,0}(z) = \frac{1}{2}\cdot(m1_{00} + m1_{10}) + \frac{1}{2}\cdot(m1_{01} + m1_{11})\cdot z = y1_{00} + y1_{01}\cdot z$$

$$Y1_{0,1}(z) = \frac{1}{2}\cdot(m1_{00} - m1_{10}) + \frac{1}{2}\cdot(m1_{01} - m1_{11})\cdot z = y1_{10} + y1_{11}\cdot z$$

This term can be multiplied by the reconstruction polynomial $S_{11}(z_2)$ but that will be left until later.

The reduction modulo $(z^2 - 1)$ of the polynomials gives:

$$H1_{1,0}(z) = (h0_{02} + h0_{22}) + (h0_{12} + h0_{32})\cdot z = h1_{20} + h1_{21}\cdot z$$
$$H1_{1,1}(z) = (h0_{03} + h0_{23}) + (h0_{13} + h0_{33})\cdot z = h1_{30} + h1_{31}\cdot z$$

$$X1_{1,0}(z) = (x0_{02} + x0_{22}) + (x0_{12} + x0_{32})\cdot z = x1_{20} + x1_{21}\cdot z$$
$$X1_{1,1}(z) = (x0_{03} + x0_{23}) + (x0_{13} + x0_{33})\cdot z = x1_{30} + x1_{31}\cdot z$$

The problem now is a two-dimensional cyclic convolution between:

$$\begin{matrix} h1_{20} & h1_{21} \\ h1_{30} & h1_{31} \end{matrix} \qquad \text{and} \qquad \begin{matrix} x1_{20} & x1_{21} \\ x1_{30} & x1_{31} \end{matrix}$$

This could be done by a direct multiplication. That is:

$$(H1_{1,0}(z_1) + H1_{1,1}(z_1)\cdot z_2)\cdot(X1_{1,0}(z_1) + X1_{1,1}(z_1)\cdot z_2)$$

A number of methods can be employed to do this. An approach more in keeping with the remainder of the algorithm and equally efficient, is to use the methods outlined at the beginning of this section. We note that the convolution can be expressed as a convolution of polynomial elements and that gives:

$$Y2_{1,0}(z) = \;<\; H1_{1,0}(z)\cdot X1_{1,0}(z) + H1_{1,1}(z)\cdot X1_{1,1}(z) \;>_{z^2-1}$$
$$Y2_{1,1}(z) = \;<\; H1_{1,0}(z)\cdot X1_{1,1}(z) + H1_{1,1}(z)\cdot X1_{1,0}(z) \;>_{z^2-1}$$

Chinese remainder theorem reconstruction can be used to compute each of these. That is:

$$Y2_{1,0}(z) = S_{13}(z)\cdot<\; Y2_{0,0}(z) \;>_{z+1} + S_{23}(z)\cdot<\; Y2_{0,1}(z) \;>_{z-1}$$
$$Y2_{1,1}(z) = S_{13}(z)\cdot<\; Y2_{1,0}(z) \;>_{z+1} + S_{23}(z)\cdot<\; Y2_{1,1}(z) \;>_{z-1}$$

where $\qquad S_{13}(z) = -\dfrac{1}{2}\cdot(z_1-1)$ $\qquad\qquad\qquad\qquad\qquad S_{23}(z) = -\dfrac{1}{2}\cdot(z_1+1)$

The reductions lead to a set of scalars, namely:

$$
\begin{aligned}
y2_{00} &= H1_{1,0}(-1)\cdot X1_{1,0}(-1) + H1_{1,1}(-1)\cdot X1_{1,1}(-1) \\
&= (\,h1_{20}-h1_{21}\,)\cdot(\,x1_{20}-x1_{21}\,) + (\,h1_{30}-h1_{31}\,)\cdot(\,x1_{30}-x1_{31}\,) \\[4pt]
y2_{10} &= H1_{1,0}(-1)\cdot X1_{1,1}(-1) + H1_{1,1}(-1)\cdot X1_{1,0}(-1) \\
&= (\,h1_{20}-h1_{21}\,)\cdot(\,x1_{30}-x1_{31}\,) + (\,h1_{30}-h1_{31}\,)\cdot(\,x1_{20}-x1_{21}\,) \\[4pt]
y2_{01} &= H1_{1,0}(1)\cdot X1_{1,0}(1) + H1_{1,1}(1)\cdot X1_{1,1}(1) \\
&= (\,h1_{20}+h1_{21}\,)\cdot(\,x1_{20}+x1_{21}\,) + (\,h1_{30}+h1_{31}\,)\cdot(\,x1_{30}+x1_{31}\,) \\[4pt]
y2_{11} &= H1_{1,0}(1)\cdot X1_{1,1}(1) + H1_{1,1}(1)\cdot X1_{1,0}(1) \\
&= (\,h1_{20}+h1_{21}\,)\cdot(\,x1_{30}+x1_{31}\,) + (\,h1_{30}+h1_{31}\,)\cdot(\,x1_{20}+x1_{21}\,)
\end{aligned}
$$

Thus:

$$
Y1_{1,0}(z) = \frac{1}{2}\cdot(\,y2_{00}+y2_{01}\,) + \frac{1}{2}\cdot(\,y2_{00}-y2_{01}\,)\cdot z = y1_{20}+y1_{21}\cdot z
$$

$$
Y1_{1,1}(z) = \frac{1}{2}\cdot(\,y2_{10}+y2_{11}\,) + \frac{1}{2}\cdot(\,y2_{10}-y2_{11}\,)\cdot z = y1_{30}+y1_{31}\cdot z
$$

All terms needed to compute the final result are now available. All that remains is the different Chinese remainder theorem reconstructions which involves multiplication by the different reconstruction polynomials. First, the formation of the result for the second stage. The computation is:

$$
\begin{aligned}
Y1_0(z) &= -\frac{1}{2}\cdot(z^2-1)\cdot Y1_{0,0}(z) + \frac{1}{2}\cdot(z^2+1)\cdot Y1_{01}(z) \\
&= \frac{1}{2}\cdot(y1_{20}+y1_{00}) + \frac{1}{2}\cdot(y1_{21}+y1_{01})\cdot z + \frac{1}{2}\cdot(y1_{20}-y1_{00})\cdot z^2 + \frac{1}{2}\cdot(y1_{21}-y1_{01})\cdot z^3 \\
&= y1_{02}+y1_{12}\cdot z + y1_{22}\cdot z^2 + y1_{32}\cdot z^3
\end{aligned}
$$

$$
\begin{aligned}
Y1_1(z) &= -\frac{1}{2}\cdot(z^2-1)\cdot Y1_{0,1}(z) + \frac{1}{2}\cdot(z^2+1)\cdot Y1_{11}(z) \\
&= \frac{1}{2}\cdot(y1_{30}+y1_{10}) + \frac{1}{2}\cdot(y1_{31}+y1_{11})\cdot z + \frac{1}{2}\cdot(y1_{30}-y1_{10})\cdot z^2 + \frac{1}{2}\cdot(y1_{31}-y1_{11})\cdot z^3 \\
&= y1_{03}+y1_{13}\cdot z + y1_{23}\cdot z^2 + y1_{33}\cdot z^3
\end{aligned}
$$

These terms are transformed back to the z_1 domain, which creates:

$$
\begin{aligned}
Y0_{1,0}(z) &= y1_{02}+y1_{03}\cdot z \\
Y0_{1,1}(z) &= y1_{12}+y1_{13}\cdot z \\
Y0_{1,2}(z) &= y1_{22}+y1_{23}\cdot z \\
Y0_{1,3}(z) &= y1_{32}+y1_{33}\cdot z
\end{aligned}
$$

The final set of reconstructions is:

$$Y_i(z) = -\frac{1}{2} \cdot (z^2 - 1) \cdot Y0_{0,i}(z) + \frac{1}{2} \cdot (z^2 + 1) \cdot Y0_{1,i}(z) \qquad\qquad i = 0, 1, 2, 3$$

which is:

$$Y_0(z) = \frac{1}{2} \cdot (y0_{02} + y0_{00}) + \frac{1}{2} \cdot (y0_{03} + y0_{01}) \cdot z + \frac{1}{2} \cdot (y0_{02} - y0_{00}) \cdot z^2 + \frac{1}{2} \cdot (y0_{03} - y0_{01}) \cdot z^3$$

$$= y_{00} + y_{01} \cdot z + y_{02} \cdot z^2 + y_{03} \cdot z^3$$

$$Y_1(z) = \frac{1}{2} \cdot (y0_{12} + y0_{10}) + \frac{1}{2} \cdot (y0_{13} + y0_{11}) \cdot z + \frac{1}{2} \cdot (y0_{12} - y0_{10}) \cdot z^2 + \frac{1}{2} \cdot (y0_{13} - y0_{11}) \cdot z^3$$

$$= y_{10} + y_{11} \cdot z + y_{12} \cdot z^2 + y_{13} \cdot z^3$$

$$Y_2(z) = \frac{1}{2} \cdot (y0_{22} + y0_{20}) + \frac{1}{2} \cdot (y0_{23} + y0_{21}) \cdot z + \frac{1}{2} \cdot (y0_{22} - y0_{20}) \cdot z^2 + \frac{1}{2} \cdot (y0_{23} - y0_{21}) \cdot z^3$$

$$= y_{20} + y_{21} \cdot z + y_{22} \cdot z^2 + y_{23} \cdot z^3$$

$$Y_3(z) = \frac{1}{2} \cdot (y0_{32} + y0_{30}) + \frac{1}{2} \cdot (y0_{33} + y0_{31}) \cdot z + \frac{1}{2} \cdot (y0_{32} - y0_{30}) \cdot z^2 + \frac{1}{2} \cdot (y0_{33} - y0_{31}) \cdot z^3$$

$$= y_{30} + y_{31} \cdot z + y_{32} \cdot z^2 + y_{33} \cdot z^3$$

The required cyclic convolution is the array:

$$
\begin{array}{cccc}
y_{00} & y_{01} & y_{02} & y_{03} \\
y_{10} & y_{11} & y_{12} & y_{13} \\
y_{20} & y_{21} & y_{22} & y_{23} \\
y_{30} & y_{31} & y_{32} & y_{33}
\end{array}
$$

The number of operations required for an N×N convolution can be found by generalizing this example. The first stage produces an $\frac{N}{2} \times \frac{N}{2}$ convolution plus a series of other terms. The number of polynomial multiplications is $\frac{3N}{2}$ and is determined by the number of products of polynomial transforms. The first stage has N of these and the second $\frac{N}{2}$, both modulo $(z^{\frac{N}{2}} + 1)$. The two reductions modulo $(z^{\frac{N}{2}} + 1)$ and $(z^{\frac{N}{2}} - 1)$ require $N \cdot \frac{N}{2}$ additions each and Chinese remainder theorem reconstruction of the final terms $N \cdot N$. The initial polynomial transform plus the inverse transform each require $\frac{1}{2} \cdot N^2 \cdot \log_2 N$ additions. The second stage of polynomial reductions involves half as many terms and so each requires $(\frac{N}{2})^2$ additions. Chinese remainder reconstruction of these second stage terms requires $N \cdot (\frac{N}{2})$ additions and the two polynomial transforms involved each require $\frac{1}{4} \cdot N^2 \cdot \log_2 N$ additions. That is, for the first iteration a total of $\frac{3}{2} \cdot N^2 \cdot (\log_2 N + 2)$ additions.

If an optimal algorithm is used for each of the polynomial multiplications, then they will require $(N - 1)$ scalar multiplications. In total, $\frac{3}{2} \cdot N \cdot (N - 1)$ scalar multiplications are needed. If N is 2^α, there will be α stages in the algorithm, ending in a 2×2 convolution which, using the technique outlined, requires eight scalar multiplications and four additions. The total number of scalar multiplications therefore, is ($2N^2 - 3N + 9$). For large N, this is just two general multiplications per point.

The total number of additions is more difficult to compute. Apart from those involved in the polynomial multiplications, there are ($\frac{2}{3} \cdot N^2 \cdot (3\log_2 N + 5) - \frac{52}{3}$) others. The additions in the polynomial multiplications are proportional to N^2, hence the number of additions per point is proportional to $\log_2 N$. Then in terms of operations this algorithm differs from the separable transform approach only by a scalar multiple. However, there are no transcendental functions employed and scaling is always by real values.

Polynomial transforms and the DFT

Polynomial transforms provide another means of computing multi-dimensional discrete Fourier transforms. They can be used to efficiently map multi-dimensional DFTs into one-dimensional and achieve a reduction in the number of operations computing those DFTs. Polynomial transform techniques involve fewer multiplications than the usual Fourier transform techniques but have a larger number of additions. Polynomial transforms are computationally more attractive though, as they do not involve sinusoids or similar transcendental functions. However, they are more complex to implement.

The objective here is to express the DFT in such a way that an efficient computational technique arises naturally. Since it will be based on polynomial transforms, that means first expressing the DFT in polynomial form then arranging those polynomials in some suitable way. Given their importance, we shall concentrate here on two-dimensional DFTs. The development will closely follow that of the last section considering first N prime and then the general case.

A two-dimensional DFT is given by:

$$X(k_1, k_2) = \sum_{n_1=0}^{N-1} \sum_{n_2=0}^{N-1} x(n_1, n_2) \cdot W_N^{k_1 n_1} \cdot W_N^{k_2 n_2} \qquad k_1, k_2 = 0, 1, \cdots, N-1$$

where W is $\exp\left(\frac{j2\pi}{N}\right)$. A polynomial expression for the values $x(n_1, n_2)$ is:

$$X(z_1, z_2) = \sum_{n_1=0}^{N-1} \sum_{n_2=0}^{N-1} x(n_1, n_2) \cdot z^{n_1} \cdot z^{n_2}$$

Hence we can express the DFT as:

$$X(k_1, k_2) = \; \ll X(z_1, z_2) \gg_{z_1 - W^{k_1}} \gg_{z_2 - W^{k_2}}$$

To apply polynomial transform techniques, we will keep this structure for one dimension but revert to the polynomial transform structure for the other. Begin by expressing $X(z_1, z_2)$ as:

$$X(z_1, z_2) = \sum_{n_1=0}^{N-1} X_{n_1}(z_2) \cdot z_1^{n_1}$$

where the polynomials $X_{n_1}(z_2)$ are given by:

$$X_{n_1}(z_2) = \sum_{n_2=0}^{N-1} x_{n_1}(n_1, n_2) \cdot z_2^{n_1} \qquad\qquad n_1 = 0, 1, \cdots, N-1$$

Link these sets of equations to the two-dimensional DFT with the equation:

$$X(k_1, k_2) = \; < X_{k_1}(z_2) >_{z_2 - W^{k_2}}$$

where

$$X_{k_1}(z_2) = \left\langle \sum_{n_1=0}^{N-1} X_{n_1}(z_2) \cdot W^{k_1 n_1} \right\rangle_{z_2^N - 1}$$

These equations are in polynomial form in the z_2 dimension but DFT form along the z_1. The reduction modulo ($z_2^N - 1$) in this last equation is unnecessary but important for the development of the polynomial transform. The subscripts for z can (and will) be dropped for clarity as it will always be obvious which applies.

Consider the case of N prime. Since:

$$z^P - 1 = (z-1) \cdot (z^{P-1} + z^{P-2} + \cdots + z + 1) = (z-1) \cdot Q(z)$$

$X_{k_1}(z)$ can be computed using Chinese remainder theorem reconstruction and that will involve two polynomials reduced modulo ($z-1$) and Q(z). For the former:

$$X(k_1, 0) = \sum_{n_1=0}^{p-1} \sum_{n_2=0}^{p-1} x(n_1, n_2) \cdot W_N^{k_1 n_1} \qquad\qquad k_1 = 0, 1, \cdots, p-1$$

and this scalar term is a one-dimensional DFT.

Now consider the polynomial obtained from the reduction modulo Q(z). As:

$$Q(z) = \prod_{n=1}^{p-1} \left(z - W^n \right)$$

and as Q(z) is a factor of ($z^N - 1$), then the DFT may be expressed as:

$$X(k_1, k_2) = \; <<< X_{k_1}(z_2) >_{z^P - 1} >_{Q(z)} >_{z - W^{k_2}}$$

From this, determine the polynomial required by computing:

$$X_{n_1}(z) = \left\langle \sum_{n_2=0}^{p-1} x(n_1, n_2) \cdot z^{n_2} \right\rangle_{Q(z)} \qquad\qquad n_1 = 0, 1, \cdots, N-1$$

$$X_{k_1}(z) = \left\langle \sum_{n_1=0}^{p-1} X_{n_1}(z) \cdot W^{k_1 n_1} \right\rangle_{Q(z)}$$

$$X(k_1, k_2) = \langle X_{k_1}(z) \rangle_{z - W^{k_2}}$$

Consider the first of these. The reduction modulo $Q(z)$ means this becomes:

$$X_{n_1}(z) = \sum_{n_2=0}^{p-2} \left(x(n_1, n_2) - x(n_1, p-1) \right) \cdot z^{n_2} \qquad n_1 = 0, 1, \cdots, N-1$$

These equations give a slight reduction in the number of operations but they can be further developed. Since k_2 is a non-zero parameter, $\langle k_1 \cdot k_2 \rangle_p$ is an isomorphism of k_1. Therefore, using this mapping in place of k_1 in the second of the above equations:

$$X_{k_1}(z) = X_{k_1 k_2}(z) = \left\langle \sum_{n_1=0}^{p-1} X_{n_1}(z) \cdot W^{k_1 k_2 n_1} \right\rangle_{Q(z)}$$

and:

$$X(k_1, k_2) = X(k_1 \cdot k_2, k_2) = \langle X_{k_1 k_2}(z) \rangle_{z - W^{k_2}}$$

But, given that the second equation is a reduction modulo $(z - W^{k_2})$ and that W^{k_2} now exists in the first equation, it may be expressed as:

$$X_{k_1 k_2}(z) = \left\langle \sum_{n_1=0}^{p-1} X_{n_1}(z) \cdot W^{k_1 n_1} \right\rangle_{Q(z)}$$

This equation is now independent of k_2 and it is also a polynomial transform.

The reduction modulo $(z-1)$ produces a length p DFT. The reduction modulo $Q(z)$ leads to a polynomial transform and that requires no multiplications. It also produces a set of polynomials of order $(p - 2)$, the order of $Q(z)$. Let that result be described by:

$$X_{k_1 k_2}(z) = \sum_{i=0}^{p-2} a(k_1, i) \cdot z^i$$

To form the required DFT, reduce this modulo $(z - W^{k_2})$, which gives:

$$X(k_1 \cdot k_2, k_2) = \sum_{i=0}^{p-2} a(k_1, i) \cdot W^{k_2 i} \qquad k_2 = 1, 2, \cdots, p-1$$

This describes a set of p p-point DFTs corresponding to the p values of n_1. Thus the polynomial transform approach requires $(p + 1)$ p-point DFTs. In contrast, the normal row and column approach requires $2p$, hence this approach gains an advantage of approximately 2 for large p.

While p-point DFTs can be used, that would be wasteful. The summation would only be over $(p - 1)$ points which means the first computed output value would never be used and as n_2 ranges from 1 to $(p - 1)$, similarly, the last input value is never used. In their current form though, advantage cannot be taken of this redundancy but that is easily overcome. Given p prime and $k > 0$, then:

$$\sum_{i=0}^{p-1} W^{ki} = 0 = 1 + \sum_{i=1}^{p-1} W^{ki}$$

Thus the last equation becomes:

$$X(k_i \cdot k_2, k_2) = \sum_{i=1}^{p-2} b(k_1, i) \cdot W^{k_2 i} \qquad\qquad k_2 = 1, 2, \cdots, p-1$$

where
$$b(k_1, i) = a(k_1, i) - a(k_1, 0) \qquad\qquad i = 1, 2, \cdots, p-2$$
$$b(k_1, p-1) = -a(k_1, 0)$$

These equations are not in the conventional DFT form but have been encountered before as part of the development of Rader's algorithm. Hence they can be computed as a cyclic correlation via many of the algorithms discussed in this Part. An advantage of using these polynomial techniques is that some of the polynomial terms can be included with the polynomial transform stage and that can create an attractive algorithm.

The general case of arbitrary N follows in almost exactly the same fashion as this last section. The most important case though, is for N equal to 2^α. Once again, we rely on the fact that $(z^N - 1)$ has factors $(z^{\frac{N}{2}} + 1)$ and $(z^{\frac{N}{2}} - 1)$. An algorithm can therefore be framed using Chinese remainder theorem reconstruction based on these terms. This factorization also neatly divides all the complex roots of the terms. For the first stage, the complex factors W^n of $(z^{\frac{N}{2}} + 1)$ correspond to all the odd values of n, while the factors of $(z^{\frac{N}{2}} - 1)$ are all the even. As we will show, the term involving $(z^{\frac{N}{2}} + 1)$ can be computed by a polynomial transform. The term involving $(z^{\frac{N}{2}} - 1)$ can be rearranged to involve a polynomial transform and as well, an $\frac{N}{2} \times \frac{N}{2}$ point DFT. Hence the procedure is repeated over $(\alpha - 1)$ stages to complete the algorithm.

Consider the original equations. Separating out the terms corresponding to k_2 odd means they may be expressed as:

$$X_{n_1}(z) = \sum_{n_2=0}^{N-1} \left(x(n_1, n_2) - x\left(n_1, n_2 + \frac{N}{2} \right) \right) \cdot z^{n_2}$$

which is the result of a reduction modulo $(z^{\frac{N}{2}} + 1)$ and:

$$X_{k_1}(z) = \left\langle \sum_{n_1=0}^{N-1} X_{n_1}(z) \cdot W^{k_1 n_1} \right\rangle_{z^{\frac{N}{2}}+1}$$

$$X(k_1, k_2) = < X_{k_1}(z) >_{z - W^{k_2}}$$

With k_2 odd and N equal to a power of 2, $< k_1 \cdot k_2 >_N$ is an isomorphism and we can substitute as before. This leads to:

$$X_{k_1 k_2}(z) = \left\langle \sum_{n_1=0}^{N-1} X_{n_1}(z) \cdot z^{k_1 n_1} \right\rangle_{z^{\frac{N}{2}}+1}$$

which is a length N polynomial transform. The output from this term can be described by:

$$X_{k_1 k_2}(z) = \sum_{i=0}^{\frac{N}{2}-1} a(k_1, i) \cdot z^i$$

which gives:

$$X(k_1 \cdot k_2, k_2) = \sum_{i=0}^{\frac{N}{2}-1} a(k_1, i) \cdot W^{k_2 i} \qquad\qquad k_2 \text{ odd}$$

This describes a set of N DFTs, each of $\frac{N}{2}$ points. They are a little unusual in that k_2 must be odd. If k_2 is expressed as $(2k+1)$ with k ranging over 0 to $\frac{N}{2}$, then this equation describes the DFT of the sequence $\{ a(k_1, i) \cdot W^i \}$.

The second term in this initial stage deals with all the even values of k_2. As they are the factors of $(z^{\frac{N}{2}} - 1)$, which in turn is a factor of $(z^N - 1)$, the problem reduces to:

$$X(k_1, k_2) = \sum_{n_1=0}^{N-1} \sum_{n_2=0}^{\frac{N}{2}-1} \left(x(n_1, n_2) + x\left(n_1, n_2 + \frac{N}{2}\right) \right) \cdot W^{k_1 n_1} \cdot W^{k_2 n_2} \qquad k_2 \text{ even}$$

This is a DFT of size $N \times \frac{N}{2}$. In similar fashion to the last section, the problem may be reversed and expressed as the set of equations:

$$X_{n_2}(z) = \sum_{n_1=0}^{N-1} \left(x(n_1, n_2) + x\left(n_1, n_2 + \frac{N}{2}\right) \right) \cdot z^{n_1}$$

$$X_{k_2}(z) = \left\langle \sum_{n_2=0}^{\frac{N}{2}-1} X_{n_2}(z) \cdot W^{k_2 n_2} \right\rangle_{z^N+1} \qquad\qquad k_2 \text{ even}$$

$$X(k_1, k_2) = \langle X_{k_2}(z) \rangle_{z - W^{k_1}}$$

Exactly the same approach as above can be applied to separate this into two terms. One will be modulo $(z^{\frac{N}{2}} + 1)$ and will lead to a polynomial transform. The other will be modulo $(z^{\frac{N}{2}} - 1)$ and leads to a $\frac{N}{2} \times \frac{N}{2}$ point DFT. The entire procedure is repeated.

Example

Consider the case of N equal to 4. The algorithm will have only one stage and reduce to a 2×2 point DFT. The array to be transformed is:

$$
\begin{array}{cccc}
x(0,0) & x(0,1) & x(0,2) & x(0,3) \\
x(1,0) & x(1,1) & x(1,2) & x(1,3) \\
x(2,0) & x(2,1) & x(2,2) & x(2,3) \\
x(3,0) & x(3,1) & x(3,2) & x(3,3)
\end{array}
$$

To begin the example, these are expressed as the set of polynomials:

$$
\begin{aligned}
x_0(z) &= x(0,0) + x(0,1)\cdot z + x(0,2)\cdot z^2 + x(0,3)\cdot z^3 \\
x_1(z) &= x(1,0) + x(1,1)\cdot z + x(1,2)\cdot z^2 + x(1,3)\cdot z^3 \\
x_2(z) &= x(2,0) + x(2,1)\cdot z + x(2,2)\cdot z^2 + x(2,3)\cdot z^3 \\
x_3(z) &= x(3,0) + x(3,1)\cdot z + x(3,2)\cdot z^2 + x(3,3)\cdot z^3
\end{aligned}
$$

The first stage of the algorithm is to reduce modulo ($z^2 + 1$) and to take polynomial transforms. The reduction gives:

$$
\begin{aligned}
\mathbb{x}_{0,0}(z) &= (\,x(0,0) - x(0,2)\,) + (\,x(0,1) - x(0,3)\,)\cdot z = x_{00} + x_{01}\cdot z \\
\mathbb{x}_{0,1}(z) &= (\,x(1,0) - x(1,2)\,) + (\,x(1,1) - x(1,3)\,)\cdot z = x_{10} + x_{11}\cdot z \\
\mathbb{x}_{0,2}(z) &= (\,x(2,0) - x(2,2)\,) + (\,x(2,1) - x(2,3)\,)\cdot z = x_{20} + x_{21}\cdot z \\
\mathbb{x}_{0,3}(z) &= (\,x(3,0) - x(3,2)\,) + (\,x(3,1) - x(3,3)\,)\cdot z = x_{30} + x_{31}\cdot z
\end{aligned}
$$

The polynomial transform is modulo ($z^2 + 1$) and uses a root of z. Thus:

$$
\begin{aligned}
\mathbb{X}_{00}(z) &= ((\,x_{00} + x_{20}) + (\,x_{10} + x_{30})) + ((\,x_{01} + x_{21}) + (\,x_{11} + x_{31}))\cdot z \\
&= m_{00} + m_{01}\cdot z \\
\mathbb{X}_{01}(z) &= ((\,x_{00} - x_{20}) - (\,x_{11} - x_{31})) + ((\,x_{01} - x_{21}) + (\,x_{10} - x_{30}))\cdot z \\
&= m_{10} + m_{11}\cdot z \\
\mathbb{X}_{02}(z) &= ((\,x_{00} + x_{20}) - (\,x_{10} + x_{30})) + ((\,x_{01} + x_{21}) - (\,x_{11} + x_{31}))\cdot z \\
&= m_{20} + m_{21}\cdot z \\
\mathbb{X}_{03}(z) &= ((\,x_{00} - x_{20}) + (\,x_{11} - x_{31})) + ((\,x_{01} - x_{21}) - (\,x_{10} - x_{30}))\cdot z \\
&= m_{30} + m_{31}\cdot z
\end{aligned}
$$

These values are now mapped into the output array. The mapping is for k_2 odd, which in this case is the two values 1 and 3, and for k_1 equal to { 0, 1, 2, 3 }. Then:

$$
\begin{aligned}
k_2 &= 1 \qquad < k_1\cdot k_2 >_4 = 0, 1, 2, 3 \qquad \text{for } k_1 = 0, 1, 2, 3 \\
k_2 &= 3 \qquad < k_1\cdot k_2 >_4 = 0, 3, 2, 1 \qquad \text{for } k_1 = 0, 1, 2, 3
\end{aligned}
$$

The DFT values are formed from:

$$
X(< k_1\cdot k_2 >_4, k_2) = <\mathbb{X}_{0<k_1 k_2>_4}(z) >_{z-W^{k_2}} = m_{(k_1)0} + m_{(k_1)1}\cdot W^{k_2}
$$

Thus:

1. $k_1 = 0$ $X(0,1) = m_{00} + m_{01} \cdot W = m_{00} - j \cdot m_{01}$
$X(0,3) = m_{00} + m_{01} \cdot W^3 = m_{00} + j \cdot m_{01}$

2. $k_1 = 1$ $X(1,1) = m_{10} + m_{11} \cdot W = m_{10} - j \cdot m_{11}$
$X(3,3) = m_{10} + m_{11} \cdot W^3 = m_{10} + j \cdot m_{11}$

3. $k_1 = 2$ $X(2,1) = m_{20} + m_{21} \cdot W = m_{20} - j \cdot m_{21}$
$X(2,3) = m_{20} + m_{21} \cdot W^3 = m_{20} + j \cdot m_{21}$

4. $k_1 = 3$ $X(3,1) = m_{30} + m_{31} \cdot W = m_{30} - j \cdot m_{31}$
$X(1,3) = m_{30} + m_{31} \cdot W^3 = m_{30} + j \cdot m_{31}$

This step has derived half the output terms we require.

Reducing the initial set of polynomials modulo $(z^2 - 1)$ provides the base for the next stage of the algorithm. The reduction gives:

$$x_{1,0}(z) = (x(0,0) + x(0,2)) + (x(0,1) + x(0,3)) \cdot z = x2_{00} + x2_{01} \cdot z$$
$$x_{1,1}(z) = (x(1,0) + x(1,2)) + (x(1,1) + x(1,3)) \cdot z = x2_{10} + x2_{11} \cdot z$$
$$x_{1,2}(z) = (x(2,0) + x(2,2)) + (x(2,1) + x(2,3)) \cdot z = x2_{20} + x2_{21} \cdot z$$
$$x_{1,3}(z) = (x(3,0) + x(3,2)) + (x(3,1) + x(3,3)) \cdot z = x2_{30} + x2_{31} \cdot z$$

To begin the stage, these are re-expressed as:

$$x2_0(z) = x2_{00} + x2_{10} \cdot z + x2_{20} \cdot z^2 + x2_{30} \cdot z^3$$
$$x2_1(z) = x2_{01} + x2_{11} \cdot z + x2_{21} \cdot z^2 + x2_{31} \cdot z^3$$

Now reduce these polynomials modulo $(z^2 + 1)$ and this gives:

$$x2_0(z) = (x2_{00} - x2_{20}) + (x2_{10} - x2_{30}) \cdot z = x3_{00} + x3_{01} \cdot z$$
$$x2_1(z) = (x2_{01} - x2_{21}) + (x2_{11} - x2_{31}) \cdot z = x3_{10} + x3_{11} \cdot z$$

These are polynomial transformed using a root of z^2. Thus:

$$\mathbb{X}2_0(z) = (x3_{00} + x3_{10}) + (x3_{01} + x3_{11}) \cdot z = m2_{00} + m2_{01} \cdot z$$
$$\mathbb{X}2_1(z) = (x3_{00} - x3_{10}) + (x3_{01} - x3_{11}) \cdot z = m2_{10} + m2_{11} \cdot z$$

The polynomial transforms of this stage derive values for the output array corresponding to the even values of k_2, namely 0 and 2. These polynomial transforms, though, are derived from the reductions modulo $(z^2 + 1)$ and so are the basis for the output terms of the odd values of k_1 for these values of k_2, namely 1 and 3. Those output terms may be described by:

$$X(k_1 < k_1 \cdot k_2 >_4) = < \mathbb{X}2_{0<k_1 \cdot k_2>_4}(z) >_{z-W^{k_1}} = m2_{(k_1)0} + m2_{(k_1)1} \cdot W^{k_1}$$

The mappings here, given k_2 is 0 and 2 and k_1 is 1 and 3, are:

$$k_1 = 1 \qquad < k_2 >_4 \; = 0, 2 \qquad \text{for } k_2 = 0, 2$$
$$k_1 = 3 \qquad < 3k_2 >_4 = 0, 2 \qquad \text{for } k_2 = 0, 2$$

Thus:

1. $k_2 = 0$
    $$X(1,0) = m1_{00} + W \cdot m1_{01} = m1_{00} - j \cdot m1_{01}$$
    $$X(3,0) = m1_{00} + W^3 \cdot m1_{01} = m1_{00} + j \cdot m1_{01}$$
2. $k_2 = 2$
    $$X(1,2) = m1_{10} + W \cdot m1_{11} = m1_{10} - j \cdot m1_{11}$$
    $$X(3,2) = m1_{10} + W^3 \cdot m1_{11} = m1_{10} + j \cdot m1_{11}$$

Now reduce the input polynomials of this stage modulo ($z^2 - 1$). This gives:

$$x3_{1,0}(z) = (x2_{00} + x2_{20}) + (x2_{10} + x2_{30}) \cdot z = x3_{00} + x3_{01} \cdot z$$
$$x3_{1,1}(z) = (x2_{01} + x2_{21}) + (x2_{11} + x2_{31}) \cdot z = x3_{10} + x3_{11} \cdot z$$

Rearrange these to:

$$x4_0(z) = x3_{00} + x3_{10} \cdot z$$
$$x3_1(z) = x3_{01} + x3_{11} \cdot z$$

What is required now is a two-dimensional Fourier transform of these points to complete the problem. The results represent the values so far not calculated, namely $X(k_1,k_2)$ for k_1 even and k_2 even. In this case, the answer can be determined directly, thus:

$$X(0,0) = (x3_{00} + x3_{10}) + (x3_{01} + x3_{11}) = X_{00}$$
$$X(2,0) = (x3_{00} - x3_{10}) + (x3_{01} - x3_{11}) = X_{20}$$
$$X(0,2) = (x3_{00} + x3_{10}) - (x3_{01} + x3_{11}) = X_{02}$$
$$X(2,2) = (x3_{00} - x3_{10}) - (x3_{01} - x3_{11}) = X_{22}$$

The final DFT is therefore the array:

X_{00}	$m_{00} - j \cdot m_{01}$	X_{02}	$m_{00} + j \cdot m_{01}$
$m1_{00} - j \cdot m1_{01}$	$m_{10} - j \cdot m_{11}$	$m1_{10} - j \cdot m1_{11}$	$m_{30} + j \cdot m_{31}$
X_{20}	$m_{20} - j \cdot m_{21}$	X_{22}	$m_{20} + j \cdot m_{21}$
$m1_{00} + j \cdot m1_{01}$	$m_{30} - j \cdot m_{31}$	$m1_{10} + j \cdot m1_{11}$	$m_{10} + j \cdot m_{11}$

That completes the problem.

SUMMARY

This Part has focused attention on the role of polynomial structures within digital convolution. In TOOLS, we examined the concept of a polynomial and showed it could be given a wide interpretation. It can be described as a tuple ($a_0, a_1, a_2, \cdots, a_{n-1}$) or, in what might be called the traditional form, as:

$$H(z) = a_0 + a_1 \cdot z + a_2 \cdot z^2 + \cdots + a_{n-1} \cdot z^{n-1}$$

In either case, what is important is the ordering of the elements concerned.

Given a definition of polynomials, it becomes possible to define a polynomial algebra. The algebra discussed here follows the principles of the earlier development of number theory and discusses polynomial congruences and related concepts. This work enables polynomial rings and fields to be defined. In these, the polynomial operations are drawn from polynomial addition where the elements are combined as per the field $GF(p)$ and polynomial multiplication where operations are modulo some given polynomial. If that polynomial is irreducible, then a polynomial field is possible. If not, then the structure can at most be a polynomial ring. In either case, a condition on the existence of a finite structure is that there can only be p^k field or ring elements and that the individual elements within each of these polynomials be drawn from $GF(p)$.

One of the most important results shown in the development of the polynomial algebra is the polynomial Chinese remainder theorem. Throughout this part, it repeatedly appears as the cornerstone of various theoretical developments. In particular, it is via this theorem that a most important result can be proved — Winograd's minimal complexity theorem. Winograd's theorem states that the product of two Nth degree polynomials modulo a third $P(z)$ requires a minimum of $(2N-k)$ multiplications where k is the number of irreducible factors over the polynomial field. While it refers only to essential multiplications and not operations between constants of the field, it shows the importance of choosing the field for any convolution computation. Practical application requires a broader perspective, though, and so much of this Part has been concerned with the rational field rather than the theoretically more attractive complex.

Winograd's theorem suggests a structure for minimal algorithms which in matrix terms is:

$$\mathbf{y} = \mathbf{C} \cdot (\mathbf{Bh} \bullet \mathbf{Ax})$$

If **h**, for example, is pre-computed, then **Bh** becomes a diagonal matrix D and:

$$\mathbf{y} = \mathbf{C} \cdot \mathbf{D} \cdot \mathbf{Ax}$$

Winograd was also able to show a dual algorithm exists given by:

$$\mathbf{y}' = \mathbf{B}^T \cdot (\mathbf{Ch} \bullet \mathbf{Ax})$$

but this defines a cyclic correlation between the input vectors **h** and **x**. A cyclic correlation is easily transformed to a cyclic convolution by either reversing one of the input sequences or permuting the rows of C and the columns of B. If the latter course is pursued, then a second expression for a minimal convolution is obtained given by:

$$\mathbf{y} = \mathbf{B}_s^T \cdot (\mathbf{C}_s^T \mathbf{h} \bullet \mathbf{Ax})$$

Since matrices A and B have very simple factors for a rational field while matrix C has more complex, this is a more appealing form for implementation. Again, if one input sequence is known, it reduces to:

$$y = B_s^T \cdot D \cdot Ax$$

Provided N has no factor of 105 or higher, then for the rational field, matrix A (and so matrix B) consists only of elements drawn from $\{ 0, -1, +1 \}$. Thus A merely defines additions and so the general multiplications are entirely due to matrix D. This is the attraction of the rational field as while it may have slightly more general multiplications than others, it has far fewer among the field constants. Practical issues, then, are more concerned with minimizing the number of additions.

Winograd's algorithm is very useful for short-length convolutions. From Rader's theorem, that enables an efficient set of prime order DFT algorithms to be evolved. Unfortunately, as N grows the number of additions in these algorithms rises rapidly and they become impractical beyond a value of N of about 20. Hence the interest in minimizing additions.

Algorithms for long-length convolutions are formed on two general principles. One is to map the convolution into a multi-dimensional structure where each dimension is now a short-length convolution. Restrictions on the approach and implementation problems limit its value to sequence lengths below about 256. The other approach is to create a nested algorithm where the convolution is transformed to a nested set of subproblems, each having exactly the same mathematical structure as the original problem. However, operations within the outer subproblems are now performed upon polynomials rather than scalars. Again, implementation problems limit its value to sequence lengths below about 256. However, a variant of nesting — recursive nesting — can, in some circumstances, be comparable to any other technique. Those circumstances are a value of N equal to 2^α where α is 2^β and each of the nested stages has 2^β elements.

A multi-dimensional DFT can be computed via the separability property by repeatedly applying a one-dimensional DFT. There are alternatives. The most interesting group uses polynomial theory to identify symmetries in these higher dimensional structures and so more efficiently decomposes them into one-dimensional DFTs. Auslander, Feig, and Winograd's method applies to k-dimensional structures having a total of p^k points where p is an odd prime. It involves a series of transformations which convert the DFT into a set of $\frac{(p^k - 1)}{(p - 1)}$ p-point convolutions. Gertner's method uses a geometric approach which results in a two-dimensional DFT being described by a set of N-point DFTs of N-point discrete Radon transforms of the input data. It applies to a range of N but a separate algorithm is required for each. It can be extended to higher dimensional problems but there are practical difficulties in doing so.

Given the existence of polynomial structures, then a logical extension of work in the other Parts is to define a transform for those structures. Polynomial transforms naturally share the characteristics of other transforms with the cyclic transform property but they have one important distinction. They are inherently two-dimensional rather than one and as a consequence, are generally more appealing for multi-dimensional convolutions. In turn, that makes them useful for multi-dimensional DFTs. The result is very efficient algorithms comparable to Auslander, Feig, and Winograd's and to Gertner's approaches.

DISCUSSION

Two concerns of abstract algebra are of importance to this Part. One is the study of symmetries within mathematical structures. The other is concerned with identifying when different mathematical structures are merely transformations of one another. For the practical concerns of this Part, the first of these offers a means of avoiding or limiting the use of transcendental functions. The second opens avenues for mapping problems into more amenable mathematical structures.

The most striking picture which emerges from the concerns of this part is the expression of a digital convolution in the matrix form:

$$\mathbf{y} = \mathbf{C} \cdot (\mathbf{Bh} \bullet \mathbf{Ax})$$

or, when the input sequence h is known and the term Bh can be pre-computed, in the form:

$$\mathbf{y} = \mathbf{C} \cdot \mathbf{D} \cdot \mathbf{Ax}$$

Matrix D accounts for all the general multiplications of the algorithm in the latter case. Variations between the techniques discussed center mainly on the construction of the matrices C and A .

For most of the methods discussed, matrix A (for most practical values of N) in these equations has terms purely drawn from the set $\{\,0, -1, +1\,\}$. It therefore contributes additions only. Matrix C is more complex and that makes the transpose algorithm more interesting for practical implementation. The transpose form is:

$$\mathbf{y} = \mathbf{B}_s^{\mathrm{T}} \cdot (\mathbf{C}_s^{\mathrm{T}} \cdot \mathbf{h} \bullet \mathbf{Ax})$$

or:

$$\mathbf{y} = \mathbf{A}_s^{\mathrm{T}} \cdot \mathbf{D} \cdot \mathbf{Ax}$$

and here there is both pre- and post-addition of terms. Although the exact form of the matrices certainly varies between the different methods and, naturally, there must be some interest in this, there is a common concern among all methods. That is, minimizing the number of additions by effective factoring of the matrices.

In this Part, some approaches to this problem have been discussed and the different techniques themselves usually suggest various factorizations. If some technique generates a matrix of the form:

$$\mathbf{y} = \mathbf{P}_1 \cdot \mathbf{P}_2 \cdot \, \cdots \, \cdot \mathbf{P}_n \cdot \mathbf{D} \cdot \mathbf{Q}_n \cdot \, \cdots \, \cdot \mathbf{Q}_2 \cdot \mathbf{Q}_1 \mathbf{x}$$

where the matrices $\mathbf{P}_i$ and $\mathbf{Q}_i$ are simple matrices resulting in additions only and D is some diagonal matrix producing all the general multiplications, then two interesting questions arise. First, is it possible to rearrange the order of these matrices to minimize the total number of operations? From earlier work, we know some problems have a forward transformation matrix which is a tensor product of the transformation matrices of the prime factors. The second question then, is whether the distribution property of tensor products over matrix products can be used to minimize the number of operations?

So far, these questions have only been addressed by Johnson and Burrus in any detail. What they have found was that the prime factors and Winograd DFT algorithms are essentially the same, differing only in the order of these elementary matrices. They were also able to show that some rearrangement of the matrices can reduce the number of operations, so answering these two questions affirmatively. However, they showed this empirically via a dynamic programming method to systematically search through the possibilities, not theoretically.

FURTHER STUDY

References

Many of the books referred to in the last Part are also applicable to this Part. In particular, the following cover much of the material of this Part.

J.H. McClellan and C.M. Rader, *Number Theory in Digital Signal Processing*, Prentice Hall, 1979

H.J. Nussbaumer, *Fast Fourier Transform and Convolution Algorithms*, Springer-Verlag, 1981

D.F. Elliot and K.R. Rao, *Fast Transforms; Algorithms, Analyses, Applications*, Academic Press, 1982

Another excellent reference is:

R.E. Blahut, *Fast Algorithms for Digital Signal Processing*, Addison Wesley, 1985

A recent book of considerable value to this Part is:

M.T. Heideman, *Multiplicative Complexity, Convolution and the DFT*, Springer-Verlag, 1988

Heideman examines the minimum number of essential multiplications needed in a number of the algorithms discussed in this book. The book is quite advanced and requires a good knowledge of linear algebra to appreciate its finer points.

An important source of reference material for this Part are the technical journals. Papers with particular relevance are the following:

B. Arembepola and P.J.W. Rayner, Discrete Fourier Transforms over Residue Class Polynomial Rings with Applications in Computing Multi-dimensional Convolutions, *I.E.E.E. Transactions on Acoustics, Speech and Signal Processing*, V 28 N 4, August 1980, pp 407–14

L. Auslander, E. Feig, and S. Winograd, New Algorithms for the Multi-dimensional Discrete Fourier Transform, *I.E.E.E. Transactions on Acoustics, Speech and Signal Processing*, V 31 N 2, April 1983, pp 388–403

P.C. BALLA, A. ANTONIOU, and S.D. MORGERA, Higher Radix Aperiodic Convolution Algorithms, *I.E.E.E. Transactions on Acoustics, Speech and Signal Processing*, V 34 N 1, February 1986, pp 60–8

G. BRUUN, Z Transform DFT Filters and FFTs, *I.E.E.E. Transactions on Acoustics, Speech and Signal Processing*, V 26, N 1, February 1978, pp 56–63

A. GUESSOUM and R.M. MERSEREAU, Fast Algorithms for the Multi-dimensional Discrete Fourier Transform, *I.E.E.E. Transactions on Acoustics, Speech and Signal Processing*, V 34 N 4, August 1986, pp 937–43

P. IONNAS and M.G. STRINTZIS, Multi-dimensional Cyclic Convolution Algorithms with Minimum Multiplicative Complexity, *I.E.E.E. Transactions on Acoustics, Speech and Signal Processing*, V 35 N 3, March 1987, pp 384–90

H.W. JOHNSON and C.S. BURRUS, The Design of Optimal DFT Algorithms using Dynamic Programming, *I.E.E.E. Transactions on Acoustics, Speech and Signal Processing*, V 31 N 2, April 1983, pp 378–87

H.W. JOHNSON and C.S. BURRUS, On the Structure of Efficient DFT Algorithms, *I.E.E.E. Transactions on Acoustics, Speech and Signal Processing*, V 33 N 1, February 1985, pp 248–54

J.B. MARTENS, Recursive Cyclic Factorization: A New Algorithm for Complex, Real and Real-symmetric Data, *I.E.E.E. Transactions on Acoustics, Speech and Signal Processing*, V 32 N 4, August 1984, pp 750–61

H.J. NUSSBAUMER, Fast Polynomial Transform Algorithms for Digital Convolution, *I.E.E.E. Transactions on Acoustics, Speech and Signal Processing*, V 28 N 2, August 1980, pp 205–15

I. PITAS and M.G. STRINTZIS, General In-place Calculation of Discrete Fourier Transforms of Multi-dimensional Sequences, *I.E.E.E. Transactions on Acoustics, Speech and Signal Processing*, V 34 N 3, June 1986, pp 565–72

S. WINOGRAD, Some Bilinear Forms whose Multiplicative Complexity depends on the Field of Constants, *Mathematical Systems Theory*, V 10, 1977, pp 169–80

M.T. HEIDEMAN and C.S. BURRUS, On the Number of Multiplications Necessary to Compute a Length 2^n DFT, *I.E.E.E. Transactions on Acoustics, Speech and Signal Processing*, V 34 N 1, February 1986, pp 91–5

I. GERTNER, A New Efficient Algorithm to Compute the Two-Dimensional Discrete Fourier Transform, *I.E.E.E. Transactions on Acoustics, Speech and Signal Processing*, V 36 N 7, July 1988, pp 1036–50

Exercises

1. Derive the optimal cyclic convolution algorithm for fourth order polynomials in the Gaussian field.

2. Derive the optimal 6th order cyclic convolution algorithm for the Eisenstein field.

3. Show that reversing the rows of matrix C, except the first, and reversing the rows of matrix B, except the first, makes a convolution of the transpose of:

$$y = C^T \cdot (Bh \bullet Ax)$$

4. If a convolution is given by the matrix equation:

$$y = A^T FAx$$

   is matrix F diagonal? If not, what is it?

5. Derive the optimal short length sixth order cyclic convolution algorithm. Comment on the matrices A and B.

6. Derive a 7-point DFT algorithm using Winograd's method.

7. Why does the factorization described for Winograd's DFT method minimize additions?

8. Derive a recursive nesting algorithm for N equal to 6.

9. Show how to derive the base-2 FFT by polynomial reduction.

10. Show how to derive the base-4 FFT by polynomial reduction.

11. In the text with reference to recursive nesting, it states 'Provided M is greater than L $\cdots$'. Why?

12. Derive the polynomial transform convolution algorithm for N equal to 6.

13. Derive a 6-point DFT algorithm by polynomial transform techniques.

14. Derive a 6-point DFT algorithm using the PFA algorithm described in this Part.

15. Prove:

$$Det(A \otimes B) = (Det\ A)^M \cdot (Det\ B)^N$$

    where A is N×N, B is M×M, and $\otimes$ is the tensor product.

16. Show $GF(p^M)$ is a subfield of $GF(p^N)$ if $(p^M - 1)$ divides $(p^N - 1)$.

17. Prove that if P(z) is $(z^P - 1)$ where p is an odd prime, the polynomial Chinese remainder theorem reconstruction polynomials are always:

$$S_1(z) = \frac{1}{p} \cdot Q(z) \qquad \text{and} \quad S_2(z) = 1 - \frac{1}{p} \cdot Q(z)$$

18. Winograd's transpose algorithm only holds true if the initial matrices are Toeplitz. Why?

19. Prove:

$$< P(z) >_{D(z)} \; = \; < P(z) >_{k\,D(z)}$$

where k is some scalar.

20. Are there matrices other than the counter identity with the property that:

$$A^T \; = \; B \cdot A \cdot B$$

where A is Toeplitz? Give an example.

21. Express the Rader-Brenner algorithm in polynomial form.

22. Derive a polynomial form for the DFT when factorization is into polynomials with Gaussian constants. Is there any practical advantage in this algorithm?

23. Derive an algorithm based on Agarwal and Cooley's nesting procedure for N equal to 30.

24. How does the Bruun algorithm change if the data is real? Does this give it any practical advantages in implementing convolutions of real-values sequences?

25. Derive a 25-point DFT algorithm using Auslander, Feig, and Winograd's method.

26. Derive a split nesting algorithm for N equal to 15.

27. Determine an algorithm for a 21×21 point DFT using Gertner's method.

28. Express Bruun's algorithm in matrix form and show it is a diagonalising procedure. Given the form of this matrix expression, there are grounds for believing Bruun's algorithm is a minimal additions algorithm. Comment on this.

PROBLEMS

1. Derive an approximate formula (or an exact one if you believe you can) for the number of additions in the Winograd short convolution algorithm.

2. Derive an algorithm based on Auslander, Feig, and Winograd's method for a DFT of sequence length $2p^k$ where p is an odd prime.

3. Show that if N is p^2, then an array of size N×N can be polynomial transformed on a ring $\dfrac{(z^p - 1)}{(z^N - 1)}$ with a root of z and a sequence length p^2.

4. Discuss convolution in the field $GF(2^q)$ where q is 8 or 16.

5. The methods discussed in this Part all concentrate on the forward DFT. Applications such as convolution require the inverse DFT as well. Discuss how this might be done with these methods.

6. Quaternions are mathematical structures consisting of four real numbers (a, b, c, d) and given by $Q = a + bi + cj + dk$. Addition of two quaternions is defined as:

$$Q_1 + Q_2 = (a_1 + a_2) + (b_1 + b_2)i + (c_1 + c_2)j + (d_1 + d_2)k$$

Multiplication is defined as:

$$Q_1 \cdot Q_2 = (a_1 a_2 - b_1 b_2 - c_1 c_2 - d_1 d_2) + (b_1 a_2 + a_1 b_2 - d_1 c_2 + c_1 a_2)i$$
$$+ (c_1 a_1 + a_1 b_2 + a_1 c_2 - b_1 d_2)j + (d_1 a_2 - c_1 b_2 + c_1 c_2 + a_1 d_2)k$$

Devise an efficient algorithm for multiplication of quaternions.

7. The basic DFT computational cell algorithm of Part 2 can be described in the form:

$$\mathbf{y} = (C_M^T \otimes C_L^T)((A_M \otimes A_L)\mathbf{x} \bullet (B_M \otimes B_L)\mathbf{h})$$

where
$$\mathbf{y_0} = C_M^T \cdot (A_M \mathbf{x} \bullet B_M \mathbf{h}) \quad \text{describes an M-point DFT}$$
$$\mathbf{y_1} = C_L^T \cdot (A_L \mathbf{x} \bullet B_L \mathbf{h}) \quad \text{describes an L-point DFT}$$

and $\otimes$ is the tensor product. Derive an alternative formulation and comment on its value for implementation.

8. Given a convolution or DFT algorithm of the form $\mathbf{y} = C^T \cdot (A\mathbf{x} \bullet B\mathbf{h})$ and a pair of matrices U and V with the property UV is the identity, then:

$$\mathbf{y} = (C^T U)(A\mathbf{x} \bullet (VB)\mathbf{h})$$

Can this be used to improve the heuristic method discussed? How should U and V be determined?

9. Modern supercomputers are predominantly vector machines. An array of processing elements can add, subtract, multiply or divide two N-element vectors in one machine cycle. They can also add to sets of numbers in a time proportional to $Log_2 N$ cycles.

   (a) What is important in a DFT/convolution algorithm to minimize vector operations?
   (b) Which of the algorithms discussed in this Part is best suited to vector operations and why?

(c) What speed-up factor is achieved by these algorithms? Does this suggest vector machines would have an advantage for these calculations (other than that due to their greater speed)?

10. A listening room, be it a major auditorium or the living room of a home, is a complex acoustic environment. Its parameters are defined by the shape of the chamber, the acoustic properties of the chamber's surfaces and the nature of the objects within the chamber. To illustrate, consider a simple one-dimensional model consisting of a chamber of length L and a single source S of sound.

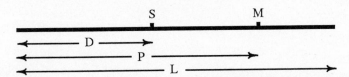

Source S is at a position D in this 'room' and a measurement system M is at a position P. Now L will be the multiple of numerous sound frequencies, hence there can be many standing waves. Also, the 'walls' will have a frequency dependent absorption.

An important property of a chamber is its reverberation. A wavefront launched at S will reflect off the left end of the room and be measured by M some $\frac{(D+P)}{V}$ seconds later where V is the velocity of sound. Exactly $\frac{2L}{V}$ seconds later, that same wavefront will be measured again from the left, and so on until the wavefront is totally absorbed. Let the two walls have some reflection coefficient ρ. The left reverberation component can be crudely modeled by the discrete equation:

$$y_R(n) = \rho x(n - d_2) + \sum_{i=e}^{\infty} \rho^i \cdot x(n - i \cdot d_3)$$

where d_0 is $\frac{(D+P)}{V}$ and d_1 is $\frac{2L}{V}$. The right reverberation component can be modeled by:

$$y_L(n) = \rho x(n - d_0) + \sum_{i=e}^{\infty} \rho^i \cdot x(n - i \cdot d_1)$$

where d_2 is $\frac{2(L-P)}{V}$ and d_3 is $\frac{2L}{V}$. The total reverberation can therefore be modeled as:

$$y(n) = (y_L(n) + y_R(n) + x(n - d_4))$$

where d_4 is $\frac{(D+P)}{V}$. The third component is the direct wave measured at M.

Reverberation is extremely important in setting the acoustic character of any chamber. Therefore, if an audio processor is required to allow simulation of real concert halls, in a large measure it must simulate their reverberation characteristics.

(a) Consider some auditorium you know. By considering it as a set of reflecting surfaces, estimate the number of reverberation filters of the simple type discussed above that you would need to model this hall.

(b) Given that an audio processor must have a real-time response and that simulating reverberation is a convolution process, what method would you recommend for implementing that convolution for your auditorium? Discuss.

11. Derive an algorithm using Gertner's method for the case of N equal to $p_1 \cdot p_2{}^\alpha$ where p_1 and p_2 are prime and α is any integer.

Appendices

Appendices

APPENDIX 1

Arithmetic modulo Mersenne and Fermat numbers

Arithmetic modulo a Mersenne number is arithmetic modulo ($2^Q - 1$) and that is one's complement arithmetic. For purposes of illustration, let Q be 8, thus:

$$M = 2^8 - 1 = 255_{10} = 1111\ 1111_2$$

Also, consider two numbers 'a' and 'b' where:

$$a = 135_{10} = 1000\ 0111_2 \qquad b = 192_{10} = 1100\ 0000_2$$

Then:

1. Negation
   A number $-a$ is interpreted as meaning ($M - a$). Thus:

   $$-a = -135_{10} = (255 - 135)_{10} = 120_{10} = 0111\ 1000_2$$

   In binary terms, negation is just the reversal of all bits.

2. Addition
   In decimal:

   $$(a + b) = <\ 135 + 192\ >_{255} = <\ 327\ >_{255} = 72_{10}$$

To add two numbers modulo a Mersenne number in binary just requires adding any overflow bit to the least significant digit. Thus:

```
         1000 0111
  +      1100 0000
  1      0100 0111
  ↑              1
         0100 1000   = 72₁₀
```

3. Subtraction
   This is just unary negation followed by addition.

4. Multiplication
   In decimal:

   $$<\ a{\cdot}b\ >_M = <\ 135{\cdot}192\ >_{255} = <\ 25920\ >_{255} = 165_{10}$$

In binary, the product is formed and the high order bits are added to the low order bits. Thus:

$$
\begin{array}{r}
1000\ 0111 \\
\text{x} \qquad\qquad 1100\ 0000 \\
\hline
0010\ 0001 \qquad 1100\ 0000 \\
0100\ 0011 \qquad 1000\ 0000 \\
0110\ 0101 \qquad 0100\ 0000 \\
\uparrow \qquad\qquad 0110\ 0101 \\
\hline
1010\ 0101 = 165_{10}
\end{array}
$$

All of these are very simple operations, very easily implemented in digital electronics, and all based on integers.

Fermat numbers are of the form ($2^q + 1$). To illustrate arithmetic modulo a Fermat number, we shall use the same example numbers as before though M now becomes the number ($2^8 + 1$), which is 257_{10} or $1\ 0000\ 0001_2$ in binary. Then:

1. Negation

   We need to consider two cases. First, all numbers up to 2^q. Let A be one of these where:

$$
A = \sum_{i=0}^{q-1} a_i \cdot 2^i \qquad\qquad\qquad a_i = 0 \text{ or } 1
$$

A number $-A$ is interpreted as meaning ($2^q + 1 - A$). Thus:

$$
-A = 2^q + 1 - \sum_{i=0}^{q-1} a_i \cdot 2^i
$$

$$
= 2^q + 1 - \left(2^q - 1\right) + \sum_{i=0}^{q} \left(2 - a_i\right) \cdot 2^i
$$

$$
= 2 + \sum_{i=0}^{q} \left(2 - a_i\right) \cdot 2^i
$$

That is, these numbers are negated by complementing the lower q bits and adding 2.

Second, the single number 2^q. Its inverse is one. That can be mechanized by complementing the ($q + 1$)th bit and adding 1 to the result.

As an example of negation, consider:

$$
< -135_{10} >_{257} = 122_{10}
$$

In binary:

$$
\begin{aligned}
-0\ 1000\ 0111_2 \quad &= 0\ 0111\ 1000_2 + 10_2 \\
&= 0\ 0111\ 1010_2 \\
&= 122_{10}
\end{aligned}
$$

2. Addition

   If two binary numbers A and B are added, the result C is:

$$C = \sum_{i=0}^{q-1} c_i \cdot 2^i + c_q \cdot 2^q + c_{q+1} \cdot 2^{q+1} \qquad\qquad c_i = 0 \text{ or } 1$$

Then:

$$\langle C \rangle_M = \sum_{i=0}^{q-1} c_i \cdot 2^i + \left\langle c_q \cdot 2^q + c_{q+1} \cdot 2^{q+1} \right\rangle_M$$

Since M is ($2^q + 1$):

$$< 2^q >_M = -1 \qquad\qquad \text{and} \qquad\qquad < 2^{q+1} >_M = -2$$

Now c_{q+1} can only be 1 if both A and B are 2^q. In this case:

$$< C >_M = -2 = 2^q - 1$$

In all other cases, addition means adding the digits of the two numbers and then subtracting the (q + 1)th bit from the result.

Consider an example:

$$< 135_{10} + 192_{10} >_{257} = < 327_{10} >_{257} = 70_{10}$$

In binary:

```
      0   1000 0111
  +   0   1100 0000
      1   0100 0111
      ↑              1
      0   0100 0110  = 70₁₀
```

$$\begin{array}{r} 0 \quad 1000\ 0111 \\ +\quad 0 \quad 1100\ 0000 \\ \hline 1 \quad 0100\ 0111 \\ \uparrow \qquad\qquad 1 \\ \hline 0 \quad 0100\ 0110 \ = 70_{10} \end{array}$$

3. Subtraction
    Subtraction involves negating one number and adding.

4. Multiplication
    Multiplication can be treated as the sum of a set of multiplications by factors 2^d. Again, we need to consider two cases. For numbers less than 2^q, each product is:

$$C = \sum_{i=0}^{q-1-d} a_i \cdot 2^{i+d} + 2^q \sum_{i=q-d}^{q-1} a_i \cdot 2^{i+q-d}$$

Since $< 2^q >_M$ is -1, for this M:

$$\langle C \rangle_M = \sum_{i=0}^{q-1-d} a_i \cdot 2^{i+d} - \sum_{i=q-d}^{q-1} a_i \cdot 2^{i+q-d}$$

From this, we deduce that if both numbers to be multiplied are less than 2^q, the procedure is simply to use normal binary multiplication and then subtract the high order q bits from the lower.

If one number is 2^q, multiplying by 2^d gives -2^d modulo ($2^q + 1$). The multiplication as a whole reduces to negation. If both numbers are 2^q, the result is just -1.

Clearly, arithmetic modulo a Fermat number is complex. Further, as only one number ever requires 2^q bits, the representation is inefficient. One approach to circumvent both problems is to assume that 2^q will simply never occur. If q is large, that is not unreasonable. While there is the risk of serious error, the possibility is easily calculated. The result of this action is to greatly simplify operations as reasonably conventional arithmetic is used, plus the word size is only q bits.

A quite different approach is to encode the numbers. It is impossible to express the set of numbers $\{\, 0, 1, 2, \cdots, 2^q \,\}$ in only q bits but they can be encoded such that zero is the only number requiring (q + 1) bits. For example, if 1 is subtracted from every number, then:

zero,	which is	0 000 $\cdots$ 0000	becomes 1 000 $\cdots$ 0000
one,	which is	0 000 $\cdots$ 0001	becomes 0 000 $\cdots$ 0000
two,	which is	0 000 $\cdots$ 0010	becomes 0 000 $\cdots$ 0001
.			
.			
.			
2q,	which is	1 000 $\cdots$ 0000	becomes 0 111 $\cdots$ 1111

This does not reduce the need for numbers to be stored with the extra bit but since only zero uses it, then it can be employed simply as an inhibit bit. That is, when detected, there is no need to add as the result is the other operand and in multiplication, the result must be zero itself.

Arithmetic becomes greatly simplified in this system, apart from becoming q-bit rather than (q + 1). It is easily shown:

1. negation is achieved by complementing all bits;
2. addition requires conventional q-bit binary addition followed by the addition of the complement of the carry bit to the sum;
3. subtraction involves negation followed by addition;
4. multiplication requires q-bit binary multiplication of the two operands, also addition of those operands as described in 2., then addition as per 2. of these results and finally subtraction of the high order q bits from the low order according to 3.

Although certainly much simpler, this scheme is still quite complex and that is a problem. Another is the need to encode input data, then decode the output. Even though modern ROMs make this easy, it is an added cost.

APPENDIX 2

The first 64 Mersenne numbers

Q	$2^Q - 1$	Factors
1	1	-
2	3	-
3	7	-
4	15	3×5
5	31	-
6	63	3×3×7
7	127	-
8	255	3×5×17
9	511	7×73
10	1,023	3×11×31
11	2,047	23×89
12	4,095	3×3×5×7×13
13	8,191	-
14	16,383	3×43×127
15	32,767	7×31×151
16	65,535	3×5×17×257
17	131,071	-
18	262,143	3×3×3×7×19×73
19	524,287	-
20	1,048,575	3×5×5×11×31×41
21	2,097,151	7×7×127×337
22	4,194,303	3×23×89×683
23	8,388,607	47×178,481
24	16,777,215	3×3×5×7×13×17×241
25	33,554,431	31×601×1,801
26	67,108,863	3×2,731×8,191
27	134,217,727	7×73×262,657
28	268,435,455	3×5×29×43×113×127
29	536,870,911	233×1,103×2,089
30	1,073,741,823	3×3×7×11×31×151×331
31	2,147,483,647	-
32	4,294,967,295	3×5×17×257×65,537
33	8,589,934,591	7×23×89×599,479
34	17,179,869,183	3×43,691×131,071
35	34,359,738,367	31×71×127×122,921
36	68,719,476,735	3×3×3×5×7×13×19×37×73×109
37	137,438,953,471	223×616,318,177
38	274,877,906,943	3×174,763×524,287
39	549,755,813,887	7×79×8,191×121,369

Q	$2^Q - 1$	*Factors*
40	1, 099, 511, 627, 775	3×5×5×11×17×31×41×61, 681
41	2, 199, 023, 255, 551	13, 367×164, 511, 353
42	4, 398, 046, 511, 103	3×3×7×7×43×127×337×5, 419
43	8, 796, 093, 022, 207	431×9, 719×2, 099, 863
44	17, 592, 186, 044, 415	3×5×23×89×397×683×2, 113
45	35, 184, 372, 088, 831	7×31×73×151×631×23, 311
46	70, 368, 744, 177, 663	3×47×178, 481×2, 796, 203
47	140, 737, 488, 355, 327	2, 351×4, 513×13, 264, 529
48	281, 474, 976, 710, 655	3×3×5×7×13×17×97×241×257×673
49	562, 949, 953, 421, 311	127×4, 432, 676, 798, 593
50	1, 125, 899, 906, 842, 623	3×11×31×251×601×1, 801×4, 051
51	2, 251, 799, 813, 685, 247	7×103×2, 143×11, 119×131, 071
52	4, 503, 599, 627, 370, 495	3×5×53×157×1, 613×2, 731×8, 191
53	9, 007, 199, 254, 740, 991	6, 361×69, 431×20, 394, 401
54	18, 014, 398, 509, 481, 983	3×3×3×3×7×19×73×87, 211×262, 657
55	36, 028, 797, 018, 963, 967	23×31×89×881×3, 191×201, 961
56	72, 057, 594, 037, 927, 935	3×5×17×29×43×113×127×15, 790, 321
57	144, 115, 188, 075, 855, 871	7×32, 377×524, 287×1, 212, 847
58	288, 230, 376, 151, 711, 743	3×59×233×1, 103×2, 089×3, 033, 169
59	576, 460, 752, 303, 423, 487	179, 951×3, 203, 431, 780, 337
60	1, 152, 921, 504, 606, 846, 975	3×3×5×5×7×11×13×31×41×61×151× 331×1, 321
61	2, 305, 843, 009, 213, 693, 951	-
62	4, 611, 686, 018, 427, 387, 903	3×2, 147, 483, 647×715, 827, 883
63	9, 223, 372, 036, 854, 775, 807	7×7×73×127×337×92, 737×649, 657
64	18, 446, 744, 073, 709, 551, 615	3×5×17×257×641×65, 537×6, 700, 417

APPENDIX 3

The first 64 Fermat numbers

Q	$2^Q + 1$	Factors
1	3	-
2	5	-
3	9	3×3
4	17	-
5	33	3×11
6	65	5×13
7	129	3×43
8	257	-
9	513	3×3×3×19
10	1,025	5×5×41
11	2,049	3×683
12	4,097	17×241
13	8,193	3×2,731
14	16,385	5×29×113
15	32,769	3×3×11×331
16	65,537	-
17	131,073	3×43,691
18	262,145	5×13×37×109
19	524,289	3×174,763
20	1,048,577	17×61,681
21	2,097,153	3×3×43×5,419
22	4,194,305	5×397×2,113
23	8,388,609	3×2,796,203
24	16,777,217	97×257×673
25	33,554,433	3×11×251×4,051
26	67,108,865	5×53×157×1,613
27	134,217,729	3×3×3×3×19×87,211
28	268,435,457	17×15,790,321
29	536,870,913	3×59×3,033,169
30	1,073,741,825	5×5×13×41×61×1,321
31	2,147,483,649	3×715,827,883
32	4,294,967,297	641×6,700,417
33	8,589,934,593	3×3×67×683×20,857
34	17,179,869,185	5×137×953×26,317
35	34,359,738,369	3×11×43×281×86,171
36	68,719,476,737	17×241×433×38,737
37	137,438,953,473	3×1,777×25,781,083
38	274,877,906,945	5×229×457×525,313
39	549,755,813,889	3×3×2,731×22,366,891

Q	$2^Q + 1$	*Factors*
40	1, 099, 511, 627, 777	257×4, 278, 255, 361
41	2, 199, 023, 255, 553	3×83×8, 831, 418, 697
42	4, 398, 046, 511, 105	5×13×29×113×1, 429×14, 449
43	8, 796, 093, 022, 209	3×2, 932, 031, 007, 403
44	17, 592, 186, 044, 417	17×353×2, 931, 542, 417
45	35, 184, 372, 088, 833	3×3×3×11×19×331×18, 837, 001
46	70, 368, 744, 177, 665	5×277×1, 013×1, 657×30, 269
47	140, 737, 488, 355, 329	3×283×165, 768, 537, 521
48	281, 474, 976, 710, 657	193×65, 537×22, 253, 377
49	562, 949, 953, 421, 313	3×43×4, 363, 953, 127, 297
50	1, 125, 899, 906, 842, 625	5×5×5×41×101×8, 101×268, 501
51	2, 251, 799, 813, 685, 249	3×3×307×2, 857×6, 529×43, 691
52	4, 503, 599, 627, 370, 497	17×858, 001×308, 761, 441
53	9, 007, 199, 254, 740, 993	3×107×28, 059, 810, 762, 433
54	18, 014, 398, 509, 481, 985	5×13×37×109×246, 241×279, 073
55	36, 028, 797, 018, 963, 969	3×11×11×683×2, 971×48, 912, 491
56	72, 057, 594, 037, 927, 937	257×5, 153×54, 410, 972, 897
57	144, 115, 188, 075, 855, 873	3×3×571×174, 763×160, 465, 489
58	288, 230, 376, 151, 711, 745	5×107, 367, 629×536, 903, 681
59	576, 460, 752, 303, 423, 489	3×2, 833×37, 171×1, 824, 726, 041
60	1, 152, 921, 504, 606, 846, 977	17×241×61, 681×4, 562, 284, 561
61	2, 305, 843, 009, 213, 693, 953	3×768, 614, 336, 404, 564, 651
62	4, 611, 686, 018, 427, 387, 905	5×5, 581×8, 681×49, 477×384, 773
63	9, 223, 372, 036, 854, 775, 809	3×3×3×19×43×5, 419×77, 158, 673, 929
64	18, 446, 744, 073, 709, 551, 617	274, 177×67, 280, 421, 310, 721

APPENDIX 4

The first 100 cyclotomic polynomials

$C_1(z)$ $= z - 1$

$C_2(z)$ $= z + 1$

$C_3(z)$ $= z^2 + z + 1$

$C_4(z)$ $= z^2 + 1$

$C_5(z)$ $= z^4 + z^3 + z^2 + z + 1$

$C_6(z)$ $= z^2 - z + 1$

$C_7(z)$ $= z^6 + z^5 + z^4 + z^3 + z^2 + z + 1$

$C_8(z)$ $= z^4 + 1$

$C_9(z)$ $= z^6 + z^3 + 1$

$C_{10}(z)$ $= z^4 - z^3 + z^2 - z + 1$

$C_{11}(z)$ $= z^{10} + z^9 + z^8 + z^7 + z^6 + z^5 + z^4 + z^3 + z^2 + z + 1$

$C_{12}(z)$ $= z^4 - z^2 + 1$

$C_{13}(z)$ $= z^{12} + z^{11} + z^{10} + z^9 + z^8 + z^7 + z^6 + z^5 + z^4 + z^3 + z^2 + z + 1$

$C_{14}(z)$ $= z^6 - z^5 + z^4 - z^3 + z^2 - z + 1$

$C_{15}(z)$ $= z^8 - z^7 + z^5 - z^4 + z^3 - z + 1$

$C_{16}(z)$ $= z^8 + 1$

$C_{17}(z)$ $= z^{16} + z^{15} + z^{14} + z^{13} + z^{12} + z^{11} + z^{10} + z^9 + z^8 + z^7 + z^6$
$+ z^5 + z^4 + z^3 + z^2 + z + 1$

$C_{18}(z)$ $= z^6 - z^3 + 1$

$C_{19}(z)$ $= z^{18} + z^{17} + z^{16} + z^{15} + z^{14} + z^{13} + z^{12} + z^{11} + z^{10} + z^9 + z^8$
$+ z^7 + z^6 + z^5 + z^4 + z^3 + z^2 + z + 1$

$C_{20}(z)$ $= z^8 - z^6 + z^4 - z^2 + 1$

$C_{21}(z)$ $= z^{12} - z^{11} + z^9 - z^8 + z^6 - z^4 + z^3 - z + 1$

$C_{22}(z)$ $= z^{10} - z^9 + z^8 - z^7 + z^6 - z^5 + z^4 - z^3 + z^2 - z + 1$

$C_{23}(z)$ $= z^{22} + z^{21} + z^{20} + z^{19} + z^{18} + z^{17} + z^{16} + z^{15} + z^{14} + z^{13} + z^{12}$
$+ z^{11} + z^{10} + z^9 + z^8 + z^7 + z^6 + z^5 + z^4 + z^3 + z^2 + z + 1$

$C_{24}(z)$ $= z^8 - z^4 + 1$

$C_{25}(z)$ $= z^{20} + z^{15} + z^{10} + z^5 + 1$

$C_{26}(z)$ $= z^{12} - z^{11} + z^{10} - z^9 + z^8 - z^7 + z^6 - z^5 + z^4 - z^3 + z^2 - z + 1$

$C_{27}(z)$ $= z^{18} + z^9 + 1$

$C_{28}(z)$ $= z^{12} - z^{10} + z^8 - z^6 + z^4 - z^2 + 1$

$C_{29}(z)$ $= z^{28} + z^{27} + z^{26} + z^{25} + z^{24} + z^{23} + z^{22} + z^{21} + z^{20} + z^{19} + z^{18}$
$+ z^{17} + z^{16} + z^{15} + z^{14} + z^{13} + z^{12} + z^{11} + z^{10} + z^9 + z^8$
$+ z^7 + z^6 + z^5 + z^4 + z^3 + z^2 + z + 1$

$C_{30}(z)$ $= z^8 + z^7 - z^5 - z^4 - z^3 + z + 1$

$C_{31}(z)$ $= z^{30} + z^{29} + z^{28} + z^{27} + z^{26} + z^{25} + z^{24} + z^{23} + z^{22} + z^{21}$
$+ z^{20} + z^{19} + z^{18} + z^{17} + z^{16} + z^{15} + z^{14} + z^{13} + z^{12}$
$+ z^{11} + z^{10} + z^9 + z^8 + z^7 + z^6 + z^5 + z^4 + z^3 + z^2 + z + 1$

$C_{32}(z)$ $= z^{16} + 1$

$C_{33}(z)$ $= z^{20} - z^{19} + z^{17} - z^{16} + z^{14} - z^{13} + z^{11} - z^{10} + z^9 - z^7$
$+ z^6 - z^4 + z^3 - z + 1$

$$C_{34}(z) = z^{16} - z^{15} + z^{14} - z^{13} + z^{12} - z^{11} + z^{10} - z^9 + z^8 - z^7$$
$$+ z^6 - z^5 + z^4 - z^3 + z^2 - z + 1$$

$$C_{35}(z) = z^{24} - z^{23} + z^{19} - z^{18} + z^{17} - z^{16} + z^{14} - z^{13} + z^{12} - z^{11}$$
$$+ z^{10} - z^8 + z^7 - z^6 + z^5 - z + 1$$

$$C_{36}(z) = z^{12} - z^6 + 1$$

$$C_{37}(z) = z^{36} + z^{35} + z^{34} + z^{33} + z^{32} + z^{31} + z^{30} + z^{29} + z^{28} + z^{27} + z^{26}$$
$$+ z^{25} + z^{24} + z^{23} + z^{22} + z^{21} + z^{20} + z^{19} + z^{18} + z^{17} + z^{16} + z^{15}$$
$$+ z^{14} + z^{13} + z^{12} + z^{11} + z^{10} + z^9 + z^8 + z^7 + z^6 + z^5 + z^4 + z^3$$
$$+ z^2 + z + 1$$

$$C_{38}(z) = z^{18} - z^{17} + z^{16} - z^{15} + z^{14} - z^{13} + z^{12} - z^{11} + z^{10} - z^9$$
$$+ z^8 - z^7 + z^6 - z^5 + z^4 - z^3 + z^2 - z + 1$$

$$C_{39}(z) = z^{24} - z^{23} + z^{21} - z^{20} + z^{18} - z^{17} + z^{15} - z^{14} + z^{12} - z^{10}$$
$$+ z^9 - z^7 + z^6 - z^4 + z^3 - z + 1$$

$$C_{40}(z) = z^{16} - z^{12} + z^8 - z^4 + 1$$

$$C_{41}(z) = z^{40} + z^{39} + z^{38} + z^{37} + z^{36} + z^{35} + z^{34} + z^{33} + z^{32} + z^{31}$$
$$+ z^{30} + z^{29} + z^{28} + z^{27} + z^{26} + z^{25} + z^{24} + z^{23} + z^{22} + z^{21}$$
$$+ z^{20} + z^{19} + z^{18} + z^{17} + z^{16} + z^{15} + z^{14} + z^{13} + z^{12} + z^{11}$$
$$+ z^{10} + z^9 + z^8 + z^7 + z^6 + z^5 + z^4 + z^3 + z^2 + z + 1$$

$$C_{42}(z) = z^{12} + z^{11} - z^9 - z^8 + z^6 - z^4 - z^3 + z + 1$$

$$C_{43}(z) = z^{42} + z^{41} + z^{40} + z^{39} + z^{38} + z^{37} + z^{36} + z^{35} + z^{34} + z^{33} + z^{32}$$
$$+ z^{31} + z^{30} + z^{29} + z^{28} + z^{27} + z^{26} + z^{25} + z^{24} + z^{23} + z^{22}$$
$$+ z^{21} + z^{20} + z^{19} + z^{18} + z^{17} + z^{16} + z^{15} + z^{14} + z^{13} + z^{12}$$
$$+ z^{11} + z^{10} + z^9 + z^8 + z^7 + z^6 + z^5 + z^4 + z^3 + z^2 + z + 1$$

$$C_{44}(z) = z^{20} - z^{18} + z^{16} - z^{14} + z^{12} - z^{10} + z^8 - z^6 + z^4 - z^2 + 1$$

$$C_{45}(z) = z^{24} - z^{21} + z^{15} - z^{12} + z^9 - z^3 + 1$$

$$C_{46}(z) = z^{22} - z^{21} + z^{20} - z^{19} + z^{18} - z^{17} + z^{16} - z^{15} + z^{14} - z^{13} + z^{12} - z^{11}$$
$$+ z^{10} - z^9 + z^8 - z^7 + z^6 - z^5 + z^4 - z^3 + z^2 - z + 1$$

$$C_{47}(z) = z^{46} + z^{45} + z^{44} + z^{43} + z^{42} + z^{41} + z^{40} + z^{39} + z^{38} + z^{37}$$
$$+ z^{36} + z^{35} + z^{34} + z^{33} + z^{32} + z^{31} + z^{30} + z^{29} + z^{28}$$
$$+ z^{27} + z^{26} + z^{25} + z^{24} + z^{23} + z^{22} + z^{21} + z^{20} + z^{19}$$
$$+ z^{18} + z^{17} + z^{16} + z^{15} + z^{14} + z^{13} + z^{12} + z^{11} + z^{10}$$
$$+ z^9 + z^8 + z^7 + z^6 + z^5 + z^4 + z^3 + z^2 + z + 1$$

$$C_{48}(z) = z^{16} - z^8 + 1$$

$$C_{49}(z) = z^{42} + z^{35} + z^{28} + z^{21} + z^{14} + z^7 + 1$$

$$C_{50}(z) = z^{20} - z^{15} + z^{10} - z^5 + 1$$

$$C_{51}(z) = z^{32} - z^{31} + z^{29} - z^{28} + z^{26} - z^{25} + z^{23} - z^{22} + z^{20} - z^{19} + z^{17} - z^{16}$$
$$+ z^{15} - z^{13} + z^{12} - z^{10} + z^9 - z^7 + z^6 - z^4 + z^3 - z + 1$$

$$C_{52}(z) = z^{24} - z^{22} + z^{20} - z^{18} + z^{16} - z^{14} + z^{12} - z^{10} + z^8 - z^6 + z^4 - z^2 + 1$$

$$C_{53}(z) = z^{52} + z^{51} + z^{50} + z^{49} + z^{48} + z^{47} + z^{46} + z^{45} + z^{44} + z^{43} + z^{42}$$
$$+ z^{41} + z^{40} + z^{39} + z^{38} + z^{37} + z^{36} + z^{35} + z^{34} + z^{33} + z^{32}$$
$$+ z^{31} + z^{30} + z^{29} + z^{28} + z^{27} + z^{26} + z^{25} + z^{24} + z^{23} + z^{22}$$
$$+ z^{21} + z^{20} + z^{19} + z^{18} + z^{17} + z^{16} + z^{15} + z^{14} + z^{13} + z^{12}$$
$$+ z^{11} + z^{10} + z^9 + z^8 + z^7 + z^6 + z^5 + z^4 + z^3 + z^2 + z + 1$$

$$C_{54}(z) = z^{18} - z^9 + 1$$

$$C_{55}(z) = z^{40} - z^{39} + z^{35} - z^{34} + z^{30} - z^{28} + z^{25} - z^{23} + z^{20} - z^{17}$$
$$+ z^{15} - z^{12} + z^{10} - z^6 + z^5 - z + 1$$

$C_{56}(z) = z^{24} - z^{20} + z^{16} - z^{12} + z^8 - z^4 + 1$

$C_{57}(z) = z^{36} - z^{35} + z^{33} - z^{32} + z^{30} - z^{29} + z^{27} - z^{26} + z^{24} - z^{23} + z^{21} - z^{20}$
$\qquad + z^{18} - z^{16} + z^{15} - z^{13} + z^{12} - z^{10} + z^9 - z^7 + z^6 - z^4 + z^3 - z + 1$

$C_{58}(z) = z^{28} - z^{27} + z^{26} - z^{25} + z^{24} - z^{23} + z^{22} - z^{21} + z^{20} - z^{19} + z^{18} - z^{17}$
$\qquad + z^{16} - z^{15} + z^{14} - z^{13} + z^{12} - z^{11} + z^{10} - z^9 + z^8 - z^7 + z^6 - z^5$
$\qquad + z^4 - z^3 + z^2 - z + 1$

$C_{59}(z) = z^{58} + z^{57} + z^{56} + z^{55} + z^{54} + z^{53} + z^{52} + z^{51} + z^{50} + z^{49} + z^{48}$
$\qquad + z^{47} + z^{46} + z^{45} + z^{44} + z^{43} + z^{42} + z^{41} + z^{40} + z^{39} + z^{38}$
$\qquad + z^{37} + z^{36} + z^{35} + z^{34} + z^{33} + z^{32} + z^{31} + z^{30} + z^{29} + z^{28}$
$\qquad + z^{27} + z^{26} + z^{25} + z^{24} + z^{23} + z^{22} + z^{21} + z^{20} + z^{19} + z^{18}$
$\qquad + z^{17} + z^{16} + z^{15} + z^{14} + z^{13} + z^{12} + z^{11} + z^{10} + z^9 + z^8$
$\qquad + z^7 + z^6 + z^5 + z^4 + z^3 + z^2 + z + 1$

$C_{60}(z) = z^{16} + z^{14} - z^{10} - z^8 - z^6 + z^2 + 1$

$C_{61}(z) = z^{60} + z^{59} + z^{58} + z^{57} + z^{56} + z^{55} + z^{54} + z^{53} + z^{52} + z^{51}$
$\qquad + z^{50} + z^{49} + z^{48} + z^{47} + z^{46} + z^{45} + z^{44} + z^{43} + z^{42} + z^{41}$
$\qquad + z^{40} + z^{39} + z^{38} + z^{37} + z^{36} + z^{35} + z^{34} + z^{33} + z^{32} + z^{31}$
$\qquad + z^{30} + z^{29} + z^{28} + z^{27} + z^{26} + z^{25} + z^{24} + z^{23} + z^{22} + z^{21}$
$\qquad + z^{20} + z^{19} + z^{18} + z^{17} + z^{16} + z^{15} + z^{14} + z^{13} + z^{12} + z^{11}$
$\qquad + z^{10} + z^9 + z^8 + z^7 + z^6 + z^5 + z^4 + z^3 + z^2 + z + 1$

$C_{62}(z) = z^{30} - z^{29} + z^{28} - z^{27} + z^{26} - z^{25} + z^{24} - z^{23} + z^{22} - z^{21} + z^{20} - z^{19}$
$\qquad + z^{18} - z^{17} + z^{16} - z^{15} + z^{14} - z^{13} + z^{12} - z^{11} + z^{10} - z^9 + z^8 - z^7$
$\qquad + z^6 - z^5 + z^4 - z^3 + z^2 - z + 1$

$C_{63}(z) = z^{36} - z^{33} + z^{27} - z^{24} + z^{18} - z^{12} + z^9 - z^3 + 1$

$C_{64}(z) = z^{32} + 1$

$C_{65}(z) = z^{48} - z^{47} + z^{43} - z^{42} + z^{38} - z^{37} + z^{35} - z^{34} + z^{33} - z^{32} + z^{30}$
$\qquad - z^{29} + z^{28} - z^{27} + z^{25} - z^{24} + z^{23} - z^{21} + z^{20} - z^{19} + z^{18}$
$\qquad - z^{16} + z^{15} - z^{14} + z^{13} - z^{11} + z^{10} - z^6 + z^5 - z + 1$

$C_{66}(z) = z^{20} + z^{19} - z^{17} - z^{16} + z^{14} + z^{13} - z^{11} - z^{10} - z^9 + z^7 + z^6 - z^4 - z^3$
$\qquad + z + 1$

$C_{67}(z) = z^{66} + z^{65} + z^{64} + z^{63} + z^{62} + z^{61} + z^{60} + z^{59} + z^{58} + z^{57} + z^{56}$
$\qquad + z^{55} + z^{54} + z^{53} + z^{52} + z^{51} + z^{50} + z^{49} + z^{48} + z^{47} + z^{46}$
$\qquad + z^{45} + z^{44} + z^{43} + z^{42} + z^{41} + z^{40} + z^{39} + z^{38} + z^{37} + z^{36}$
$\qquad + z^{35} + z^{34} + z^{33} + z^{32} + z^{31} + z^{30} + z^{29} + z^{28} + z^{27} + z^{26}$
$\qquad + z^{25} + z^{24} + z^{23} + z^{22} + z^{21} + z^{20} + z^{19} + z^{18} + z^{17} + z^{16}$
$\qquad + z^{15} + z^{14} + z^{13} + z^{12} + z^{11} + z^{10} + z^9 + z^8 + z^7 + z^6$
$\qquad + z^5 + z^4 + z^3 + z^2 + z + 1$

$C_{68}(z) = z^{32} - z^{30} + z^{28} - z^{26} + z^{24} - z^{22} + z^{20} - z^{18} + z^{16} - z^{14} + z^{12} - z^{10}$
$\qquad + z^8 - z^6 + z^4 - z^2 + 1$

$C_{69}(z) = z^{44} - z^{43} + z^{41} - z^{40} + z^{38} - z^{37} + z^{35} - z^{34} + z^{32} - z^{31} + z^{29} - z^{28}$
$\qquad + z^{26} - z^{25} + z^{23} - z^{22} + z^{21} - z^{19} + z^{18} - z^{16} + z^{15} - z^{13} + z^{12}$
$\qquad - z^{10} + z^9 - z^7 + z^6 - z^4 + z^3 - z + 1$

$C_{70}(z) = z^{24} + z^{23} - z^{19} - z^{18} - z^{17} - z^{16} + z^{14} + z^{13} + z^{12} + z^{11} + z^{10} - z^8$
$\qquad - z^7 - z^6 - z^5 + z + 1$

$C_{71}(z) = z^{70} + z^{69} + z^{68} + z^{67} + z^{66} + z^{65} + z^{64} + z^{63} + z^{62} + z^{61}$
$\qquad + z^{60} + z^{59} + z^{58} + z^{57} + z^{56} + z^{55} + z^{54} + z^{53} + z^{52} + z^{51}$
$\qquad + z^{50} + z^{49} + z^{48} + z^{47} + z^{46} + z^{45} + z^{44} + z^{43} + z^{42} + z^{41}$

$$+ z^{40} + z^{39} + z^{38} + z^{37} + z^{36} + z^{35} + z^{34} + z^{33} + z^{32} + z^{31}$$
$$+ z^{30} + z^{29} + z^{28} + z^{27} + z^{26} + z^{25} + z^{24} + z^{23} + z^{22} + z^{21}$$
$$+ z^{20} + z^{19} + z^{18} + z^{17} + z^{16} + z^{15} + z^{14} + z^{13} + z^{12} + z^{11}$$
$$+ z^{10} + z^9 + z^8 + z^7 + z^6 + z^5 + z^4 + z^3 + z^2 + z + 1$$

$$C_{72}(z) = z^{24} - z^{12} + 1$$

$$C_{73}(z) = z^{72} + z^{71} + z^{70} + z^{69} + z^{68} + z^{67} + z^{66} + z^{65} + z^{64} + z^{63} + z^{62}$$
$$+ z^{61} + z^{60} + z^{59} + z^{58} + z^{57} + z^{56} + z^{55} + z^{54} + z^{53} + z^{52}$$
$$+ z^{51} + z^{50} + z^{49} + z^{48} + z^{47} + z^{46} + z^{45} + z^{44} + z^{43} + z^{42}$$
$$+ z^{41} + z^{40} + z^{39} + z^{38} + z^{37} + z^{36} + z^{35} + z^{34} + z^{33} + z^{32}$$
$$+ z^{31} + z^{30} + z^{29} + z^{28} + z^{27} + z^{26} + z^{25} + z^{24} + z^{23} + z^{22}$$
$$+ z^{21} + z^{20} + z^{19} + z^{18} + z^{17} + z^{16} + z^{15} + z^{14} + z^{13} + z^{12}$$
$$+ z^{11} + z^{10} + z^9 + z^8 + z^7 + z^6 + z^5 + z^4 + z^3 + z^2 + z + 1$$

$$C_{74}(z) = z^{36} - z^{35} + z^{34} - z^{33} + z^{32} - z^{31} + z^{30} - z^{29} + z^{28} - z^{27} + z^{26} - z^{25} + z^{24}$$
$$- z^{23} + z^{22} - z^{21} + z^{20} - z^{19} + z^{18} - z^{17} + z^{16} - z^{15} + z^{14} - z^{13} + z^{12}$$
$$- z^{11} + z^{10} - z^9 + z^8 - z^7 + z^6 - z^5 + z^4 - z^3 + z^2 - z + 1$$

$$C_{75}(z) = z^{40} - z^{35} + z^{25} - z^{20} + z^{15} - z^5 + 1$$

$$C_{76}(z) = z^{36} - z^{34} + z^{32} - z^{30} + z^{28} - z^{26} + z^{24} - z^{22} + z^{20} - z^{18}$$
$$+ z^{16} - z^{14} + z^{12} - z^{10} + z^8 - z^6 + z^4 - z^2 + 1$$

$$C_{77}(z) = z^{60} - z^{59} + z^{53} \vee z^{52} + z^{49} - z^{48} + z^{46} - z^{45} + z^{42} - z^{41} + z^{39}$$
$$- z^{37} + z^{35} - z^{34} + z^{32} - z^{30} + z^{28} - z^{26} + z^{25} - z^{23} + z^{21} - z^{19}$$
$$+ z^{18} - z^{15} + z^{14} - z^{12} + z^{11} - z^8 + z^7 - z + 1$$

$$C_{78}(z) = z^{24} + z^{23} - z^{21} - z^{20} + z^{18} + z^{17} - z^{15} - z^{14} + z^{12} - z^{10}$$
$$- z^9 + z^7 + z^6 - z^4 - z^3 + z + 1$$

$$C_{79}(z) = z^{78} + z^{77} + z^{76} + z^{75} + z^{74} + z^{73} + z^{72} + z^{71} + z^{70} + z^{69}$$
$$+ z^{68} + z^{67} + z^{66} + z^{65} + z^{64} + z^{63} + z^{62} + z^{61} + z^{60} + z^{59}$$
$$+ z^{58} + z^{57} + z^{56} + z^{55} + z^{54} + z^{53} + z^{52} + z^{51} + z^{50} + z^{49}$$
$$+ z^{48} + z^{47} + z^{46} + z^{45} + z^{44} + z^{43} + z^{42} + z^{41} + z^{40} + z^{39}$$
$$+ z^{38} + z^{37} + z^{36} + z^{35} + z^{34} + z^{33} + z^{32} + z^{31} + z^{30} + z^{29}$$
$$+ z^{28} + z^{27} + z^{26} + z^{25} + z^{24} + z^{23} + z^{22} + z^{21} + z^{20} + z^{19}$$
$$+ z^{18} + z^{17} + z^{16} + z^{15} + z^{14} + z^{13} + z^{12} + z^{11} + z^{10} + z^9$$
$$+ z^8 + z^7 + z^6 + z^5 + z^4 + z^3 + z^2 + z + 1$$

$$C_{80}(z) = z^{32} - z^{24} + z^{16} - z^8 + 1$$

$$C_{81}(z) = z^{54} + z^{27} + 1$$

$$C_{82}(z) = z^{40} - z^{39} + z^{38} - z^{37} + z^{36} - z^{35} + z^{34} - z^{33} + z^{32} - z^{31}$$
$$+ z^{30} - z^{29} + z^{28} - z^{27} + z^{26} - z^{25} + z^{24} - z^{23} + z^{22} - z^{21}$$
$$+ z^{20} - z^{19} + z^{18} - z^{17} + z^{16} - z^{15} + z^{14} - z^{13} + z^{12} - z^{11}$$
$$+ z^{10} - z^9 + z^8 - z^7 + z^6 - z^5 + z^4 - z^3 + z^2 - z + 1$$

$$C_{83}(z) = z^{82} + z^{81} + z^{80} + z^{79} + z^{78} + z^{77} + z^{76} + z^{75} + z^{74} + z^{73}$$
$$+ z^{72} + z^{71} + z^{70} + z^{69} + z^{68} + z^{67} + z^{66} + z^{65} + z^{64} + z^{63}$$
$$+ z^{62} + z^{61} + z^{60} + z^{59} + z^{58} + z^{57} + z^{56} + z^{55} + z^{54} + z^{53}$$
$$+ z^{52} + z^{51} + z^{50} + z^{49} + z^{48} + z^{47} + z^{46} + z^{45} + z^{44} + z^{43}$$
$$+ z^{42} + z^{41} + z^{40} + z^{39} + z^{38} + z^{37} + z^{36} + z^{35} + z^{34} + z^{33}$$
$$+ z^{32} + z^{31} + z^{30} + z^{29} + z^{28} + z^{27} + z^{26} + z^{25} + z^{24} + z^{23}$$
$$+ z^{22} + z^{21} + z^{20} + z^{19} + z^{18} + z^{17} + z^{16} + z^{15} + z^{14} + z^{13}$$
$$+ z^{12} + z^{11} + z^{10} + z^9 + z^8 + z^7 + z^6 + z^5 + z^4 + z^3 + z^2 + z + 1$$

$$C_{84}(z) = z^{24} + z^{22} - z^{18} - z^{16} + z^{12} - z^8 - z^6 + z^2 + 1$$

$$C_{85}(z) = z^{64} - z^{63} + z^{59} - z^{58} + z^{54} - z^{53} + z^{49} - z^{48} + z^{47} - z^{46} + z^{44} - z^{43}$$
$$+ z^{42} - z^{41} + z^{39} - z^{38} + z^{37} - z^{36} + z^{34} - z^{33} + z^{32} - z^{31} + z^{30}$$
$$- z^{28} + z^{27} - z^{26} + z^{25} - z^{23} + z^{22} - z^{21} + z^{20} - z^{18} + z^{17} - z^{16}$$
$$+ z^{15} - z^{11} + z^{10} - z^{6} + z^{5} - z + 1$$

$$C_{86}(z) = z^{42} - z^{41} + z^{40} - z^{39} + z^{38} - z^{37} + z^{36} - z^{35} + z^{34} - z^{33} + z^{32}$$
$$- z^{31} + z^{30} - z^{29} + z^{28} - z^{27} + z^{26} - z^{25} + z^{24} - z^{23} + z^{22}$$
$$- z^{21} + z^{20} - z^{19} + z^{18} - z^{17} + z^{16} - z^{15} + z^{14} - z^{13} + z^{12}$$
$$- z^{11} + z^{10} - z^{9} + z^{8} - z^{7} + z^{6} - z^{5} + z^{4} - z^{3} + z^{2} - z + 1$$

$$C_{87}(z) = z^{56} - z^{55} + z^{53} - z^{52} + z^{50} - z^{49} + z^{47} - z^{46} + z^{44} - z^{43}$$
$$+ z^{41} - z^{40} + z^{38} - z^{37} + z^{35} - z^{34} + z^{32} - z^{31} + z^{29} - z^{28}$$
$$+ z^{27} - z^{25} + z^{24} - z^{22} + z^{21} - z^{19} + z^{18} - z^{16} + z^{15} - z^{13}$$
$$+ z^{12} - z^{10} + z^{9} - z^{7} + z^{6} - z^{4} + z^{3} - z + 1$$

$$C_{88}(z) = z^{40} - z^{36} + z^{32} - z^{28} + z^{24} - z^{20} + z^{16} - z^{12} + z^{8} - z^{4} + 1$$

$$C_{89}(z) = z^{88} + z^{87} + z^{86} + z^{85} + z^{84} + z^{83} + z^{82} + z^{81} + z^{80} + z^{79}$$
$$+ z^{78} + z^{77} + z^{76} + z^{75} + z^{74} + z^{73} + z^{72} + z^{71} + z^{70} + z^{69}$$
$$+ z^{68} + z^{67} + z^{66} + z^{65} + z^{64} + z^{63} + z^{62} + z^{61} + z^{60} + z^{59}$$
$$+ z^{58} + z^{57} + z^{56} + z^{55} + z^{54} + z^{53} + z^{52} + z^{51} + z^{50} + z^{49}$$
$$+ z^{48} + z^{47} + z^{46} + z^{45} + z^{44} + z^{43} + z^{42} + z^{41} + z^{40} + z^{39}$$
$$+ z^{38} + z^{37} + z^{36} + z^{35} + z^{34} + z^{33} + z^{32} + z^{31} + z^{30} + z^{29}$$
$$+ z^{28} + z^{27} + z^{26} + z^{25} + z^{24} + z^{23} + z^{22} + z^{21} + z^{20} + z^{19}$$
$$+ z^{18} + z^{17} + z^{16} + z^{15} + z^{14} + z^{13} + z^{12} + z^{11} + z^{10} + z^{9}$$
$$+ z^{8} + z^{7} + z^{6} + z^{5} + z^{4} + z^{3} + z^{2} + z + 1$$

$$C_{90}(z) = z^{24} + z^{21} - z^{15} - z^{12} - z^{9} + z^{3} + 1$$

$$C_{91}(z) = z^{72} - z^{71} + z^{65} - z^{64} + z^{59} - z^{57} + z^{52} - z^{50} + z^{46} - z^{43}$$
$$+ z^{39} - z^{36} + z^{33} - z^{29} + z^{26} - z^{22} + z^{20} - z^{15}$$
$$+ z^{13} - z^{8} + z^{7} - z + 1$$

$$C_{92}(z) = z^{44} - z^{42} + z^{40} - z^{38} + z^{36} - z^{34} + z^{32} - z^{30} + z^{28} - z^{26} + z^{24} - z^{22}$$
$$+ z^{20} - z^{18} + z^{16} - z^{14} + z^{12} - z^{10} + z^{8} - z^{6} + z^{4} - z^{2} + 1$$

$$C_{93}(z) = z^{60} - z^{59} + z^{57} - z^{56} + z^{54} - z^{53} + z^{51} - z^{50} + z^{48} - z^{47}$$
$$+ z^{45} - z^{44} + z^{42} - z^{41} + z^{39} - z^{38} + z^{36} - z^{35} + z^{33} - z^{32}$$
$$+ z^{30} - z^{28} + z^{27} - z^{25} + z^{24} - z^{22} + z^{21} - z^{19} + z^{18} - z^{16}$$
$$+ z^{15} - z^{13} + z^{12} - z^{10} + z^{9} - z^{7} + z^{6} - z^{4} + z^{3} - z + 1$$

$$C_{94}(z) = z^{46} - z^{45} + z^{44} - z^{43} + z^{42} - z^{41} + z^{40} - z^{39} + z^{38} - z^{37} + z^{36} - z^{35}$$
$$+ z^{34} - z^{33} + z^{32} - z^{31} + z^{30} - z^{29} + z^{28} - z^{27} + z^{26} - z^{25}$$
$$+ z^{24} - z^{23} + z^{22} - z^{21} + z^{20} - z^{19} + z^{18} - z^{17} + z^{16} - z^{15}$$
$$+ z^{14} - z^{13} + z^{12} - z^{11} + z^{10} - z^{9} + z^{8} - z^{7} + z^{6} - z^{5}$$
$$+ z^{4} - z^{3} + z^{2} - z + 1$$

$$C_{95}(z) = z^{72} - z^{71} + z^{67} - z^{66} + z^{62} - z^{61} + z^{57} - z^{56} + z^{53} - z^{51} + z^{48}$$
$$- z^{46} + z^{43} - z^{41} + z^{38} - z^{36} + z^{34} - z^{31} + z^{29} - z^{26} + z^{24}$$
$$- z^{21} + z^{19} - z^{16} + z^{15} - z^{11} + z^{10} - z^{6} + z^{5} - z + 1$$

$$C_{96}(z) = z^{32} - z^{16} + 1$$

$$C_{97}(z) = z^{96} + z^{95} + z^{94} + z^{93} + z^{92} + z^{91} + z^{90} + z^{89} + z^{88} + z^{87}$$
$$+ z^{86} + z^{85} + z^{84} + z^{83} + z^{82} + z^{81} + z^{80} + z^{79} + z^{78} + z^{77}$$
$$+ z^{76} + z^{75} + z^{74} + z^{73} + z^{72} + z^{71} + z^{70} + z^{69} + z^{68} + z^{67}$$
$$+ z^{66} + z^{65} + z^{64} + z^{63} + z^{62} + z^{61} + z^{60} + z^{59} + z^{58} + z^{57}$$
$$+ z^{56} + z^{55} + z^{54} + z^{53} + z^{52} + z^{51} + z^{50} + z^{49} + z^{48} + z^{47}$$

$$+ z^{46} + z^{45} + z^{44} + z^{43} + z^{42} + z^{41} + z^{40} + z^{39} + z^{38} + z^{37}$$
$$+ z^{36} + z^{35} + z^{34} + z^{33} + z^{32} + z^{31} + z^{30} + z^{29} + z^{28} + z^{27}$$
$$+ z^{26} + z^{25} + z^{24} + z^{23} + z^{22} + z^{21} + z^{20} + z^{19} + z^{18} + z^{17}$$
$$+ z^{16} + z^{15} + z^{14} + z^{13} + z^{12} + z^{11} + z^{10} + z^{9} + z^{8} + z^{7}$$
$$+ z^{6} + z^{5} + z^{4} + z^{3} + z^{2} + z + 1$$

$$C_{98}(z) = z^{42} - z^{35} + z^{28} - z^{21} + z^{14} - z^{7} + 1$$

$$C_{99}(z) = z^{60} - z^{57} + z^{51} - z^{48} + z^{42} - z^{39} + z^{33} - z^{30} + z^{27} - z^{21} + z^{18} - z^{12}$$
$$+ z^{9} - z^{3} + 1$$

$$C_{100}(z) = z^{40} - z^{30} + z^{20} - z^{10} + 1$$

INDEX